T0291822

CAMBRIDGE LIBRARY COLLECTION

Books of enduring scholarly value

Technology

The focus of this series is engineering, broadly construed. It covers technological innovation from a range of periods and cultures, but centres on the technological achievements of the industrial era in the West, particularly in the nineteenth century, as understood by their contemporaries. Infrastructure is one major focus, covering the building of railways and canals, bridges and tunnels, land drainage, the laying of submarine cables, and the construction of docks and lighthouses. Other key topics include developments in industrial and manufacturing fields such as mining technology, the production of iron and steel, the use of steam power, and chemical processes such as photography and textile dyes.

Papers, Literary, Scientific, Etc.

An electrical engineer, university teacher and wide-ranging writer, Fleeming Jenkin (1833–85) filed thirty-five British patents in the course of his career. Edited by Sidney Colvin (1845–1927) and J.A. Ewing (1855–1935) and first published in 1887, this two-volume work brings together a selection of Jenkin's varied and engaging papers. The collection ranges from notes on his voyages as a marine telegraph engineer, to a critical review of Darwin's *On the Origin of Species*, essays on literature, and thoughts on technical education. A memoir written by Robert Louis Stevenson, his former student, provides biographical context and attests to Jenkin's many interests and talents across the arts and sciences. Volume 2 includes Jenkin's papers on political economy, scientific education, and applied science, notably marine telegraphy. Abstracts of his scientific papers, along with a list of his patents, form an appendix to the volume.

Cambridge University Press has long been a pioneer in the reissuing of out-of-print titles from its own backlist, producing digital reprints of books that are still sought after by scholars and students but could not be reprinted economically using traditional technology. The Cambridge Library Collection extends this activity to a wider range of books which are still of importance to researchers and professionals, either for the source material they contain, or as landmarks in the history of their academic discipline.

Drawing from the world-renowned collections in the Cambridge University Library and other partner libraries, and guided by the advice of experts in each subject area, Cambridge University Press is using state-of-the-art scanning machines in its own Printing House to capture the content of each book selected for inclusion. The files are processed to give a consistently clear, crisp image, and the books finished to the high quality standard for which the Press is recognised around the world. The latest print-on-demand technology ensures that the books will remain available indefinitely, and that orders for single or multiple copies can quickly be supplied.

The Cambridge Library Collection brings back to life books of enduring scholarly value (including out-of-copyright works originally issued by other publishers) across a wide range of disciplines in the humanities and social sciences and in science and technology.

Papers, Literary, Scientific, Etc.

VOLUME 2

FLEEMING JENKIN
EDITED BY SIDNEY COLVIN
AND J.A. EWING

CAMBRIDGE
UNIVERSITY PRESS

University Printing House, Cambridge, CB2 8BS, United Kingdom

Published in the United States of America by Cambridge University Press, New York

Cambridge University Press is part of the University of Cambridge.
It furthers the University's mission by disseminating knowledge in the pursuit of
education, learning and research at the highest international levels of excellence.

www.cambridge.org
Information on this title: www.cambridge.org/9781108068048

This edition first published 1887
This digitally printed version 2014

ISBN 978-1-108-06804-8 Paperback

PAPERS AND MEMOIR OF FLEEMING JENKIN

VOL. II.

PRINTED BY
SPOTTISWOODE AND CO., NEW-STREET SQUARE
LONDON

PAPERS

LITERARY, SCIENTIFIC, &c.

BY THE LATE

FLEEMING JENKIN, F.R.S., LL.D.

PROFESSOR OF ENGINEERING IN THE UNIVERSITY OF EDINBURGH

EDITED BY

SIDNEY COLVIN, M.A. AND J. A. EWING, F.R.S.

WITH A MEMOIR BY ROBERT LOUIS STEVENSON

IN TWO VOLUMES

VOL. II.

LONDON

LONGMANS, GREEN, AND CO.

AND NEW YORK: 15 EAST 16th STREET

1887

CONTENTS

OF

THE SECOND VOLUME.

PAPERS BY FLEEMING JENKIN (continued).

DIAGRAMS.

POLITICAL ECONOMY

TRADE-UNIONS: HOW FAR LEGITIMATE.[1]

TRADE-UNIONS are on their trial. Large and increasing numbers of workmen have banded together to promote their common interests; they claim to represent the feelings and wishes of the artisans in each trade; they possess large resources, and have used their power with such effect that while they proclaim higher wages, lighter labour, and increased influence, as fruits of their exertions, they are denounced by numerous opponents as illegal societies, using outrage and intimidation to coerce men and masters, and as injuring commerce, without permanent benefit even to the members of these unlawful unions. The men say that they have found a plan by which they can, and do, better their condition, and they claim the right to put their plan in practice under the protection of the law. They are answered by the assertion that trade-unions are nurseries of disaffection, fatal to liberty, hostile to merit, and injurious at once to capital and labour. On all hands legislation is called for; each party awaits with impatience the report of the Royal Commission now sitting; the unions demand recognition as corporate bodies able to possess property and to recover debts at law; while their opponents press for the complete abolition of the system, or at least for such repressive and restrictive measures as shall break the power of these combinations.

The issues are of imperial importance. Mistaken legislation in one direction may involve great detriment to our commercial prosperity; and errors of an opposite kind may alienate the great body of skilled artisans to whom the suffrage has just been largely intrusted. The loyalty of these men to our constitution as hitherto worked will in a great measure depend on the justice

[1] *Review of Reports of the Commissioners appointed to inquire into the organisation and rules of Trades-Unions and other Associations.* From *The North British Review*, March 1868.

done to their demands by the expiring Parliament; and no worse effect could follow the extension of the suffrage than an attempt by workmen to use their new power to alter legislation in a sense favourable to their immediate interests, but adverse to those of the nation. The evidence received by the Royal Commission, and the questions asked by the members of that Commission, seem to show that even among those who are familiar with trade, with workmen, and with the handling of economical questions, the gravest errors are rife—errors indorsed without hesitation by the greater portion of the press. The claims and practices of unions are judged on no fixed principles and their legitimate action is condemned with almost the same rigour as is justly displayed in branding the foul crimes which they have fostered. The wishes of workmen are misunderstood, their habits are unknown, and they are pitied for hardships unheard of from the mouths of artisans, but mercifully vouched for by master builders.

The principles of political economy, though often quoted, are little understood; we propose—*first*, to discuss those principles as affecting trade-unions; *secondly*, to consider the right to combine; *thirdly*, to describe unions as they exist; and *finally*, to examine what legislative action is required. Before entering on these four subdivisions of our task, we will state briefly the general features of the case for and against trade-unions, and for the latter purpose shall draw largely from an article in the *Quarterly Review*.

This article begins with the assertion that unions are not economically beneficial to their members; that they do not, and cannot, raise wages permanently. It is not denied that wages have risen since the establishment of unions, but this rise may have been due to large profits made in trade—not to the unions at all. When profits are large the demand for labour will be great, and wages must rise. When profits are small in a given trade, capital will be driven from that trade, and wages will fall. The action of trade-unions cannot, it is said, increase the wages fund or capital out of which the workmen are to be paid, nor do they diminish the number of the recipients, though they may prevent the increase of that number by arbitrarily limiting the number of apprentices. Now, wages depend simply on the ratio between the capital employed as wages and the number of persons to be paid; and unless by augmenting the capital or by diminishing the number, in other words, by augmenting the demand or diminishing the supply, no permanent alteration in wages can be effected. The question for those who wish to raise the wages of labour is, not how to divide the existing wages fund in a manner more favourable to the working man, but how to increase competition for his labour among employers; in other words, how to increase the wages fund. Trade-unions, far from even aiming at this end, drive capital away from trade by harrassing employers, diminishing profits, and increasing risks. Therefore, in the long run they tend to diminish wages, and though for a little while they may obtain an increase from an employer working, for instance, under a penalty, the increase is only temporary, and is little, if at all, short of a theft from that employer. But while they fail to increase wages, they do increase the cost of production; they do therefore injure all consumers, themselves as well as others. By excluding competition, they may raise their own wages, but this exclusion constitutes a tyrannous monopoly which cannot be permitted for a day; and even this monopoly can never raise the wages of working men as a whole. The main aim and object of trade-unions being to raise wages, the above arguments lead to the conclusion that this object is a delusion based on an obvious fallacy, that unions are, so far even as concerns the interest of their members, an enormous blunder. But worse than this, they are injurious to the country at large, and their existence is irreconcilable with public policy. They injure the quality of articles produced, by diminishing competition among artisans; they are hostile to excellence among workmen, discouraging piece-work and over-time, by which the skilful man may hope to better his condition; they oppose machinery, and foster dissension between employers and employed; they limit the quantity of wealth produced by limiting the number of producers;—by all these means, without benefit to themselves, they banish trade, and increase the cost of produce to consumers. Worse still, they are not even honest, nor do they represent the true feelings and wishes of workmen; they are governed by glib democrats, who resort to force and outrage to establish their power; they are secret societies, and therefore odious; they have been established by fraud, on the pretence of being benefit societies for which purpose they are even now bankrupt; the savings which should have been invested to provide for the benefits have been squandered in futile strikes, and even had every sixpence been profitably invested, the subscriptions are inadequate to provide for the payments promised. It is really a comfort to think that such monstrous organisations are even by the present law illegal, and we must readily grant that the only remedy practicable is total abolition. Here and there we have reinforced the *Quarterly* argument by extracts from the evidence of Mr. Mault and others; and assuredly this impeachment, supported by evidence of murder, theft, and outrage, can be met with no light denial.

Let us now hear what men in unions claim to have accomplished, what objects they avow, and how they answer the accusations against them. As to wages, the men say:—'We *have* raised wages; if political economy says that this is impossible, so much the worse for political economy; we know that unions do raise wages, and our employers know it, and this is one reason why they are hostile to unions. Our opinion is no conjecture, but based on evidence collected for years from all parts of England—evidence which we lay before you. The cry used always to be that strikes could not raise wages; now it often is that wages have by unions and strikes been raised so high that trade is banished to other countries. Not only do we raise wages, but by the establishment of working rules, by the collection of information as to the want and excess of labour in different towns, by the selection of good and exclusion of bad workmen, by discouraging piece-work and over-time, noxious practices both, we have greatly benefited our members, and at the same time we have benefited both our employers and the consumers of the wealth we produce. By the establishment of organised bodies with whom employers can treat and argue, we have diminished the number of strikes, and facilitated arbitration; our unions supervise the conduct of their members, and we have notably raised the social position of the artisan; by our benefit funds we encourage frugality and have banished pauperism from among us. Your calculations as to our bankruptcy are based on a misconception of our rules; we do not discourage excellence; we do not oppose machinery; we are not governed by democrats; we no more injure trade by refusing to work for less than 36*s.* per week than a capitalist injures trade by refusing to invest his money for less than 10 per cent. The unions are popular even among non-members, and are recognised by the whole working class of the country as acting in their interests, and we so love our unions that we will emigrate or starve rather than abandon them. We admit that the great power given by unions has been abused by the ignorant in certain trades; we admit that even the best unions have from time to time made mistakes, and that the worst have incited men to murder and outrage; we will second every endeavour to prevent the recurrence of such crimes, but we contend that such great power has never yet been wielded by single men or by large bodies with less abuse of that power; we claim to have our rights recognised by law; we will cheerfully submit to those restrictions of our power which are required for the general good, but if you determine on abolition we will use our whole political power to reverse your decision.'

The answer reads tamely after the accusation. Here and there it involves direct contradiction as to facts. It does not meet the case as to limiting the number of competitors, and it could only be honestly delivered by members of the best unions; but before examining any of the minor contradictions, we must endeavour to settle the first question at issue, Can or cannot unions raise wages? This really is a fundamental question. If unions cannot raise wages, it is futile to discuss whether they should be permitted to try; they certainly cause great annoyance and loss by their endeavours, and also suffer much themselves; if they cannot raise wages, neither can they obtain other indirect benefits, such as shorter hours, equivalent to increased pay. The one argument in favour of permitting combinations of workmen to bargain with their employers is that these combinations do enable men to make a more advantageous bargain with the capitalist. If this be not true, the policy of allowing an apparent but unreal privilege would be dishonest to the workman and unjust to the capitalist. It could only be palliated on the ground that we dare not interfere with the ignorance of the workmen, and must deceive them to keep them quiet. Let us, then, examine closely the arguments in favour of the proposition dinned daily into our ears, that no combination of workmen or of masters can alter the rate of wages.

These arguments take two forms, different in wording, but the same in essence, and are enounced as the doctrine of the Wages Fund, and the Law of Demand and Supply. Mr. Mill writes of the wages fund as follows:—

Wages depend, then, on the proportion between the number of the labouring population and the capital or other funds devoted to the purchase of labour; we will say, for shortness, the capital. If wages are higher at one time or place than at another, if the subsistence and comfort of the class of hired labourers are more ample, it is and can be for no other reason than because capital bears a greater proportion to population. It is not the absolute amount of accommodation or of production that is of importance to the labouring class; it is not the amount even of the funds destined for distribution among the labourers: it is the proportion between those funds and the numbers among whom they are shared. The condition of the class can be bettered in no other way than by altering that proportion to their advantage; and every scheme for their benefit which does not proceed on this as a foundation, is, for all permanent purposes, a delusion.

Very clear and very true. When you know the number of recipients and the sum to be paid them, divide the number of shillings by the number of men, and you obtain the mean wages. If 1,000 eggs are sold daily, and 1,000 pence are daily spent on eggs, the mean price of eggs that day will be a penny a piece. The price of eggs depends on the egg fund, it seems. Diminish the number of men, diminish your divisor, says Mill, and your quotient will be larger. Sell only 500 eggs, and the price will be twopence, if the egg fund remains the same, which it will not. Still, we do not deny that by restricting the number of eggs for sale, and of labourers applying for employment, the price of eggs and rate of wages will rise though the wages and egg funds will fall. But we seem now to be leaving the clear and beaten path of simple division: apparently this same wages fund is not a constant quantity. It may diminish, it may increase. This becomes interesting to our labourers, who cannot readily diminish their numbers. Cannot this same wages fund be persuaded to increase for their benefit? How does it happen to be exactly the amount it is? What will make it rise, what will make it fall? The stock answer is, 'My poor fellows, do not delude yourselves; the wages fund depends on the profits of capital; if profits are large the fund may increase, but everything tending to diminish the profits diminishes the wages fund, so if you or some of you for a little while get increased wages, diminishing our profits, the fund to be divided among you next year will be smaller; and so, however much we may regret it, you will infallibly get less than you do now; what you are now getting is the market price of your labour—the laws of political economy say so.' Workmen do not always believe this, and sometimes do get an increase of wages; but the argument of the economists is elastic—they say the wages fund has increased; your new wages are now the market price of labour; you would have got it without asking. But all workmen are not quite sure that this is true, nor are we. The fallacy lies in the premiss that everything which diminishes profits diminishes the wages fund, or the saving which the capitalist applies to the purchase of labour. Of course the tendency in that direction must be admitted, but the motion of a body is not determined by one force only; to deduce its motion by calculation from the forces in action, we must take all the forces into account, all the tendencies; and we venture to say that in a large number of cases diminished profits on capital may cause an increase in the saving applied to the purposes of production. Take a concrete case first. A manufacturer having a large fixed capital in the form of a factory, has for some years cleared as gross receipts 100,000*l.*; he has paid as wages 80,000*l.* per annum; for simplicity's sake we may assume that he pays for his raw material and tools in wages only; he has spent 20,000*l.* per annum on his personal establishment. Under pressure from trade-unions he has thought it wise to give an increase of wages to his workmen for one year, though they have neither diminished in numbers nor have his profits increased; he would rather not face the loss entailed by a strike. Such things do happen. That year he pays his workmen 90,000*l.*, and finds he has only 10,000*l.* clear profits. What will this man do? Will he next year pay a smaller number of workmen, employing say only 50,000*l.* in wages at his original mill, and diverting the balance of 50,000*l.* to increase other investments or his personal expenditure, or will he curtail his expenditure, and provide 90,000*l.* as wages for his workmen next year too? As a matter of observation, many manufacturers will continue the production to the extent of 100,000*l.* per annum, and will increase the amount paid as wages. Their gains do not represent the profits on the wages fund alone, but on the fixed capital as well. If they diminish the rate of production in their factory by investing a large portion of their gross annual receipts elsewhere, they greatly diminish the returns on their fixed capital, so much indeed as to outweigh any moderate advantage they can obtain by investing annual receipts more profitably. This consideration will often lead a manufacturer to continue his business with increased wages and diminished profits. No manufacturer has come before the Commission to say, 'I have always, or generally, diminished my business whenever I have had to give increased wages;' and yet, whenever a man does continue his business at the original rate of production, with increased wages and constant, or nearly constant, receipts, he is increasing the wages fund or his circulating capital in the face of a diminished profit. Should he obtain an increase of price from the consumer, our argument is strengthened.

The greatest portion of the circulating capital of a country constituting its wages fund is of this nature. Year by year the savings of other classes add to this fund, but it is mainly composed of the price received by the manufacturer for his produce, a portion of which he habitually re-invests in the payment of labour without any conscious effort to save. We now assert that the proportion which he does so re-invest is not necessarily smaller because wages are larger or profits smaller. A manufacturer will generally work his mill or factory to the utmost so long as he does obtain a profit; he does not voluntarily set aside a certain sum for wages, diminishing and increasing that sum according to profits, but he employs as many men as he can, and pays them what he must. How this 'must' is determined, shall be considered further on. Obviously there is a limit to action of this sort. Any conscious savings he will generally invest in other undertakings, and if his profits fall below a certain point, he will endeavour in all ways to divert the use of his fixed capital to other objects. He may be unable to control his personal expenditure; and so, if wages rise and profits fall beyond a certain point, his contribution to the wages fund will diminish, and possibly disappear, owing to his ruin. But how is this certain point to be determined? Are all manufacturers habitually carrying on their business at such profits, that should these diminish they will diminish their annual payments in wages? We think not, and if not, the wages fund may increase in the face of diminished profits.

Let us turn to the second class of capitalists, men who, not being manufacturers, save to invest money, with the object of obtaining an assured income as the reward of their saving. A portion of each new saving will find its way into the wages fund. Will these savings be increased or diminished as the rate of interest is high or low? On the other hand, a high rate of interest is a greater temptation to investment than a low one; but then with a low rate of interest, a much larger sum must be invested to return a given income, and a given ideal income, of say 1,000*l.* per annum, is generally the object of investors of this class. Is it clear that the saving for investment in countries with a high average rate of interest is greater in proportion to the incomes than in countries where a low rate of interest obtains? We doubt it extremely; indeed, we entirely disbelieve that saving increases in proportion to the rate of interest to be obtained. We do not believe that men determine if they can get 10 per cent. that they will invest 1,000*l.*, but if they can only get 5 per cent. they will spend it. The contrary proposition is more nearly true: if men can get 10 per cent. for their money they will consider they have made a sufficient provision for their family by investing 10,000*l.*; if they can only get 5 per cent. they feel compelled to invest 20,000*l.* before retiring from business. In fine, both with the manufacturer re-investing old savings in his own business, and the professional man investing new savings, diminished profits on capital lead to diminished expenditure, and not always or generally to diminished saving. The reason why this fact is very generally denied may probably be a confusion between profits and the fund out of which savings are taken. If, it is said, the profits from which savings are made for re-investment diminish, how can we expect savings to increase? But the renewed savings for re-investment which constitute the great bulk of the wages fund do not come out of profits, but out of gross receipts. Our typical manufacturer does not annually obtain the 80,000*l.* or 90,000*l.* used for wages out of his profits of 10,000*l.* or 20,000*l.*, but out of his gross annual receipts of 100,000*l.* So that diminished profits do not entail a diminution in the fund from which renewed savings are made. They do diminish the fund from which new savings are drawn, and no one will deny that in face of falling profits and rising wages new investments in a particular trade will be checked.

Hitherto we have assumed that notwithstanding the increased cost of production consumers would pay no more for the produce. If they do pay more, the fund from which the renewed savings are drawn will increase also, and the wages fund will or may be increased still further. We took the other case first, as more unfavourable to our views.

But, it may be urged, the argument as to a possible increase of the wages fund, notwithstanding diminished profits, cannot hold good in those employments which require but little fixed capital. Undoubtedly the more easily capital can be transferred from one business to another, the sooner will any possible increase of wages fund applicable to that business reach its limit; but this same transfer of capital is by no means an easy matter, as any manufacturer or man of business will tell us.

Some persons in speaking of the wages fund seem to imagine that there would be a material as well as a moral difficulty in paying increased wages. They reason as if wages were limited by the amount of cash which manufacturers hold. It is of no use, they say, that the manufacturer may be willing to pay increased wages, if he has not already saved money enough for the purpose. A man cannot give what he has not got. A man who will open his eyes and will look at the way in which wages are paid in practice, will never be deceived by this fallacy. There is a very considerable available sum in the hands of all solvent persons and manufacturers, used to provide against irregularities in receipts and payments. This fund in money, or in assets easily convertible into money, forms a kind of distributing reservoir, and might be called the reservoir fund. If it were not for a fund of this kind the richest man might be in continual straits for a few pounds because receipts do not arrive daily, but at intermittent and at more or less uncertain times. No solvent manufacturer (except in times of panic) would have any difficulty in doubling the wages of all his workmen next week and for many weeks following (though he might ultimately be ruined by the process), any more than a solvent consumer would have any difficulty in doubling his weekly expenditure, though he might ultimately leave himself without a penny. All money received by a manufacturer is first paid into the reservoir fund; from that fund it may pass into four distinct channels: it may be spent, it may be invested in fixed capital unproductively, in fixed capital productively, or finally, in circulating capital out of which wages are paid. If the manufacture is profitable, the receipts paid into the reservoir fund continually exceed the sum returned into the circulating channel; an increase in the wages of the workmen increases the sum to be returned, and diminishes the sums flowing into the three other channels. Even if the trade is not profitable, wages may be increased, but only by drawing back through the second or third channel sums previously invested, until the manufacturer is wholly ruined. To allow this last re-absorption, savings made by some other person are certainly required, and in hard times these savings may not be forthcoming, so that in common language our manufacturer cannot realise his assets. He will be all the sooner ruined in this case by unprofitable trade; but so long as trade is profitable, in order to pay increased wages he need only divert out of the reservoir fund what, up to that time, he has habitually spent or consumed as income. Thus there is no material obstacle to an increase of wages, so long as any profit whatever is made by trade.

Having sifted the wages fund argument, we find that it tells us nothing as to the possible price of labour, because it does not tell us how the wages fund itself is determined. It may increase by obtaining a larger share of the gross receipts from the sale of produce, though profits may be less; it may be swelled by the increased savings of the community made in the face of a diminished average rate of interest; it may rise by an increase in the gross receipts received by the maker from the consumer, and the want of specie opposes no obstacle to an increase of wages so long as produce will sell for more than its cost. We see that in some uncertain way the wages fund is affected by the security of property, the effective desire of accumulation, profits made on capital, the number of labourers, peace or war, but we have found no better way than mere observation of determining whether a given change of circumstances will or will not augment the fund. One class of economists believe they can give a definite rule by which the price of labour may be determined, or at least by which any permanent change in that price is regulated. This rule would, therefore, if true, allow us to calculate either the wages fund or the change in the wages fund due to altered circumstances. This rule they name the Law of Demand and Supply. We will again take our definition of the law from Mr. Mill, who says—

The idea of a ratio as between demand and supply is out of place, and has no concern in the matter; the proper mathematical analogy is that of an equation.

Demand and supply, the quantity demanded and the quantity supplied, will be made equal. If unequal at any moment, competition equalises them, and the manner in which this is done is by an adjustment of the value. If the demand increases, the value rises; if the demand diminishes, the value falls; again, if the supply falls off, the value rises; and falls if the supply is increased. The rise or the fall continues until the demand and supply are again equal to one another; and the value which a commodity will bring in any market is no other than the value which in that market gives a demand just sufficient to carry off the existing or expected supply. . . . This then is the law of value with respect to all commodities not susceptible of being multiplied at pleasure.

There are commodities of which, though capable of being increased or diminished, to a great or even unlimited extent, the value never depends on anything but demand and supply. This is the case in particular with the commodity labour.

Well, as Mill says, labour and commodities not capable of being multiplied at pleasure have their value fixed by demand

and supply alone, let us first see what that law means as applied to some given commodity.

A thousand equal diamonds are offered for sale; a thousand purchasers equally desire them: what will be the price of diamonds? A thousand eggs have been imported to a henless island; a thousand islanders would like to have them for breakfast: what will be the price of eggs? No economist has hitherto stated the law of demand and supply so as to allow this calculation to be made.

Let us examine more nearly what is meant by 'demand' and 'supply.' The word *demand* is used in two distinct senses, and the confusion arising from these two meanings lies at the bottom of much bad reasoning. *Supply* is almost always used to signify 'the quantity offered for sale,' and can be expressed or measured by a number. Thus the supply of eggs is the number of eggs; the supply of land the number of acres in the market. When the demand is said to be equal to the supply, men mean that all of the commodity offered for sale is bought, and that no more would have been bought had it been offered. In this sense demand also means a quantity measured by a number; as Mill says, 'A ratio between demand and supply is only intelligible if by demand we mean the quantity demanded.' But the word *demand*, as popularly used, signifies a desire; and when 100 eggs are sold each day at 2*d.* each, instead of 1*d.*, the demand is said to have increased; and so correct would this language be in any other than in a highly technical sense, that correct reasoning will be impossible so long as this ambiguous word is used to signify the quantity demanded. The quantity demanded depends on the price at which the goods can be purchased; and the demand in the sense of a desire may oe measured by value as expressed in money. Thus, in a place where 1,000 eggs per diem are sold at 2*d.* each, the desire for eggs may be said to be twice as strong as in a place where 1,000 eggs will fetch daily 1*d.* a piece only. Again, the number supplied, popularly called the supply, must not be confounded with the readiness to sell the commodity in question. We may say that the people who sell the 1,000 eggs at 1*d.* are twice as ready to part with or supply eggs as those who sell them at 2*d.* The equality between demand and supply means equality between the number demanded and the number supplied at a given price; and to signify these numbers we shall use these words, and not the words demand and supply. No equality or ratio can be said to exist between the desire to buy and the readiness to sell. When our 1,000 eggs are sold at 2*d.*, the desire to buy was clearly greater than when they were sold at 1*d.*, but the readiness to sell was less in the former case than in the latter. A high price indicates a great demand and a small supply, in the sense of readiness to sell. If the desire be measured by the product of the number sold and their price—in other words, by the whole sum spent—the readiness of a community to sell, being inversely proportional to the price, might be measured by what is called the reciprocal of that number, or by the quotient of the number supplied by the money spent. Measured thus there is no equality or constant ratio between desire to buy and readiness to sell. The two may increase together, as when a larger number are sold at a constant price, or either may increase while the other diminishes, or both may decrease together. There is, indeed, an equality between the wish to buy and the reluctance to sell each individual thing, but this means no more than that the purchaser and seller must agree on one price before a transaction can take place. Still, as reluctance to sell is measured by the price demanded, we might state that when prices are constant, the desire to purchase is equal to the reluctance to sell, measuring one by the money spent and the other by the money received. These two equations, first between two numbers, and secondly between two values, are both true, and can one be deduced from the other; but unfortunately, because the number demanded has an effect on the reluctance to sell, and *vice versa*, people speak as if the equation lay between the number demanded and the readiness or perhaps the reluctance to sell—which is nonsense. Any increase in the number demanded at a given price indicates an increase at the time in the whole desire for the thing wanted; but it is not true that an increased total desire for the thing wanted necessarily indicates an increase in the number demanded. At one time in a given town 1000 workmen may be wanted at 20s. per week, and at another time only 800 workmen at 30s. If the value of money has remained constant with respect to other commodities, the total desire of the community for that particular kind of labour may be said to be greater in the second case than in the first, though the number of labourers wanted is less. Again, if at one time 1,000 are willing to work at 20*s.* and at another time none, or say only 100, will work at 20*s.*, while 900 are willing to work for 25*s.*, the readiness to supply labour will have diminished, though the number of labourers remains the same. To avoid confusion, we will avoid the equivocal words *demand* and *supply* altogether, and speak only of the number or quantity demanded and supplied as one pair of corresponding ideas, and the desire to purchase and reluctance to sell as a second pair of comparable magnitudes.[1]

[1] We may now try to write the equation indicated by Mr. Mill. Let the quantity demanded be called D, and the variable price x. We know that D is affected by the price, diminishing as the price increases, and may therefore write $D = f(\frac{1}{x})$, where f is not a simple factor, but is a mere symbol, indicating that D increases as the price diminishes, and is affected by no other circumstance, an assumption which on any given market-day may be true. Next, let S be the number which at the price x will be supplied during the same time that the quantity D is bought. S will also vary with the price, but it will increase as the price increases. We may therefore write $S = F(x)$, expressing the assumption that S is a function of the price, and is affected by no other circumstance. When D is equal to S, we have the equation $f(\frac{1}{x}) = F(x)$, by which the price x could be calculated, and would be determined, if the quantities demanded and supplied varied according to any constant law, and merely in consequence of variation of price. There would then be only one natural and invariable value or price for each article. But this equation does not express all that Mill says. If the desire for the article increases, the value tends to rise. The quantity demanded then is not a mere function of the price. D must therefore be considered equal to some more complicated expression, such as $f(A + \frac{1}{x})$, where A is some unknown variable quantity. Again the readiness to sell at a given price may diminish, and so diminish the quantity supplied, which is therefore not a mere function of price. To express this we write $S = F(B + x)$, where B again is an unknown variable quantity; thus when D is equal to S, we have the new equation, $f(A + \frac{1}{x}) = F(B + x)$,—an equation in which, so long as A and B and f and F were all constant in value and form, x would remain constant, and would be fixed in terms of these magnitudes. If x were to rise by what we may term an accident for a day or two above the value determined by the equation, the first number would be smaller than the second, the quantity supplied would be in excess of that required, competition would therefore at once lower the price to its true value, as determined by the above equation, so that all goods supplied might be sold. The consequence of a fall in

We assert that the number of things bought and sold may remain perfectly constant and yet a considerable change of price take place. Not only may the number remain equal at very different prices—this no one denies; but a thousand transactions may take place this year at one price and a thousand transactions may take place next year at double the price without any variation whatever having taken place in the demand or supply, as measured by the number of goods supplied and sold. What is necessary for this result is simply that while a disinclination to supply the article at the old price arises, an inclination on the part of purchasers to buy at a higher price shall also arise. So long as the total desire for the article, and reluctance or readiness to sell it, are unaltered, the price of the commodity remains fixed. Competition between both buyers and purchasers brings back the price to this fixed amount whenever any accidental deviation occurs. This is the law of demand and supply, as usually understood. The price is no more fixed by competition than a weight is fixed by a balance and scales; but the balance and scales serve to measure weights, and competition brings the price to the amount fixed by other considerations, which, in the case of a limited article, may be infinite in number, including

price diminishing the second member would be to raise the first. The increased quantity wanted would bring back x to its true market value. Again, suppose that the desire to possess the goods increases by an increase of A; if B remains constant then x must rise to maintain our equation. If the readiness to sell increases by an increase of B, x must fall. Our equation thus expresses every relation between value, demand, and supply, which Mill states as expressing the law of value with respect to all commodities not susceptible of being multiplied at pleasure. But there is nothing in this law to prevent A, f, F, and B from varying any day or any hour, from motives of the most opposite kind.

When the quantity demanded at a fixed price increases, A is increased. If B varies at the same time to a corresponding extent, we may have x the same as before, but a brisk trade instead of a slow one. If, on the other hand, B diminishes while A is constant, the price will rise while the number of transactions will become more limited, but if A rises while B falls, we may have a new and higher value of x, with a constant number of transactions; this is the conclusion to which we especially wish to draw attention. A diminished supply conveys to the minds of most persons the idea of absolutely fewer things for sale; but when an exact definition is sought of the number of things for sale, the idea of price is necessarily added; the things must be for sale at a given price, and an increase in the price at which a given number will be supplied produces many of the effects due to a diminution in the number supplied at a constant price.

everything capable of increasing the desire for the commodity, or the reluctance to part with it. The action of the law as usually described is true, but partial; the effect due to a disinclination on the part of purchasers to sell at the old price is admitted as a virtual diminution of supply, but this increased price will, it is said, diminish the demand, meaning the number demanded, and so the price may rise, but the number demanded, as well as the number supplied, will fall; on the other hand, the demand, meaning either number demanded or desire to purchase, may rise; this they say will increase the number supplied under the stimulus of an increased price, and the price will rise with an increased number of transactions. No one has ever denied these two actions, both tending to an increase of price—but one with an increased trade, the other with a diminished trade. How is it that we are not equally familiar with the third case, where the demand, meaning the desire to purchase, increases, and the supply, meaning readiness to sell, diminishes at the same time, so that as before we have an increased price, but this time with neither an increase nor a decrease in the number of transactions?[1] A change of price at any time may be due to increased desire for possession, or increased reluctance to sell; the increased reluctance to sell may increase the desire for possession, or it may diminish this desire; the action in any one case can only be determined by experiment. If the holders of a thousand diamonds refuse to sell except at an average increase of 20 per cent. in price, no one can tell except by experiment whether more or less money will be spent in diamonds next year, nor even whether more or fewer diamonds will be bought.

To apply this reasoning to labour:—Wages, it is said, can only increase by an increase in the demand or by a decrease in the supply; and decrease in the supply is always interpreted to mean decrease in the number of men in want of employment.

[1] The omission of this case from consideration tends to obscure the fact, that a great change in price may accompany a very small change in the number of transactions, and indeed that change of price has no invariable connection with a change in the number of transactions, unless on the assumption that the quantities A, B, f, and F, all remain constant, which will be sensibly true during any short period. These quantities in the long run all may and do vary, for every commodity.

Now, an equivalent effect to that produced by a decrease in the number supplied is produced whenever a given number of men who were yesterday willing to work at 30*s*. per week are to-day unwilling to work at that price, and require 31*s*. instead of 30*s*. If while the readiness to sell labour is decreased the desire to purchase it does not increase, we allow that to re-establish equality between the number demanded and the number supplied, the number demanded or employed must fall as wages rise; but if the diminished readiness to work be accompanied by an increased wish for labourers, wages may rise, and the number employed remain the same, though the demand and supply, as measured by the number demanded and supplied, would remain constant. Really it seems ridiculous to take so much pains to prove the self-evident proposition that if men want higher wages, and masters see that it is their interest to give those wages, the transaction may occur and all the men remain employed.

A second effect which may follow, and perhaps most generally does follow, the unwillingness of men to work except at increased wages, is this: the number employed may actually diminish, and yet the desire for labour, as measured by the total fund spent for labour, may increase; so that the reduced number, with augmented wages, may receive more than the larger number at lower wages; in this case it may be the interest of the workman to support his fellows out of work by a contribution from his gains, rather than by a reduction in his own requirements, to allow them to find employment. We have reasoned so far on the assumption that the workmen act as one body, as is sensibly the case where unions are strong. We have therefore neglected the effect of competition among workmen. Where competition can occur, it weakens the effect which an increased reluctance to sell their labour on the part of some workmen can produce in increasing the total desire for their work. The smaller the united body which refuses the low wages, the less their power; but whatever their size and importance, the tendency of their action remains the same.

It may here be argued, that the increased desire or demand on the part of the masters would have given a rise of wages independently of any action on the part of the men; but it by no means follows that without the diminished willingness to work on the part of the men the increased desire would ever have arisen. The master builders of London want for their present work 2,000 men. They are paying them 30*s.* a week; there may be no reason why they should want an increased number, and still less reason why *proprio motu* they should wish to give them 36*s.*; but let the men decline to work for less than 36*s.*, the masters, if making a good profit, will still want 2,000 men to do their work, and may therefore agree to advance the wages. The demand in the sense of desire for labour may thus be said to have increased, but it has increased solely in consequence of the diminished willingness to sell. On the other hand, if trade is bad and the workmen are unwilling to work, the masters will not care to give 36*s.*, and so the diminished readiness to sell labour may diminish instead of increasing the desire for it; and if the men are obstinate, some may get employment at 36*s.* for urgent matters, but the whole desire for labour and number demanded will both diminish.

An antagonist might still urge this argument: When trade is so good that masters can afford the advance of wages, they would naturally extend their business, and would want more hands; it is this potential increase in the number of hands wanted that really determines the increase of wages—not the refusal of the men to work for less than 36*s.* This need not always or even generally be true, but even in this case the action of the men in demanding more wages determines a rise of wages instead of extension of employment.[1]

Our argument is briefly this:—Wages, like the price of all other limited commodities, depend on a conflict between the desire for the commodity and the reluctance to sell it. Anything affecting either feeling as to labour will alter wages. The total desire measured by the total sum paid for wages may increase in consequence of large profits leading men to wish for an extension of trade, but it may also increase owing to increased

[1] To return to our equation: Under the influence of a good trade A may rise and B fall, raising the value of x, and leaving the numerical value of each side of the equation unaltered. Or, on the other hand, A and B may both rise, while x remains constant. The action of the men determines the former change, corresponding to an increase of wages, in distinction to the latter change, indicating an increased number employed.

reluctance on the part of the labourers to sell, leading the purchasers of labour and produce, one or both, to pay more, lest they should lose wholly, or in part, their profits, or the enjoyment of the produce. Competition is the process by which the price is ascertained at which the desire for the commodity and the reluctance to sell it are equal, but in no way can be said to determine the price.

We have come to a point where the identity of the wages fund argument with the demand and supply argument is obvious. The wages fund is the desire for labour, as measured by the total sum paid for it. That desire may increase or decrease in consequence of the increased reluctance of men to sell their labour. The increase of the fund invested by the capitalist may be due to increased payments he receives or expects to receive from his customers, or it may be directly due to a relinquishment of profits. It is wholly impossible to say when this will or will not be the case; it is impossible to fix any one given rate of average return on capital which may be taken as a kind of standard towards which, in all times and places, the profits tend. By the joint action of capital and labour, profits are made; that is to say, produce results from their action which exceeds the value of produce consumed by them in the process. Each claims a share in the profits; each must have some share, or each will refuse his aid. How much must each have? in what proportion shall the profits be divided? We apprehend that this is purely a question of bargain, and that the share each receives will vary, and may legitimately vary, within very wide limits. The capitalist may not force the labourer to work; the labourer may not force the capitalist to invest savings productively; each must tempt the other, and it is entirely a question of experiment how much temptation will in each case be required. It is quite possible that the temptation which was sufficient yesterday will not be sufficient to-day. Those who misapply the doctrine of demand and supply, or the wages fund argument, assume that the sum available to pay the workmen is fixed beforehand, or, if not fixed, must be diminished by any increase of wages. To assume this is to beg the question. Every effect which is distinctly seen to follow on changes in demand and supply, as popularly understood, will follow without this wholly arbitrary assumption of fixed wages for a fixed supply of workmen; and these known effects are not inconsistent with the fact that workmen, by bargaining, may in certain cases raise their wages. When more workmen are wanted than can be found, undoubtedly wages will rise without any bargaining; the competition among masters for workmen in that case indicates the increased desire, it does not create it; and when more workmen want work than are wanted, wages will fall in spite of bargaining; the competition among workmen indicates their increased readiness to sell; but when the number wanted and the number able to work are not very different, bargaining may raise wages or prevent a fall; and in the two other cases it may increase a rise and diminish a fall—a conclusion surely not far removed from common-sense. The contrary view, that somehow wages or prices are fixed by a law, is something like the idea that the strength of a beam is fixed by an equation. We can imagine a party of wiseacres who should meet the proposal of an engineer to cheapen their bridges by saying, 'Pray, don't be so foolish; you ought to know that the strength of a beam is determined by mathematics;' and our primitive engineer, guiltless of algebra, might say, 'So much the worse for mathematics; I know I can make beams lighter and stronger and cheaper, and I've done it.' At first this would be shortly denied; but at last one of the party would find out that the mathematics were all right after all, the equations for the strength of a beam perfectly correct, only, that as some of the terms were variable, it was quite consistent with algebra that beams should be made stronger by a better distribution of material. Even so economists who know that the equation exists, determining prices, should remember that there are other variables in the equation besides prices, and that the law only determines the price in terms of these variables.

If it be granted that bargaining does affect wages, it will readily be allowed that an association with savings enables its members to bargain more advantageously than isolated workmen could do. If the alternative before the labourer is work at the wages offered or starvation, he will be much less resolute in his views as to his worth, than when the alternative lies between work at high wages and mere privation; and a large mass, acting in concert, finds support in the mutual approval of its members. Joint action also causes greater inconvenience to the capitalists, and forces them to make up their mind at one given time. This point requires no elaboration. Many persons think the unions ought not to be allowed to exercise the powers they possess, but few, if any, will deny that if wages can be altered by bargaining, unions can drive the harder bargain.

We have so far, with Mill, assumed that labour is on the same footing as to value as commodities of which the quantity cannot be increased; but the grounds of that assumption should be understood. The cost of articles which *can* be multiplied at will is rightly supposed to depend ultimately on the cost of production. Why? Because there is no room for the exercise of any unwillingness to sell, such as may occur in the case of holders of a monopoly. If one set of holders will not sell without a profit above the average, new makers will produce, and by their competition soon reduce the cost to that which represents an ordinary profit on outlay.[1]

The *primâ facie* reason why labour cannot be included in this category of objects is, that the quantity for sale cannot be increased or diminished quickly enough. The cost of manufacture of labour is (neglecting previous outlay on education) the cost of the weekly sustenance of the labourer, who has to go on producing himself, and however small his profits on his absolutely necessary outlay may be, he is forced to sell or die; but then he has the great advantage that by eating twice as much he cannot do twice as much work, so that at any time when he is all wanted, he gets the benefit of being a limited article,

[1] Returning to our equation, there is no room for B; the number supplied can only depend on $F(x)$; but if the variable B disappears, A, as an independent variable, disappears too; for A could only vary in our equation without a change in $f(x)$ or $F(x)$ by the variation of B. We then find that the price of a commodity such as this is absolutely fixed once for all, so long as the relation between f and F remains constant. Now, the number supplied at a given price will increase precisely as the number wanted increases, so long as the profit remains unaltered—in other words, $F=fz$, where z is a factor or function depending on the cost of manufacture—hence the relation between F and f can vary only from a variation in the cost of manufacture, in which one possible variation is a change in the average rate of profit expected by the maker.

and may get more than his prime cost; but when he is not all wanted, and must sell his labour, he may be driven to cheapen his prime cost to starvation wages, or wages at which he can barely exist; whereas other articles, if the profit falls too low, are simply not produced. Mill points out, very justly, that if time be given for adjustment, the labourer comes into the category of unlimited articles; for though he will not avoid daily producing himself by eating, he may avoid reproducing himself in children, and will avoid doing so if his profit as a labourer be below a certain amount. This certain amount depends on what the labourer considers the minimum at which it is worth while to exist. The natural price of labour is fixed in this manner quite as definitely as the natural price of any unlimited commodity is fixed by the cost of its production, including in that cost the current rate of profit in trade. So far, therefore, it would appear that, after all, granting time, we might bring wages into the second category, in which bargaining avails nothing; but there are here one or two remarks to be made. If the standard of comfort be so raised that our labourer positively will not work unless he has more food and better clothes than last year, his prime cost is raised; but, considering the objection that men have to starvation and the workhouse, it is impossible that his standard should rise, unless he has some saving or fund to prevent his starving or to allow of emigration. This increase in the standard of comfort held by the labourer is analogous to the rate of profits expected by the manufacturer. If manufacturers, as a body, determine that it really is not worth while to produce goods except at an increased profit, the prime cost of their produce will be increased. Manufacturers could not act up to this determination unless they had savings —unless they combined, and unless they could prevent competition. The workman can only raise his price on precisely the same conditions, but he is fortunate so far, that competitors cannot readily be produced for any skilled employment. Ultimately, whenever population increases at such rate that competitors are practically unlimited, and where this population can flow without check into any skilled employment, wages must fall to such a point that no further competitors will enter the lists. But this increase in the number of competitors, and fall in the standard required as an inducement, both depend on man's own choice and on the standard of comfort once established. Whether, therefore, we look on wages as determined at any given time by the law of demand and supply, or as determined in the long-run by the cost of living, we find that the standard of comfort expected by the men, and the possession of savings sufficient to ward off starvation, may exercise great influence on the value of labour.

Much has been said on the identity of the interests of capital and labour. Well-meaning but fruitless attempts are made to teach workmen that capital and labour are never in antagonism. No one can deny that each needs the other; but they have a common interest only as the horse and his rider have a common interest. If the horse starves the rider must walk; if the horse jibs he must go to the knacker. So the rider feeds the horse, and the horse carries the rider. So far they have common interests; but it is none the less true that they have opposed interests, inasmuch as the horse would like to eat plenty of corn and do as little work as possible; while the rider, on the contrary, would be better pleased the less his horse ate and the farther he trotted. Workmen are all the less likely to see the common interest, if they hear the antagonism persistently denied with what seems to them hypocrisy. They think it monstrous that one of two parties to a bargain should be told to shut his eyes, and open his hands and take the wages fixed by Political Economy, which allegorical personage looks very like an employer on pay-day. On this ground of a common interest the workmen might as well require that all profits should be paid to them, and that employers should thankfully accept the share Political Economy, in the shape of a union secretary, might think fit to award them.

Let us openly face the fact, that wages and profits on capital are matters of bargain between men and master, and then we shall be prepared to consider under what conditions that bargain may be most advantageously made in the interests of the whole community. Revising our argument, and confining its application to a stationary community, we find that annually, in addition to fixed capital employed in production, a certain circulating capital is employed in the payment of wages to productive labourers. Annually the capital and labour produce wealth of more value than the circulating capital. This new wealth may be divided in an infinite variety of ways. The same sum as before may be spent as circulating capital in the shape of wages, and the whole excess of the wealth produced be consumed by the capitalist as his reward for saving. But this constancy need not be maintained; it is equally possible that the wealth may be divided in other proportions. What does in practice determine the proportions? We answer, the will of the capitalist and of the labourer. The exercise of this free will is subject to the heaviest penalties. The capitalist may refuse to re-invest so large a proportion of his wealth, and may diminish wages. His penalty may be that fewer workmen will work —perhaps none; his profits next year may diminish instead of increasing, he may find no profits, and have to live by consuming the capital he would fain invest. Again, if the workmen demand a larger share, they do it at their peril; they may get a smaller share, they may get none. Thus a perpetual and inevitable strife arises as to the distribution of the wealth produced by the conjunction of labour and capital; each party declares their share to be the smallest they can possibly accept. 'I will starve or emigrate rather than take less than 36*s.* per week,' says the workman. 'I will spend my wealth, or invest it abroad or in non-productive investments unless I get 15 per cent.,' says the capitalist.[1] The sincerity of the two parties to the bargain cannot be tested except by the practical test of a refusal to work, or a refusal to employ workmen. It cannot be contended that the proportions of distribution once fixed will be constant, or that any natural proportion whatever does exist. No man by reasoning beforehand can discover what rate of profit will reward a man for saving, or in other words, what is the natural interest on capital. No man by reasoning *à priori* can determine what food, lodging, raiment, amusement will be sufficient for an artisan, or in other words, what are the natural wages of that artisan. Both the necessary reward to induce saving, and the standard of comfort, will vary immensely with

[1] There is no means in any one case of knowing what profit an employer really does make; the workman openly states his claim for so much a week; the employer does not state his profit, nor can he be expected to do so.

custom, education, government, climate, and indeed with every circumstance which affects man's desire for wealth.

In the assertion of their determination, the capitalists stand at a great advantage when compared with an individual labourer who has no accumulated savings. He must work or starve, or break into open rebellion. When he has saved money, he may emigrate or change his occupation, since he will have time at his command. If many workmen at once determine not to work below a certain standard, and if they have accumulated funds, they stand more or less on an equal footing with the capitalist. They can wait, and he can wait; they suffer, and he suffers; the force of their determination is tested by the time during which each will endure the loss entailed. The capitalist sees opportunities of profits lost, he sees rivals supplanting him in trade; if his capital has been borrowed, he may see ruin impending. The workman sees his savings vanish, he endures privation at his home, he sees a rival workman at his bench, he must face unknown changes, starve, or live on charity.

This torture soon settles whether really the capitalist will be content with 14 per cent., or the workman with 30*s*. Nor is there any other test by which the proportion required to induce investment and to induce work can be settled. Workmen will continue to think it outrageous that the capitalist will not be content with less than 15 per cent. Masters will continue to think it monstrous that workmen who live uncommonly well on 36*s*. will not work for less than 40*s*. In truth, the master has no moral obligation to save or invest capital in consideration of any particular rate of interest, nor is it the duty of the workman to work at any given rate of wages. Capital and labour *are* antagonists, they must fight for the spoil, but they fight under this singular condition, which should put buttons on the foils—if one kills the other, the victor cannot long survive; nay, each feels every wound he gives his foe.

We have now completed the first branch of our inquiry, and, assuming that trade-unions can and do materially increase wages, will proceed to consider whether combination for this and analogous purposes ought to be permitted, and if permitted, under what restrictions, both as to the objects sought and the means employed to compass those objects; in brief, what are or what ought to be the rights of trade-unions, taking for our guide the interest of the community and the laws of positive morality.

Writers who admit that unions do and can raise wages, rarely contend that any legal restriction should be put on what they call the *right* to combine for the purpose of raising wages. Even the *Quarterly*, before venturing to recommend the abolition of unions, undertakes to prove that they do not benefit the workman by increasing his pay. Workmen generally hold the most decided belief that they have a *right* to combine with this object. So they have, while the law remains unaltered, but (we are almost afraid to write such heresy) they do not come into the world clothed with any natural right to combine, and the utility of these combinations to the nation is not so clear as they think. Granting that the law forces no man to sell his labour except on such terms as suit him (with exceptions which do not vitiate the reasoning), it does not follow that the law must and ought to grant a right of combination. How that poor word 'right' is misused! It is perhaps hopeless to try to explain in a few words to those who do not know it already, that a 'right' has any other meaning than something which is thought nice by the person using the word. We will, however, quote a passage from Mr. Austin's work on the Province of Jurisprudence :—

Every right supposes a duty incumbent on a party or parties other than the party entitled. Through the imposition of that corresponding duty, the right was conferred. Through the continuance of that corresponding duty, the right continues to exist. If that corresponding duty be the creature of a law imperative, the right is a right properly so called. If that corresponding duty be the creature of a law improper, the right is styled a right by an analogical extension of the term. Consequently, a right existing through a duty imposed by the law of God, or a right existing through a duty imposed by positive law, is a right properly so called. Where the duty is the creature of a positive moral rule, the nature of the corresponding right depends upon the nature of the rule. If the rule imposing the duty be a law imperative and proper, the right is a right properly so called. If the rule imposing the duty be a law set by opinion, the right is styled a *right* through an analogical extension of the term. Rights conferred by the law of

God, or rights existing through duties imposed by the law of God, may be styled *Divine*. Rights conferred by positive law, or rights existing through duties imposed by positive law, may be styled emphatically *legal*. Or it may be said of rights conferred by positive law, that they are sanctioned or protected *legally*. The rights, proper or improper, which are conferred by positive morality, may be styled *moral*. Or it may be said of rights conferred by positive morality, that they are sanctioned or protected *morally*.

No one will contend that Divine law enforces the duty of permitting or aiding trade-unions. Positive law may or may not, as it pleases Parliament. The whole question then as to the right of combination depends on the question whether there is a positive moral law imposing the duty of allowing or sanctioning trade-unions. Positive morality is unfortunately less well defined than Divine and positive law. We, for our part, cannot admit that any positive morality sanctions such combinations if they are injurious to the country, but will freely grant that so far as they are beneficent to the community they have a sanction. What we wish workmen would understand is, that they have no rights other than are sanctioned by Divine law, the law of their country, and positive morality; and that whether a supposed right has or has not the sanction of positive morality is a fair matter for argument, not to be settled by doggedly repeating a set phrase that every man has a right to vote, or a right to combine, or a right to be comfortable, etc. etc., but to be proved by showing that the exercise of this right benefits the community. Especially this right to combine is no clear matter, and always has been and ought to be conferred with great caution by positive law. For instance, almost every man has a right to walk up and down in the streets of London, but it would be intolerable that any 500 men should be allowed to combine, and all walk one way, blocking up the street; when the right to combine is granted, as to Volunteers, the right of walking about in any direction they please is restricted. Any one may go into Trafalgar Square; but a right to combine, even to hold a meeting, is quite another matter. And one may carry on a trade, but if several people combine to carry on trade, the right to combine, whether as partners, as a joint-stock or limited liability company, is conferred with restrictions devised in the interest of the community. People may think the laws affecting the joint-stock companies bad, and may wish to change them, but no one complains that the great powers of those companies are regulated by positive law. If joint-stock companies were clearly injurious to the community, they might be and ought to be abolished to-morrow, for there is no positive moral right to combine for the purpose of trading, nor is there any positive moral right to combine for the purpose of selling labour. Those who support trade-unions must therefore argue thus: These unions raise wages; they so far benefit the community by benefiting that section of it which is most numerous and least well off. Diminished profits to capital cause an evil which does not outweigh the good of increased wages, especially as there is a limit beyond which, if wages rise, the whole payment to the working classes will diminish, so that they will learn by experience at what point consistently with the good of the community their wages must cease for the time to rise. Their opponents, granting, as some do, that the unions raise wages, contend that by doing so they injure the consumer, first, by the direct increase of cost of the goods which he buys; and secondly, by the indirect decrease of production likely to result from diminished profits to capital. Unions raise prices and restrict trade. If the prices of produce rise in all trades, the purchasing power of the wages will remain the same, and the nominal benefit to workmen will confer no real benefit, while the loss to capitalists and annuitants will be doubled. It must, we think, be admitted that if unions become very general and the wages of the whole working classes rise, the purchasing power of the wages will not increase so much as the nominal value of the wages. But as the cost of produce does not wholly depend on the wages paid in this country, nor wholly on wages paid anywhere, but partly on the profits of capital, it must equally be admitted that the purchasing power of the wages will rise with their nominal amount, though not equally, and there will result, therefore, a tangible gain to the workman, and a loss to capitalists and annuitants. Looking at the relative position of the rich and poor, we do not think that the permission to combine should be withheld because it tends to diminish the present inequality of condition. Great inequality is necessary and desirable, but it is at present great enough to admit of some reduction. The accusation that unions do restrict trade is also well founded. No rationally conducted combination will so restrict trade as to diminish the total wages fund, but a rational combination may diminish the rapidity of its extension, by diminishing the profits of capital. The inducement to save, and the fund out of which new savings are made, are both diminished: and though other reasons, such as the desire for a given income, may tend to increase capital, still observation seems to show that trade will extend faster with large profits and small wages than with small profits and large wages. Is the rapid extension of trade a permanent good? Is it better that there shall be a working population of twenty-five millions with small wages, much pauperism, and great total wealth, or a population of twenty millions, less total wealth, but good wages, and little pauperism? To put the question is to answer it. If unions raise wages and the standard of comfort, the mere restriction to an increased trade will be no evil, provided the increased standard of comfort leads to a corresponding restriction of the increase of population. If it do not, then indeed the temporary gain to the fathers will be fatal to the children.

At one and the same time to diminish the increase in the production of wealth, and increase the number among whom the wealth is to be divided, is to insure a future generation of paupers. Trade-unions may for once increase the share of the workman in the profits on production, but they can only do it once, and so soon as the limit has been reached beyond which the wages fund under their action will decrease instead of increasing, they can no longer benefit the workman further than by maintaining the good they have won. When that wages fund has reached its maximum ratio to the total produce of the country, then every word said by Mill on the subject of the necessary limitation to population is applicable. Trade-unions could not maintain themselves in the face of paupers clamouring for employment, and perhaps the clear perception which those unions produce of the necessity of limited competition to the wellbeing of competitors for bread, may lead even the English workman to act on the precepts of Mill, as well as to vote for him and cheer him. Meanwhile, simple restriction of the extension of trade is not *per se* an evil, and none of the pleas against trade-unions founded upon it will hold water. When the Bank of England raises its rate of discount to 6, 7, 10 per cent. it restricts trade—unsound trade, you say; but is not trade unsound which requires for its success that the workmen shall be *quasi* paupers? The laws on joint-stock companies, the standing orders of the House of Commons, the determination of any board of directors not to invest money in an undertaking which promises to return less than 5 per cent., taxes, wars, Factory Acts—all these things are restrictions on trade, some wise, some inevitable; thus, we cannot forbid actions simply because they restrict trade, and we can see no reason why combinations of capitalists should be permitted to fix the rate of interest at which they will invest their money, and combinations of workmen forbidden to fix the rate at which they will sell their labour. They no more restrict trade by demanding high wages than capitalists do by demanding high profits. The same reasoning answers the allegation that trade-unions drive away trade. Unquestionably, if the workmen are sufficiently foolish to persist in their demands for wages which the trade cannot afford, they may drive away the trade; but again, if capitalists are so foolish as not to sell unless at a profit so great as to prevent successful competition with other countries, they may lose their business, ruining themselves and their workmen. We do not, therefore, prescribe a given rate of profit as a maximum, but trust to self-interest as the strongest of motives to prevent such suicidal action. Workmen in practice may be found less sensible than employers, but there is much evidence in the Blue-Books to show that the unions do look very keenly into the possibility of foreign competition; and in an ideal union it is clear that information among the men that the trade was being lost would lead them to abate their demands.

An odd fallacy has been mooted lately, chiefly by Americans, to the effect that free-trade and high wages are incompatible—that, in effect, free-trade tends to lower wages, and that if the unions raise wages free-trade must be abandoned. The effect of free-trade at any place is to reduce the price of articles which cannot advantageously be made there, but it increases the price of articles which can be advantageously made there; and as under perfect free-trade no article would be produced anywhere but where it would be advantageously produced, it raises the price paid at each place for those articles, and raises the fund out of which wages arise. Free-trade, therefore, not only increases the purchasing power of fixed wages, but actually tends to raise wages. Thus, supposing wine can be more advantageously made in France and beer in England, under free-trade the average price of beer in the two countries will be higher than it was in England when excluded from France, and wine with free-trade will be dearer in the two countries than wine in France when excluded from England. But the average price of beer and wine in France and England will, with free-trade, be lower than without it. The Frenchman, if he sets his heart on alternate bottles of Bass and Beaune, will be able to purchase them for less than before; but the brewer of Bass and the grower of Beaune will get more money with free-trade than without it, and will be able to pay higher wages, until of course by competition his profits are brought down to the average rate. Free-trade can only depress wages of those commodities which were already made at a disadvantage in any given place. If this disadvantage be due to excessive wages, it will depress wages; but unless the manufacture can bear the average rate of wages, it ought not to be carried on in that place. The workmen have, therefore, in such a case to decide whether they prefer to abandon their trade or to work for lower wages; but here again they are simply in the same position as the employer. Free-trade tends to diminish profits on all articles which cannot be advantageously made in a place, and so a producer of such articles must either abandon his trade or be content with small profits. Free-trade is good for both capital and labour when applied to proper objects; it is inimical to capital and labour when improperly, that is to say wastefully, employed. It is found expedient to allow the capitalist to consult his own interest rather than prescribe his course of action by law: and we think it will be found equally expedient to allow the workman to consult his interest, and to make no attempt to keep down wages by preventing the combination necessary to allow workmen to make a bargain.

When the right to combine is granted, it can only be granted in the interest of the whole community, not in the interest of the members of any particular combination. Joint-stock companies are allowed not in the interest of their shareholders, but because joint-stock companies are supposed to benefit the nation. The law granting the right ought therefore to impose limits on the action of the combination wherever that action is hurtful to the community, as in the case of a company, by imposing a limit to the profits it shall divide among its shareholders. The very first limitation to the powers of a trade-union should be aimed at preventing any violent or sudden change in the labour market. A sudden refusal to work causes much greater inconvenience than a refusal to work at a future time; it may cause great suffering to the community, as when all cabs are withdrawn, or when engine-drivers strike suddenly; and it may extort wages for a time which the capitalist would never have given, if he had been aware, before entering on a certain course of action, of the demand his workmen would make. This can create no permanent rise of wages, and it does harm both to public and to employer, driving away capital without any advantage to the workman. No law could permit all the bakers one day to declare that they would not sell bread under double or treble the price charged the day before, or to declare that for the next month they would make no bread. No law could permit all the railway officials round London to declare that to-morrow they would not work. We need not, however, deny to bakers and railway officials the right to combine. Let them give six months' notice, and the public can provide against the threatened loss or inconvenience. Any employer receiving a six months' notice will be free to choose whether he will enter into new engagements; if so, on what terms: and though he may still be fettered by old engagements, a six months' notice will generally extricate him from any serious embarrassment.

This simple restriction, which apparently would be accepted readily by the unions, is far from being the only one required. A combination permitted with the object of raising wages inevitably uses its power to obtain collateral benefits, generally equivalent to increased pay, though differing in form; in fine they bargain not only as to wages, but as to all the conditions of the contract between man and master.

All arguments in favour of permitting bargains for money apply to bargains for other privileges, such as a diminution of the hours of labour, the notice to be given before dismissal, the allowance to be made for travelling, etc. But as some conditions are illegal in any contract, we are at liberty to consider what conditions shall be declared illegal in this particular class of agreement between employer and workman. We assert that the *contract must contain no provision in virtue of which the workman or the master shall undertake to injure a third person who is no party to the contract, and that all other conditions may properly be made a matter of bargain.* This principle will serve to distinguish the right from the wrong action of unions, when in the next division of our subject we consider their actual practice, as explained in the evidence before the Commissioners. Observe, we do not say that workmen must not combine to injure other people. Masters might say that by combining to make them pay high wages unions injured them. Consumers might complain of high prices as an injury. Fellow-labourers thrown out of work by a strike may complain that they suffer by the action of the combination. Yet if a bargain is to be allowed at all, these injuries must follow. We say that the workmen and employers must not be allowed to agree on terms one of which is the injury of a third person. If a contract of this form is entered into, the workman is bribing his employer to injure this third person. The employer wants work done; the workman says, 'I will do it on these conditions:—1st, You shall pay me 30*s.* a week; 2nd, My working hours shall not exceed 56 in each week; 3rd, You shall turn off John Smith.' Wherever, as in this case, one condition of the agreement is that a third person shall be injured, the agreement is contrary to the laws of positive morality, and it is and should be not only illegal, but subject to a penalty for both the parties to such an agreement. We need hardly have recourse to first principles to prove this, and shall assume it as self-evident; our only care will be to prove that, if enforced, it is sufficient to restrict the action of trade-unions within harmless limits.

To resume: We find that although combination to raise wages and guard the other interests of workmen is no natural right, it may be permitted consistently with the interests of the community, provided sudden action be prevented, which might both derange the necessary machinery of daily production and traffic and also unnecessarily harass the capitalist engaged in production; and we further declare that the legitimate field for the action of the combination in driving its bargain is defined by the principle that no injury to a third person shall form any part of that bargain.

We turn now to the description of trade-unions as they are; and assuming that the general scope and action of unions is sufficiently known, we shall forthwith discuss those rules and practices which are either certainly pernicious, or are thought so by many writers.

The atrocious outrages detected at Sheffield, and among the Manchester brickmakers, require little comment here; not, indeed, that too much can be said to show the execration in which such crimes are held: they are only possible in societies where the criminal is conscious of the support and approbation of his associates—where the opinions of men are vile, and their conscience degraded. It is therefore most necessary that the thieves and murderers should know that beyond that depraved circle they are known and loathed as simple thieves and murderers. We do not pass by these outrages quickly, as of small account, but because there is no question but that they are outrages, that they deserve the heaviest penalties, and that further legislation is desirable for their better prevention, detection, and punishment. By and by we will discuss the remedies and safeguards against these crimes; but now, when about to discuss the merits of various rules and practices, it were waste time to prove that assassination, arson, theft, and the destruction of property must remain crimes, even if committed by members of a trade-union in the interests of what they call the trade.

Unions wholly free from outrage, and whose members neither practise personal violence nor even intimidation, do nevertheless interfere with non-society men—knobsticks, as they are called by engineers. The wretched knobstick need not fear that he will be murdered or even beaten, but he is persecuted nevertheless; he is jeered at and snubbed on all possible occasions; he is betrayed to foremen for peccadilloes; he receives none of those little aids by which the other men lighten one another's labour; apprentices fetch him no beer; he is generally rather an inferior workman, and his work receives its full due of criticism; he is an outcast, a pariah, and fear of personal violence is not required to render this position a wretched one. Some societies will not allow him to work in the same shop with their members, even as though he tainted the air; and upon the whole, perhaps, these societies are the most merciful. Workmen in general cannot be brought to see the wickedness of their conduct towards the poor knobstick. They reason thus: 'If he is a competent workman, and will pay a very moderate subscription, we will receive him among us; if he is not with us he is against us; and while he acts as our enemy, he receives great part of the benefits we painfully gain for ourselves by self-denial and privation; we strike, we starve, we gain the victories, and then this fellow who fought against us shares the spoil. Our wages rise, and so do his, unless we can prevent it, as we certainly will if we can by any means within the law.' Odd as it may seem, the knobstick takes much the same view of his own position; he feels himself a sneak, who for money betrays his fellows, he looks on the union with fear and longing, but with reverence. He is unskilful, poor, weak, and a traitor; they are skilled, rich, strong, and noble; yes, even when they morally kick him; for they serve a common cause, he stands alone an outcast; he wishes he could work better, could scrape that entrance-money together, and pay the fine standing against his name. Sometimes he does and feels himself a free man at the very moment when he would generally be described as entering into slavery.

The above description is drawn from experience among the engineers. In trades where the union is weaker, non-society men may meet with less contempt, and greater facilities in joining are often held out; and again, there are unions which treat them much worse, refusing to work in a shop where a single non-society man is employed. With the engineers, every man would belong to the union if he could. In other trades there are doubtless men who disapprove of the conduct of the unions, and would much rather not belong to them or acquiesce in their proceedings, but who are nevertheless driven into the unions by the harassing conduct above described; but we believe this to be a small class. In considering the treatment which the knobsticks receive, and which is cruelly wrong, we must remember how natural that particular form of cruelty is to man, and how society is pervaded from top to bottom with a similar feeling. The knobstick is the *parvenu*—the man who has not entered the profession by the right gate. A saving clause generally exists in favour of great merit; and it does take *great* merit to overcome the barriers erected by the actual possessor of any patronage or privilege. Men can get into the Artillery or Engineers by competition; does any one think a snob could stop in those corps? We remember very well the case of a young man who had served his apprenticeship as pupil to a very eminent mechanical engineer, but who was told by a civil engineer, that whatever his merits or knowledge he must not look forward to the position of a resident engineer on a railway, a position worth from 250*l.* to 400*l.* per annum, because he had come into the profession by a wrong gate, *i.e.*, through a mechanical engineer's office, not through a civil engineer's office. Be it in law, physic, church, army, or navy, the man who does not come in at the right gate will be looked upon with an evil eye. We fear the feeling is too deep-rooted in our English nature to be met by any law to the contrary. Great merit of course overcomes the feeling, and wins regard in spite of all restrictions, and this more readily among cultivated than uncultivated men; but shall we not feel greater indignation at the physician who refuses to attend a case where a midwife has been employed, than at a workman who declines to work beside a knobstick?

While we frankly allow that there is no hope of obtaining full justice for the non-society man wherever unions are strong, we can point out one great distinction between the cases of oppression among gentlemen and those among workmen. The social indignities heaped on the victim are the same in both cases; but the workmen often go further—they make a bargain with their employer that he shall join in the persecution, that as one consideration received in return for their labour they shall be able to shut the door against their weak competitor. They thus bribe the employer to deprive workmen of the wages they could otherwise gain. This action falls distinctly within the rule which we laid down, that no compact should be allowed, one condition of which was the injury of a third party. Neither master nor man should be suffered to agree to a rule excluding the knobstick; we fear he will be excluded by the system of contumely, after all has been done that can be done for him, but we can at least protect him against positive enactments.

The case of labourers employed to do the work of skilled artisans is closely analogous but not identical. Perhaps some of our readers may not be aware of the great distinction in social status between the artisan and the labourer. In works on political economy the labourer means a man who lives by the labour of his hands, but in workmen's language a labourer means a man of wholly different and much lower standing than the skilled workman. The labourer in each trade does the work requiring comparatively little skill, but much strength and hard work, and in workshops the line between labourers and artisans or mechanics is as clear and as strongly drawn as that between employers and workmen. Not that a labourer is necessarily or generally a mere beast of burden without any skill; on the contrary, an engineer's labourer or a bricklayer's labourer requires considerable training; and so it is in each branch of trade. Custom has partitioned the work between two classes, one receiving nearly twice the wages of the other, and consisting of men with some education, men who dress in a good black coat on Sundays, and who look on the other or labouring class as one with whose members they cannot associate out of the workshop, while in the workshop the labourers are treated as servants. Labourers have sometimes foremen of their own; they have unions also, in emulation of their betters; but the labourer and mechanic are as different as the mechanic and the gentleman. This being the relation between the two classes, it is a mortal sin in a labourer to presume to encroach on the field which the mechanic arrogates to himself. Of course labourers, by seeing mechanics constantly at work, are frequently able to do the simpler parts of the work as well as they. Woe to the labourer who is caught doing the work of his betters; he will not be beaten, any more than a gentleman's servant wearing his master's clothes will be thrashed, but he will not long keep his employment. The subdivision of the work into two categories has come about in the interest of all concerned.

It has analogies in the distinctions between apothecaries, general practitioners, and physicians, between solicitors and barristers. It would be very inconvenient in any workshop if the labourers were generally looking to promotion as mechanics, nor will they ever desire it as a body; but precisely as there should be no legal impediment to a practitioner who wished to become a physician, or to a solicitor who wished to become a barrister, so there should be no legal, or rather illegal, disabilities preventing a labourer from changing his condition or work. Few of the unions, we imagine, would object to this; they object to a labourer who remains a labourer, but does odds and ends of their work. The objection will never be eradicated, but judged by our rule the men would not be justified in striking against the employment of one or more labourers in ways of which they did not approve.

There is yet one more form of interference with competition: those who will not work on certain conditions, who are on strike, bribe other men who are willing to work, not to do so. This is indefensible, according to the rule laid down. The union must not contract with any man, or body of men, to the injury of a third person—the master. If this simple rule could therefore be enforced against men and masters, it would prevent strikes against non-society men, and against the employment of labourers to do any special class of work. It would remove the disabilities which, under unions as they are, do weigh on labourers and similar competitors in the labour market, and it would abolish the system of buying off competing workmen during strikes. The coercion of non-society men by what are sometimes called moral means would remain; but against this society at large is powerless in all ranks; and we warn all those who fancy that the unions are oligarchies ruling tyrannically a disaffected multitude outside the pale, that the competition of outsiders against the societies will not be much more active than at present, when all coercion is at an end: for incredible as it may seem, trade-unions are looked up to by the mass of workmen of all grades as the champions of labour, whose rules may injure individuals here and there, but on the whole benefit the great majority.

From what precedes it is already apparent that, in bargaining as to wages, unions think it their business to settle many collateral conditions; and, in fact, no relation between employer and employed escapes their vigilance. At first sight, all men who pique themselves on being liberal, are disposed to concede to workmen the right of refusing to work unless the conditions of the employment suit them as well as the wages; but a little consideration has already shown us that we cannot allow men to stipulate for any conditions whatever. We will now point out some of those conditions which are indefensible, but which have been claimed, ay, and established, by the workmen.

Sometimes the societies choose the materials the masters shall employ, such as the size and make of bricks, or the quality of stone to be used. Sometimes they choose the place where the materials shall be prepared for use, as when they refuse to set stones worked at the quarry instead of at the building. Sometimes they refuse to allow the employment of certain machinery. Sometimes they claim the right of dismissing their own superior officers, as their foremen. Sometimes they even choose the means of transport of the materials to be used, refusing to fix bricks brought on a given canal, or by a given carter. They even claim a veto over contractors, and sometimes architects. In fine, it is hard to say in what matter affecting their employer they will not occasionally interfere, when it is their interest to do so.

These claims are selected from isolated instances in special trades; they do not represent the general conduct of unions, and it must be singularly galling to workmen to find every instance of unjust action discovered in any petty branch or trade attributed to the general policy of their societies. On the contrary, many skilled artisans will not hesitate to denounce interference in every case above cited as unjust and intolerable. The difficulty is to show this to the more ignorant workman, who replies doggedly, 'I have a right to do what I please with my own labour, and will not work if you get bricks from Jones less than four inches thick, or stones ready dressed from Robinson's quarries, or if Smith cuts them, or if Green is to be foreman, or if you use barrows instead of hods.' We answer, 'O dogged objector! you have not a right to do what you please with your own, but only to do that which is lawful, and it shall not be lawful for you to use your labour as a payment to your employer for injuries which you wish to inflict on Brown, Jones, Robinson, Green, and the customer who wants $2\frac{1}{4}$ inch bricks. You must not spend your money standing outside your grocer's door, and paying all who come there sixpence each, on condition that they shall deal elsewhere. A bargain between two people to injure a third is a conspiracy, and you shall not sell your labour on those most objectionable conditions.' The ground commonly taken against these and similar conditions, that they are contrary to free-trade and injurious to customers, though true as far as it goes, is insufficient. Our dogged bricklayer asks if he and his employer are not to be free to agree on any conditions they please, and calls that free-trade. It is true that we cannot pretend to prevent everything injurious to customers. High wages and high profits are injurious to customers; we do not interfere with these in general; but from the principle of preventing a contract between two parties, containing a condition injurious to a third party, the impropriety of all the above claims follows as a simple consequence. If our bricklayer does not see it, he must be made to see it.

The case of one trade striking in support of another, as masons in support of bricklayers, offers greater difficulty, supposing the original strike to be for a legitimate object. If the support were bought by one trade from another, the action would be illegal. The masons' union must not contract with the bricklayers' union to injure the master for a consideration obtained from the bricklayers; but if the masons received no consideration, specified or implied, we do not see that they could be prevented from supporting their colleagues. So far as their contract with their employers is concerned, they are at liberty to make an advantage to a third party one of the conditions of the contract. They may say to the employers, 'You shall pay me five shillings a day, and that mason one shilling extra, or I will not work for you.' The hours of labour, the conditions on which notice of dismissal shall be given, the regulations as to lost time, allowances for walking, for travelling, are all proper subjects for negotiation, and may fitly be included in any contract.

The rules of a workshop include all these matters, but

they include others which are uncondemned by the principle hitherto employed to distinguish good from evil. These are the rules as to over-time, piece-work, and the standard rate of wages—all vexed questions. The standard rate of wages is differently interpreted by men and masters. The men say 'it is a minimum below which no one of our members shall work, and we will take care to have no members much below the average. If we do not carefully select our members as competent workmen, they will not find employment at all at our standard rate. We shall then have to support these incompetent members out of our own funds; we are therefore bound, under a very stringent penalty, to admit none but competent workmen.' The masters, on the other hand, declare the so-called minimum to be virtually a maximum, fixed so high that they cannot afford to give the best workmen more than the standard rate, because they have to pay the inferior men more than their value. They also allege that the men interfere to prevent skilled workmen obtaining higher wages. This the men deny, we think with truth, alleging that masters are not at all in the habit of offering superior men higher wages. The masters allege that they would more often do so if they were not afraid that the rise would be made a pretext for a demand for an increased standard rate; they also express great compassion for the inferior workmen, who, not being worth the standard rate, cannot get employment at all. On this last point all grievance would cease if there were no coercion against non-society men. The inferior workman would simply not join the society, and the master could then pay him what he pleased. We doubt whether the best masters would be very anxious to see him in their shops. We think the men have here the best of the argument, and that the masters are hardly candid in speaking of bargains with individual men. Such bargains never have been common. New workmen are almost always engaged at the current rate for average workmen, and either kept or dismissed without material change in those wages, unless the rate changes. Mr. Smith in his evidence, in answer to a leading question, says, 'he' (the employer) 'would not bargain with each individual man,' but points out that if he wanted more workmen he would instruct his foreman to offer 6*d*. a day extra, that is to say, over the current rate of wages. Even Mr. Mault, who acted as a kind of advocate, with a brief against the unions, speaks of the 'current rate of wages.' The separate bargain with each man, except in extreme cases, has never obtained, and never will. No satisfactory evidence was brought showing that men of extra skill did receive higher wages without unions than with unions, and the case against unions as levelling the men broke down, though urged almost with importunity by the Commissioners.

No workmen came forward to say, 'I should have 5*s.* a week more but for the unions.' There is good evidence that highly skilled workmen in unions do receive more than less skilled workmen.[1] This point would be a small one if no coercion were used to non-society men, as these ambitious workmen would quit the union if they thought it retarded their progress; but it is certainly to be regretted that all the unions do not act upon a simple principle involved in the carpenters' and joiners' trade-rules, recently adopted at Birmingham: 'The ordinary rate of wages for skilled operatives of the various branches to be 6½*d.* per hour. Superior and inferior workmen to be rated by the employer or foreman.'

The middle-class public greatly misapprehend the question of extra skill. In such professions as the law or physic extra skill has an enormous extra value. The best advice may be worth 100 guineas, when the average advice is only worth one. The difference in skill between workmen is not at all after this manner. If the average workman be worth 35*s.*, the very best will not be worth more than 40*s.* The difference in wages usually given does not generally exceed a shilling or two per week. The great advantage given by skill is certainty of

[1] The assertion that unions wish to reduce good and bad men to one level is continually repeated, but we find no corroborative evidence anywhere. Masters say that they are afraid of giving good workmen extra wages, lest this should lead the men to expect a general rise. They also say, they cannot afford to give extra pay when the general rate is high. Neither statement bears on the workman's wish for equality. Neither with nor without unions is there any machinery of competition by which the man of extra skill can enforce full extra payment, because of the understanding between masters that one shall not entice away the other's best hands. The objection to piece-work is due to no objection to skill; if there be any such objection, it can only exist in some local or small trades, and we are really curious to know how the cry arose.

employment. The highly skilled workman is always spoken of by middle-class writers as a man anxious and likely to rise in the world. This is untrue; the men who rise do not rise in virtue of their skill as workmen, but because they possess other qualities far more valuable, and which, in fact, are rarely found in combination with extreme skill at the bench. The evidence given before the Commission has failed to show that skilled workmen think themselves aggrieved, or that the unions have prevented workmen from rising. In general, all allegations on the part of masters that unions are baneful to their members must be received with great caution. The members of unions are extremely well satisfied with them, as any one mixing with workmen will soon discover. In fine, if unions are to bargain, they can only bargain for the standard rate of wages; they may refuse to allow their own members to receive less, taking the risk of having to support incompetent members. It is wrong that skilled workmen should be prevented from gaining in unions as much as they would out of the union, and this is, in fact, not practised; if it were and if the highly skilled men chose to remain in the unions without coercion, we fail to see how we can interfere to prevent their working at less wages than they are worth. The complaint that unions made men indolent seems also based on misconception. Many masters complain that men are lazy, and declare that unions make them lazy. It is undeniable that men who are tolerably certain of employment will not work so hard as men to whom loss of employment means pauperism. If, therefore, unions have made and do make men more independent and less liable to starve, they probably do make them less industrious; but though hard work is good, we doubt the propriety of keeping men poor in order that they may work the harder. As a proof of indolence, masters cite the general dislike of over-time avowed by trade-unions.

The question of over-time is thoroughly misunderstood by the general public. By refusing to work more than ten hours, or even eight hours a day, a man may put his employer to some inconvenience; he may make less money than if he worked fourteen or sixteen hours per diem, and indirectly he may increase the cost of articles to the consumer; but surely if he can by working eight hours each day gain as much money as he requires, society has no right to ask him to work longer; when he bargains with his master that he shall not be made to work longer, this condition, so far from being directly injurious to any third party, is beneficial to his fellow-workmen, since more of them will be employed than if he worked sixteen hours each day. But, says the Press, he ought to be energetic, hard-working; he ought not to be satisfied with what he can earn in eight hours. Why not? Is contentment so great a crime? The country will never progress if our workmen become indolent, it is said. True enough; but what is indolence? Do you for the progress of the country desire that workmen shall work eight, ten, twelve, or sixteen hours each day? Is it not perhaps quite as well that 1,000 men should work hard for ten hours a day, as 800 should work for fourteen, or even 700 for sixteen hours each? 'Ah! but,' says the middle-class lawyer, 'where should I have been if I had not worked late every night for years; and what a shame it is to prevent the ambitious workman from pushing his way by hard work too; or suppose he has a large family, as I have, what an atrocious thing is this that a trade society shall tell him No, you must not work extra hours to gain enough to support and start your sons and daughters in life! These trade-unions are levellers, foes to merit and progress.' Gentlemen who reason thus know neither the objects nor the habits of workmen. If any individual who pleased could work over-time without entailing equal work on all his fellows, there would be little or no objection to over-time; but if over-time is made at all, it must be made by a large proportion of the men employed in a shop. The engine must be at work, the gas burning, the time-keeper at the gate, the foreman present; and does any one suppose this can be done for an odd man here and there, who wishes to get on, or earn extra pay? No; the rule in a shop is, that all or none work over-time. Of course, one branch of the shop, as the pattern-makers, may not be working extra hours, though the erecters are; but the work in any branch of the shop where over-time is made must be in full swing, or over-time would not pay the masters. Over-time, gentlemen, means this:—You are engaged at a salary to work in an office from 9 to 5, which most of you think long hours. One day your employer comes into the office and says to you, 'For the next six months you must all come back after dinner, and work from 7 to 10 every evening; of course you will be paid for your extra hours at an increased rate.' The consternation in the office would be great; here and there one of you would like it, but to the mass it would be intolerable. They could not go out to dinner or to the theatre; they would have to give up their reading at home; they could not see their friends; and if this sort of thing were to go on year after year, and become the rule, not the exception, most of you would look out for lighter work and less pay. Over-time, habitual over-time we mean, is due to the simplest possible cause. It allows employers to make more money, with a given fixed capital. Suppose that their works are large enough to turn out 50 locomotives per annum, with men working ten hours a day; then if the men work fourteen hours a day, the works may perhaps turn out 60 or 65 locomotives per annum. The profits on the capital invested will therefore be so much increased, that for the extra hours wages can be profitably paid at a higher rate—at time and a quarter, or time and a half, in technical language. Masters say, as Mr. Smith says in his evidence, their works are not elastic, and if they get extra orders they must work extra time. As brick walls are not elastic, they stretch flesh and blood, and it being, as we have shown, clearly their interest to keep their productive powers at a maximum, they keep flesh and blood somewhat tightly stretched, so that in many works the habitual hours are from 6 in the morning to 8 at night, and in some from 6 to 10. Unions have opposed this, and most properly so. It is better for the men and better for the country that a larger number of men should be employed for the smaller number of hours. Never mind how the men employ their leisure; we will neither assume with some that they pass it in laudable courses of study, nor with others that they pass it in the pothouse; independently of all these really irrelevant arguments, we say there is no reason why workmen as a body should not decline to work more than a given number of hours, provided in those hours they can make the wages they require. It may be inconvenient to a few of their number not to have the opportunity of making more, but it would be intolerable that a large mass of workmen should, night after night and year after year, have all of them to work till 10 o'clock, in order that 1 per cent. of their number should rise to be a master, or that even 5 per cent., with extra large families, should be more at their ease. In truth, the middle-class mind is so imbued with the one longing to *get on*, that they cannot conceive a healthy state of society in which the members are actually contented with their position. Your middle-class man must make his way and end his days with greater means, and in a higher rank of society, than his father, or he has failed. Nay, such is the struggle, that unless he steadily strives to advance he will recede, and, falling back, will come to real ruin and privation. Failing to perceive the happier condition of skilled workmen, who need not struggle at all, and who scoff at the idea that to become a draughtsman or a clerk is an advance, the middle class think, as Mr. Gladstone said at Oldham, that the best condition of things for the labouring classes is that in which it shall be easiest for the able or the diligent man to rise out of it. What a blunder is this! On the contrary, the best state for the working man is that in which he can be good, happy, and well off, remaining in that state. Not one in a thousand can ever hope to rise, and we must not legislate for this unit, but for the 999 who desire no better than to do their duty *in* the condition of life in which they were born, not *out* of it—which last is the whole aim of the educated Englishman.

We have spoken of habitual over-time; as to occasional over-time in the face of a real emergency, no union ought to object to it, and we think few do, unless there be some standing quarrel. We have known a gang of a dozen shipwrights work thirty-six hours on a stretch, ay, and work hard too, and be back at their work after one night's rest. No indolence this, nor any uncommon case; but then the men must feel that there is a real occasion calling for extra exertion.

Piece-work is no better understood than over-time. In some places, and in many trades, as at Birmingham in the hardware trades, the men *will* have piece-work, and decline to be paid by the day. There some masters deprecate piece-work. In other trades and other places, the men set their face against piece-work, and there the masters uphold it. This is no accusation against either party; what answers best for the masters does not always answer best for the men, and *vice versâ*. Unlike over-time, piece-work has its good as well as bad points. The clever, hard-working, and ingenious workman, who contracts to do a given piece of work for a fixed price, will work harder and make more money than a man working by the day. His invention is also called into play, and various clever devices in aid of work are continually invented by men working by the piece. They make tools specially adapted to the job, and getting handy, often turn out the work, done well, and with surprising quickness. A master will then often diminish the contract price; the man grumbles a little, but submits, so long as it is his interest to do so; and in good workshops, there is a kind of honourable understanding that men at piece-work shall be allowed to make time and a third; that is to say, at the end of the week their profits will not be considered excessive if they receive one-third more than men in receipt of daily wages. So far, piece-work is distinctly beneficial. Men working by the day do not like it, for it makes them seem lazy; they therefore urge against it, that it tends to make men scamp work, *i.e.*, do it only just well enough to pass, and that where work cannot be thoroughly inspected this scamping is carried very far. This is probably true, but it would apply to all contracts; and with all submission to the workmen, we do not think that their zeal for good work would lead them to oppose the practice very resolutely. No rules against using inferior kinds of iron or unseasoned wood appear to be issued by the societies, and the secretaries would no doubt say these matters rested between the employers and their customers, so that zeal against the bad work due to a particular plan of payment seems uncalled for. In truth, piece-work has some tendency to diminish wages in certain trades, and also tends to make men work harder; and as the average man dislikes low wages and hard work, he opposes piece-work, to the detriment undoubtedly of the skilful hard-working man. This is much to be regretted, and might drive the skilful man out of the unions, if there was no moral coercion keeping him in. We do not see how legislation can force a body of men to take contracts rather than wages. We can only provide legal protection for those men who prefer a contract to a salary. No general rule about piece-work can be laid down. Where articles are made by thousands, not by tens, piece-work tends to raise wages. Men become very expert, so that their labour is worth more; they do work at home also, and keep their tools and inventions secret; and the master well knows that paying by the day he would get less for his money, even if the earnings of the workmen were less per week. In these cases piece-work is the rule, not the exception; and yet it has some very bad effects. It prevents any modification in the design and pattern of the articles produced. The workmen either flatly refuse to make the new design, or finding by a few trials how much longer it takes them to make than the old form, they demand such an exorbitant price that the manufacturer prefers to keep in the old rut. Birmingham is, we believe, losing her pre-eminence in the hardware market partly if not mainly in consequence of this vicious system of payment restraining invention and progress, while both in America and on the Continent the quality of work and the patterns used have greatly improved. Thus in laying down the law about piece-work, general rules must be avoided, and the attention directed to the special customs of each trade.

As an instance of misconception due to ignorance of special customs, we may remind our readers of a paragraph which went the round of the papers, stating that the union of engineers had a rule under which a man making any extra profit by piece-work was forced to share that profit among all his mates, though they were simply receiving daily wages. What a picture this raised of a hard-working man who, before he could make one shilling for himself, had to gain a pound for twenty other idle people! The explanation turns upon a special form of contract devised for the convenience of men and masters, and applicable, for instance, to the erection or putting together of a locomotive engine, the parts of which have been prepared in the fitting-shop and boiler-shed. A gang of half a dozen men may be employed to erect the engine, and these work under a leading hand. The employer finds it his interest to let the erection of the engine as a contract to this gang, who undertake to finish the work for a fixed sum, say 50*l*. The contract is not, however, made in form between the half-dozen men, who have no corporate capacity, but with the leading hand as their representative. As it would be very inconvenient to the men to wait for the completion of the job before receiving payment, they are each paid the usual weekly wages, and the balance due to them when the work is done is paid to the leading hand for distribution. He is sometimes, indeed generally, allowed by his mates rather a larger share than he would get if the division were made strictly according to the wages at which each man is rated in the shops. It appears that some leading hands took it into their heads that they might keep the balance to themselves, as probably at law they might have a right to do. The union very properly stopped this. All the men are working by the piece, and all should make like profit. If any one of them skulked his work, the others would either force him to quit the gang, or at the least would take care never to work associated with him again. This was explained to the Commission by Mr. Allan, but it was apparently not very clearly understood.

The limitation of apprentices is a common but not universal rule among trade-unions—the object being to keep up wages by preventing competition. This condition directly injures all apprentices who are excluded under it, and we think it therefore an improper condition in the contract between master and man. It is highly valued by the men as a very powerful means of raising wages; and while they admit that this is the general scope of the rule, they defend their conduct by several arguments which deserve consideration. First, they say that they are willing to enter into a bargain to work for their masters, but not to teach; that they do not, in fact, impose this condition injurious to a third party, but simply refuse to enter into a special subsidiary contract to teach, that being no essential part of their business. This is so far a sound argument, that we think it would be unanswerable if they would allow masters to employ apprentices in distinct rooms, taught by workmen who did not share this objection to the education of competitors; but neither masters nor men will look at this as a practical issue from the difficulty. Unless, therefore, the men allow apprentices to work along with them they do exclude young men from the trade, and make their injury a condition upon which the society man will work. Another argument is, that if no limitation were imposed wages would fall so rapidly that really the benefit to those admitted into the trade would vanish, and that the union is acting kindly in preventing lads from embarking in a trade in such numbers as would prevent them from ever earning a comfortable livelihood. Specious this, but false—as most arguments are which attempt to prove that a rule devised for your own benefit really benefits the person against whom it is aimed. It would no doubt be pleasant for skilled workmen to possess a monopoly of their trades, and only to admit such numbers as would keep their wages at a comfortable rate. Administered with a little good sense, such a rule as this would insure the existence of a class of well-to-do artisans;—but how about those excluded? No monopoly can be allowed for the benefit of a privileged class of workmen who are to administer the patronage as seems good to them, regardless of the poverty of all applicants whom they refuse. Workmen compare their trades to ships, which when full can receive no more with comfort; but if a ship's crew, finding a crowd of famished creatures on an island, told them, 'Really, good people, we should be most inconveniently crowded if you came on board; why, we should have to be put on short rations, and you know you would not like that yourselves;' the answer would be—'Have pity on us; short rations are not starvation, overcrowding is not abandonment;' and the crew would deserve hanging who left the wretches behind rather than sacrifice some comfort. A low standard of comfort, implying low wages, is an evil and a great evil; but it is a worse evil to create an artificially high standard among the few, to the detriment of the many. Of course, rules which simply prevented the accumulation of an undue number of lads in one shop would be defensible enough, and educational restrictions might also be permitted, analogous to those which fence round most of the learned professions. These restrictions do limit competition; but the members of the several professions do not simply select *proprio motu* who shall and who shall not be free to enter these professions. Mr. Roebuck told at Sheffield a pitiful story of an orphan lad[1] supposed to have suffered exclusion under one of these arbitrary

[1] It so happens, the lad was not excluded, but the union did ask for his exclusion.

rules determining who may and who may not become an apprentice. It is a pity he should have used an argument so easily answered. All rules, all laws, however beneficial on the whole, work hardly in individual cases, and workmen know this as well as Mr. Roebuck ought to know it.

We have now discussed the main rules of trade-unions—some bad, some indifferent, some good. There are minor regulations about which a great fuss has been made. Here and there a rule is found that members shall not speak to employers, which simply is an endeavour to stop talebearing; there is here and there a rule against *chasing*, which means that some men have been suspected of maliciously, or for extra pay, driving their mates to work harder than was pleasant, by showing what they, the chasers, could do;—wrong, no doubt, and meaning that workmen squabble sometimes in an undignified manner, but having no reference whatever to the really skilled workman, who is honoured in and out of unions by all men. Then there are lists of black sheep here and there. Some masters copied this practice by the way, but explained that their black sheep got white in time, whereas the men's black sheep were dyed in grain; but the men explained that their black sheep would be bleached by the payment of a fine; and indeed, that these portentous lists mean that if a sinner is repentant, he must pay a fine varying say from 2*s*. 6*d*. to 2*l*., according to the enormity of his offence, before be can once again be admitted as a lamb into the fold—many of those fines being simply safeguards against the intermission of the weekly payments whenever the said sheep preferred not to pay them. It is preposterous to make a fuss about these trivial matters. Let us settle on some rational principle, how far the action of unions may extend on really important points, and leave the management of bricklayers' etiquette to bricklayers, and smile rather than frown when we hear that a man may be fined 6*d*. for tattling.

Passing from the examination of the rules, with their merits and demerits, we will say a few words as to their administration. On this point there is as great a difference between the practice of various unions and various trades as between man and man. Such societies as the Amalgamated Engineers or the Amalgamated Joiners and Carpenters fight with courtesy when they think they must fight, and enforce their rules against peccant brethren with justice and without rancour. Of course where there are opposing interests there will be disputes, and where there are disputes there will be some recrimination; but after reading Mr. Mault's attack and Mr. Applegarth's reply, we conclude that masters have little cause to blame these unions of superior workmen. The executive council and secretaries are really superior men, and prevent instead of fomenting strikes. The masons do not stand so high; bricklayers lower still; with them may be classed plasterers; and when we reach bricklayers' labourers and brickmakers we reach the realm where violence and outrage are used as the sanction of trade rules. In the better societies, moderate fines or exclusion from the society are ample securities against any infraction of the laws. It is not till we reach Sheffield and the grinding trades that we find the payment of arrears enforced by maiming and murder. The wretches do not see when they whine a complaint that they are driven to it; having no legal redress against defaulters, they pronounce the condemnation of their unions. Exclusion should be and is the bitterest punishment in the better unions, even though exclusion is followed by no necessary loss of work. The grinders dare not exclude their members. A club or an insurance office need never sue for arrears. Expulsion is a very simple remedy, entirely in their own hands; and unless expulsion be felt as a punishment, the club is of no benefit to the member. There is evidence to show that the better class of unions facilitate arbitration upon disputed points, and settle rules with the masters more easily than can be done when the workmen are disorganised. It is natural enough that masters should resent having to settle any rules at all, and having to meet the unions as their equals, with whom they are to bargain, discussing every condition of the contract, as if with a brother capitalist. They naturally regret the good old times when the workman was a servant, often a trusted and devoted servant, but still a servant who must do as he was told. That is past, and the world will not turn back, so it is useless to discuss whether or not a reverse motion would, on the whole, be profitable. The old form of good feeling as between master and dependent is gone, but it is quite possible that good feeling in a new form should grow up. We hope and expect that it will; but so long as masters try to crush the unions, and to detach men from them, this new kind feeling is impossible. The sincere attachment of men to their unions admits of no rational doubt. Over and over again employers have tried to put an end to unions by declaring that they would employ no union men; as often unions have come out of the struggle more vigorous than ever. Men will starve, they will emigrate; they have starved, they have emigrated, rather than abandon these institutions. Men trust in them, as they trust in themselves, with a thorough British self-reliance. A Frenchman clamours for work and protection from his Government, or from his master. They look for their benefits from the head of the establishment as they look for benefits at the hands of the Government. Englishmen are too self-reliant to follow any similar course of life. The English workmen ask nothing but wages and respect from their employers; and from the Government they ask leave to be allowed to manage their own affairs. They organise themselves and govern themselves on a small scale, as Great Britain at large is governed on a large scale; and, when organised, they say little about the rights of man, or communism, or principles of any kind. They want good wages, and where the shoe pinches they try to ease it. They have done so with so much success, and have had so much pleasure in managing their own affairs, that they feel a loyalty to their unions akin to that felt by the middle classes to Parliament. To deny this feeling shows ignorance, to ignore it folly. Would that the workmen felt towards our Government what they feel for the unions; they may come to feel this, and if they do England will be stronger than she is now. It is the fashion to speak of the workmen as tools in the hands of secretaries and delegates, who foment strikes to their own profit. Among the lower trades the men may be in the hands of low men, though probably even there the governors truly represent the governed. The large unions are no more in the hands of their leaders than England is in the hands of Parliamentary leaders. The unions have their Gladstones and Disraelis *in parvo*, no doubt, but these are representative men; and the constitution of a union is singularly well suited to secure an accurate representation of the feelings of a majority, and a full expression to the opinions of a minority. The officers are elected by universal suffrage, and all decrees are passed in the same manner. This, we allow, affords small guarantee for a true expression of feeling. Shareholders may all vote, but directors govern; fellows of learned societies elect, but the councils choose. But why? Because of the great difficulty in organising any opposition, in finding a nucleus round which discontented members can rally. But trade-unions are divided into many branches, each with a committee and local secretary, each holding a separate meeting, generally in a separate town, before any vote is given. Thus the carpenters have 190 branches, the engineers 308 branches; and any discontented branch can express its opposition, and can make known its feelings to all other branches, while the executive council or committee can never personally explain their motives, or personally influence more than a very few branches. No better plan could be devised against the growth of dictation; and except in small local societies we see no signs of dictatorship. In the large societies the accounts are regularly printed, distributed, and scrutinised by every branch; and each one has a direct interest in preventing a misapplication of funds by any of the others. The incomes of six of the societies concerning which evidence was given before the Commissioners, ranged from 2,700*l*. per annum to nearly 87,000*l*. per annum. Is not the collection and successful administration of these funds a very striking proof of the powers of self-government possessed by workmen?

Not an instance of malversation in these societies was brought before the Commissioners; no workmen appeared to complain that they were defrauded; no complaint was made of any difficulty in collecting the funds. The accounts appear to be well kept, and the expenses of management were not shown to be excessive. (The small local societies, such as those in Sheffield, differ *toto cœlo* from the account just given.) The monthly circulars published by the leading societies are very creditable documents. They record the votes given on all questions by all the branches. They contain the reports from all branches of the state of trade in the several districts; also the number of sick and the number out of employment in each place, with the amount of relief distributed from the funds of the union. The decisions of the executive council and resolutions of branches are also printed. A number of the circular or report issued by the Amalgamated Joiners and Carpenters, taken at hazard, contains, besides the above official matters, an account of the presentation of a testimonial to a gentleman who had rendered assistance in courts of arbitration; a suggestion that technical education might prove one of the benefits of trade-unions, with a resolution of the executive council in support of the suggestion; a report of a speech by Mr. Grenfell, M.P., on trade-unions, urging the stock doctrines of political economy; a portion of a paper on trade societies and co-operative production, by Mr. Ludlow; some short account of co-operation in America; reports of the proceedings at branch anniversaries, with the accompaniments of loyal toasts, evergreens, and allegorical designs, such as Justice trampling outrage under foot, and holding a balance with a scale on which the word 'Arbitration' is inscribed. Next comes a letter from the operative bricklayers of Burslem, who mean well, though the style of their secretary is cloudy. He says of trade-unions:—

> Although they may in some instances have exceeded the bounds of discretion, and perhaps acted tyrannically, yet, as a body of men, they must execrate the conduct of such officials as those of Sheffield and Manchester, believing that education (compulsory or otherwise) would have prevented such a state of things—as witness those trades where the greatest amount of it exists.

Inarticulate this, but good. The report concludes with an open column, containing letters from their members. One letter suggests a plan for a co-operative society; one advocates a reform in the method of voting; and one calls for a trade directory. We are tempted to give this last letter *in extenso*. The style of the joiner differs considerably from that of the bricklayer.

> Brother Members,—At a time when trade is generally in a very depressed state throughout the country, it may not altogether be out of place to consider whether we cannot afford some additional facilities to those of our members who are unfortunately compelled to search for employment.

I have heard many members state the difficulty they have experienced in finding out the workshops whenever they have ventured into a locality with which they were not well acquainted. This is not to be wondered at in London, where many of the shops are situated in some court or alley, so that a man might pass by every day for a month without once dreaming that a joiner's shop was to be found in the immediate vicinity. And I am quite sure that many of us who reside in the north of London would be nearly as much at a loss in looking for a job in Lambeth or the Borough, as we should be in Birmingham or Manchester. This state of things is not, I believe, confined to the metropolis ; it prevails also in other districts.

To supply the want which I consider at present exists, I would suggest that schedules be issued from the General Office, on which each Branch could forward a return of the names and addresses of all the building firms in the vicinity. A committee might be appointed by each Branch for the purpose of filling up the schedules, and the result of their labours might be read over to the Branch for final approval, and signed by the officers, before it is forwarded to the General Office. From these returns a trade Directory might be compiled, and issued to the Branches ; a copy might be kept with the vacant book of each Branch for the use of any member who might require it, whilst those who might desire a copy for private use could be supplied at a reasonable price. The returns could be revised and a new edition issued whenever such a course might be deemed necessary.

If this plan were adopted, I believe much time and trouble might be saved which is now needlessly expended, as a member when signing the vacant book might copy on a slip of paper the addresses of any firms he might wish to visit. A member seeking employment in a strange town would be specially benefited by such an arrangement.

The policy of our society, as I understand it, is to endeavour to remove as much as possible of our surplus labour into those districts where trade is brisk, and where it may find profitable employment. With this view we publish a monthly return of the state of trade in each town where a Branch of our Society exists. Would it not also be a step in the right direction if we published a Directory which would furnish valuable information to members on travel, and to many others in want of employment ?

The adoption of this suggestion would involve very little expense, and might easily be carried into effect by the Executive Council, should it meet with the approval of the members. I therefore take the liberty of soliciting the Branches to express their opinions thereon by resolution in the usual way.—Yours fraternally,

JOHN D. PRIOR, Islington Branch.

5 WAKELING TERRACE, BRIDE STREET, N.
January 4*th*, 1868.

Remark, that in the above report there are no leading articles, and no matter but what strictly bears on the union and the interests of its members.[1] In a society conducted upon this plan, we cannot doubt that any course of action decided upon does truly represent the wishes of the members. Yet, when men refuse to work for certain wages, a portion of the Press invariably deplores the unhappy fate of the poor men misled and duped by secretaries and delegates who are supposed to find their account in ruining the societies they serve. Lately, even the leading journals have deplored the blind obstinacy of the shipwrights at the East-end of London, who will not consent to a reduction of wages. We are told that it is intolerable that men who will not work for 6*s*. a day should be supported by the poor-rates and by charity. Only as a matter of fact, we believe that none of the union shipwrights have received anything either from charity or the poor-rates. Other papers say the strike is supported by contributions from distant branches, whose members force the Millwall men to refuse reasonable wages. The Millwall men remark that there is no strike, and that they are living on their savings, and are not supported from the union funds. Probably this assertion would require some qualification before it expressed all the facts; but we believe the Millwall men to have been hitherto quite as much in favour of refusing to submit to any reduction of wages as the other branches or unions. They may be wise or foolish; it may be better for them that few or no ships should be built at Millwall, or it may be a great loss. If, owing to the dearness of provisions and cost of transport, ships are built in the Thames at a disadvantage, it will be better for the whole country, in the long run, that shipbuilding should not be practised there.

[1] The Annual Reports of large societies contain detailed statements of expenditure, receipts, funds, etc. The Engineers' Report for 1867 has 429 pages that of the Carpenters 159 pages.

That is the true free-trade principle. But whatever be thought of these questions, we cannot refuse men the right to decline 20*s*. a day, so long as they support themselves or one another, and do not hinder competition. But 'think of the distress they occasion among the labourers, and other trades who would take lower wages, but who cannot work without shipwrights.' Poor fellows! they do suffer sadly, but to force shipwrights to work at wages they will not voluntarily accept is equivalent to confiscation of property. Vast misery is caused when a capitalist, finding that he can invest his money more profitably elsewhere, closes a mill. We do not compel him to be content with 2 per cent., when he will not invest without the profit of 10. People are amazed when they hear a man declare that he cannot bring up his family if he has less than 7*s*. a day, and point to labourers who support large families on 3*s*. a day. The shipwright may very properly plead that his standard of comfort and education is wholly different from that of the labourer, and that what he means is just what a gentleman means who says he can't marry under five hundred a year. A high standard is very far from an unmixed evil; it is almost unmixed good.

There is much discrepancy between the various estimates of the proportion of men in each trade who have hitherto joined unions. Mr. Mault, for the building trades, puts the number as low as 10 per cent., and tries to convince us that these 10 per cent., being organized, do lead and govern 90 per cent. disorganized; though the latter are backed by the masters and Colonel Maude. Mr. Applegarth thinks about half the men in the building trade belong to the unions, and that in large towns this proportion is far exceeded. Mr. Mault includes, as in the trade, the boys, the labourers, and all the little country workmen, taking his gross numbers from the census; his estimate is, therefore, obviously very incorrect, and we do not think many masters will endorse his estimate from practical experience. According to one estimate 700,000 men are now enrolled in trade-unions. The large societies are increasing very rapidly; most of them increased by about one-fourth during last year. The Engineers' Society, with 33,600 members, an income of 86,885*l*., a reserve fund of 140,000*l*., and 308 branches, stands far ahead of all others, but it increased by only 3,300 members last year. If, as some think, it already includes 90 per cent. of all the men working at the trade, no further very rapid increase is possible. No masters came forward to give evidence against this society. Nor did Mr. Allan, the secretary, complain of the masters. Such disputes as have occurred in this trade of late years seem to have been of a very trifling character. The engineers did not go to Geneva, nor take part in the great trade conference with which Mr. Potter was connected.

The Amalgamated Society of Carpenters and Joiners, numbering 8,261 members, with an income of 10,487*l.*, 8,320*l.*[1] in hand, and 190 branches, is very similar in organization and in general conduct to the Engineers' Society. Both of these unions are benefit or mutual insurance societies as well as trade societies. They have an allowance for the sick, a superannuation allowance, a payment for burial expenses, and they give 100*l.* to any member who is so disabled by an accident that he cannot follow his trade. Most of the unions have some benefits, and partake in some degree of the nature of friendly societies, but the superannuation payment is generally omitted. These benefits are sometimes most unjustly described as mere traps, to entice prudent men into unions. It is far more true that the trade-unions have taught their members to be provident. The benefits, great or small, are so unmixed a good, that the opponents of unions have endeavoured to show that after all they are, as benefit societies, mere swindling concerns, that the subscriptions from members are quite insufficient to provide for the benefits promised, even in the great Engineers' Society, with its 140,000*l.* of capital. These enemies to unions have got an actuary, Mr. Tucker, to come and pronounce the curse of bankruptcy on unions from this point of view; and he has been generally acknowledged a true prophet by writers. Mr. Allan holds up facts in the face of Mr. Tucker's calculations which he does not attempt to reconcile with his deductions, and the facts seem to contradict the figures.

The following table shows the payments which, according to Mr. Tucker, would be required to provide for the engineers' benefits:—

[1] In 1868 the fund is 14,171*l.*

Age of Entry	Monthly Payments			Total
	To provide 10*s.* per Week in Sickness, up to 65	To provide Superannuation Allowance	To provide for a Payment of 12*l.* at Death	
	s. *d.*	*s.* *d.*	*s.* *d.*	*s.* *d.*
25	1 1	2 2	0 4¾	3 7¾
30	1 3	2 10	0 5¼	4 6¼
35	1 3½	3 10	0 6¼	5 7¾
40	1 5	5 5	0 7¾	7 5¾
45	1 7¼	6 10	0 9¾	9 3

Now, as the actual subscriptions of the members amount to only 4*s.* 4*d.* per month, it seems clear that the society, spending about half its income on trade purposes and management, must be bankrupt.

Mr. Allan in reply says, 'We have paid all calls upon us for sixteen years, and our funds in hand increase rapidly. We had ten years' experience in an older society, and may therefore count twenty-six years' experience against your calculations. We have also many sources of income that you do not count.' Mr. Tucker rejoins, saying, 'Your members have increased so rapidly that your soundness has never been put to the test.' Mr. Allan hands in statements showing for each year the payment under each head, and points out that one-third of the members leave before dying or receiving superannuation allowance, and Mr. Glen Finlaison has been called in to advise the Commissioners further.

On looking over the figures it is clear that the statistics on which Mr. Tucker reasoned do not apply to trade societies. Thus, from 1858 to 1866, the amount paid by the Engineers for sick benefits amounted on the average to 8¼*d.* per month per member, and this payment per member was sensibly constant during those nine years, being 8¼*d.* for 1858 and 7¾*d.* in 1866. During the seven preceding years the benefit was a little smaller, and the average per month per member was 6¾*d.*, presenting the same stationary character. These amounts are about half what Mr. Tucker would make a man of thirty pay. The difference is partly accounted for by the fact that the 10*s.* per week is reduced to 5*s.* after twenty-six weeks' illness; but it must be chiefly due to the supervision under which every such member lives— a supervision of great service to the really sick, but fatal to malingering. Mr. Allan in fact here completely refutes the calculations, and the constancy of the payments year by year proves that their small amount does not depend on any exceptional youth on the part of the members. The superannuation allowance, on the other hand, has not reached its maximum among the engineers; it has increased from ¾*d.* per month per member in 1852 (and 0 in 1851) to 3*d.* in 1865 and 1866—still very far from the 2*s.* 2*d.* which Mr. Tucker would relentlessly exact from every subscriber. This enormous discrepancy is due to three causes:—

1. The maximum has not yet been reached.
2. No man has a right to the superannuation allowance at any given age, but must continue to work so long as the society can find employment for him, so that a very large proportion of the men work till they die.
3. One-third of the members fall off before becoming entitled to the allowance.

Mr. Glen Finlaison will in course of time tell us how much all these circumstances ought theoretically to diminish Mr. Tucker's estimate.

The payment per month per member for the burial benefit shows a gradual increase, rising from 1½*d.* to 3*d.* in the sixteen years, but during the last nine years the increase has been very slow, being 2¾*d.* in 1858, and 3*d.* in 1866. Out of every 100 men in the society at a given time, 33 do not die at all, but retire; this ought therefore to diminish Mr. Tucker's estimate by one-third; but these men who never receive the funeral benefit contribute to the fund from which the others are paid, and diminish by so much their contributions. The longer they stop in the society the greater is this action; without any very complex calculation, we see that from this one cause Mr. Tucker's estimate must be diminished by considerably more than one-third, nearly by one-half—in which case the actual payments of the engineers will, even from Mr. Tucker's table, have nearly reached their maximum. If the average age of members, as would appear from this, has reached a constant maximum, in which case 8¼*d.* for sickness, with 3*d.* for superannuation, and

3*d*. for funeral, in all 14*d*. a month, will really be sufficient to provide for benefits which would cost on the actuary's estimate 3*s*. 7¾*d*., even on a preposterously favourable assumption as to the youth of members when they join the society. The question turns chiefly on the superannuation allowance. We wait with curiosity for Mr. Glen Finlaison's report; but even this gentleman can have no statistics as to the number of mechanics who are unable to support themselves in old age by their craft, and how long infirm men live after stopping work. In the absence of data, assumptions are quite worthless. Meanwhile, we venture to remind Mr. Tucker that, for trade purposes, and expenses of management combined, the Engineers' Society does not spend so large a proportion of its receipts as the St. Patrick's and some other friendly societies spend on management alone. Moreover, if the funds do fall short of the calls upon them, trade-societies can call upon their members for extra payments.[1] Sir Daniel Gooch and others suggest that these calls will not be met. No such case has yet arisen, nor in a mutual insurance society do we think it likely to arise, but of course with long practice workmen may emulate the financial morality even of railway directors. Meanwhile, let it be well understood that not a single case of repudiation has been discovered among any of the larger societies. Even assurance companies have met their liabilities with less certainty than trade-unions; between 1844 and 1866, 308 assurance companies have ceased to exist; of these, 59 are winding up in the Court of Chancery. In 1867, the total number of companies was 204, so that the failures form a considerable percentage of the whole number.

Reviewing, as a whole, the conduct of trade-unions, we find that they differ one from another as man differs from man. Among small unions of ignorant uneducated men we find organised villainy of the grossest stamp. In larger unions of better workmen we find narrow views enforced with blind selfishness, but without violence. In the largest unions, formed by the most skilled artisans, we find few objectionable rules, and few disputes between master and man; while the struggles that do occur are carried on with little bitterness and

[1] The ironfounders are now being severely tested, but they have survived many tests during the last fifty-seven years.

absolutely no violence. These last unions comprise benefit societies of great value. In all cases we find an intense attachment of workmen to the union, joined with dislike of those who cannot or will not join the society,—a dislike which in the better trades involves social discomfort, and in Sheffield the risk of assassination. We find the cries about piece-work and over-time to be founded on ignorance; that the indolence complained of arises not from unions, but from the natural slackening which results from increased comfort and diminished risk of want. We find the accusation of levelling unsupported by the evidence of any levelled workman, and wholly denied by the unions. We find that the government of unions does truly represent the wishes of their members, that they do assure those members against want, and that they do increase the wages of the working classes. Let us not reason of an imaginary working man, ground down by the tyranny of a secretary, secretly loathing his oppressor, losing the substance of wages while grasping at the shadow, and using violence to coerce a majority whom he cannot convince, and with whom he secretly agrees. Let us not seek with middle-class complacency to patronize the oppressed being, and deliver him from his thraldom. So doing, we shall seem but wretched hypocrites to workmen blindly unable to comprehend our blindness. No; it is the wearer who knows where the shoe pinches. Masters hate unions; workmen love them. Let those who feel them to be adversaries destroy them if they can; the workmen will fight hard, but in the great trades they have used and will use no foul weapons, and will feel little bitterness to open opponents.

We prophesy no dismal revolution, no war of fustian with broadcloth, no violence of any kind, if the attempt be made to abolish unions; we only expect then shortly to see candidates of the highest respectability on the hustings swallowing unwholesome pledges to support the worst rules trade-unions have yet devised. Now is our opportunity. If we show that we can govern wisely, workmen may consent to be governed. If we act with folly we must soon learn to follow our new masters. Educated Englishmen have hitherto known how to lead, and we therefore dismiss the question whether workmen shall still be permitted to combine, and consider only what remedies shall be applied to the gangrene spots found here and there, and what restrictions are really necessary in the interests of society and in protection of the rights of the minority.

Admitting that total abolition is out of the question as impolitic, undeserved, and impossible, we must insist that the great power granted to the bodies of workmen shall be administered under stringent regulations, clearly defining the rights and duties of the workmen, securing masters against extortion, independent workmen against coercion, and individual members of the unions against fraud or oppression by the majority. The better unions may complain that they have deserved no penal enactments, but laws are made for good and bad alike, the good man differing from the bad, not as living under a different law, but as never incurring its penalties. In treating of the legal action required, we have to consider simply how to prevent these crimes and misdemeanours which some unions have been shown to foster.

First, it will be necessary to give the unions a corporate existence, enabling them to sue and be sued; not but that the better unions are almost indifferent upon this point, finding expulsion an ample remedy against defaulting subscribers, as is the case with clubs; nor yet is a corporate existence necessary to allow the unions securely to possess property—the device of trustees would meet, and has met this want. Giving a legal remedy against debtors will remove that shadow of justification which has been quite falsely pleaded in extenuation of rattening (*i.e.* coercion by theft), in the case of the grinding trades; but removing the excuse will not prevent the crime. We advise legalizing not on the above grounds, but in order that the whole body of workmen may be responsible for their conduct to individual members and to society; in order that any benefits promised may be secured; in order that no unjust expulsion or illegal levy of funds may be enforced by an irresponsible body, and in order that the unions may suffer as a body when they transgress the law. There is so clear an agreement between all parties on this point that arguments in support of legalizing are unnecessary, and we need only discuss the conditions under which a corporate capacity may be granted.

The conditions on which unlimited joint-stock companies

are allowed to exist need not be very widely departed from in the case of unions. The Government ought no more to interfere as to the sufficiency of the payments by members to meet the benefits promised, than they ought to declare publicly whether a given joint-stock company is sound or unsound; but they may properly insist that the liability of the members of the association shall be unlimited, so that no member subscribing on the faith of mutual assurance need be without a legal remedy against the body and the individuals for any sums which may become due to him. The names of all members should be made public, and every change of membership, by death, expulsion, or withdrawal should also be published, with the cause of the change, and a legal appeal against expulsion should be established. The rules of the union should be the articles of association, providing for their own modification, and for the passing of bye-laws within certain limits. The duty of the registrar should be confined to certifying whether the articles of the association contained any illegal provisions, and no society should be permitted to exist except in the form now sketched. We would leave the widest possible scope to legal societies, and would forbid secret societies under heavy penalties. Of course the accounts of these societies should be audited, but we attribute little virtue to the system of audits. We do not see how an auditor, even if he examine a voucher for every payment, is to discover whether or no the voucher be forged; if a dozen men, when a branch wishes to misapply funds, sign receipts, say for payments during sickness, and the secretary duly enters these payments in his book, how can any auditor, however appointed, discover that these men were not sick, but that the funds have been misapplied? In cases of crime paid for out of union funds, a bungle in the accounts might assist detection, but simple misappropriation of funds will not be detected by auditing. Again, suppose it to have been detected, the auditor refuses to pass the accounts. If the union approve the misappropriation they have only to subscribe the amount, recoup the peccant branch, and the accounts must pass. This is no punishment for misappropriation. If 10*l.* are misapplied without detection the union will possess a balance of say 90*l.* If detected they will have to subscribe 10*l.*, but they do not lose this, they simply raise their balance to 100*l*. This form of punishment would be as sensible as though a judge were to condemn a prisoner to pay out of his pocket a fine of 40*s*. to his own bankers. The simple refusal of an auditor to pass accounts will be no punishment, and will not even cause temporary inconvenience, unless the misappropriation has been very large. What can follow a refusal to pass the accounts? In a joint-stock company no dividend can be declared; but are we prepared to say that in a union no benefits shall be paid while the accounts remain uncertified? No; there is no magic either in the word audit or in the thing, and if the auditor is to have any power to enforce correct accounts, he must have the power of inflicting penalties for non-compliance with the regulations. He will be of little use as a protection against the action of unions, but may be useful in protecting the interests of members defrauded by their officers.

What rules shall be legal, what rules shall be illegal? We propose that the union should be treated as a single body, existing for the purpose of contracting for the sale of labour, and that no contract shall be allowed which, by any of its conditions, requires the injury of a third person or body, not a party to the contract. No rules permitting or enforcing such contracts should therefore be legal, and we see no other restriction which has been shown by evidence to be necessary. This principle would render illegal,—strikes against outsiders; against machinery; against any special materials, any given contractor; against the limitation of apprentices. It would leave the union the fullest scope to determine the conditions on which its members would sell their labour, so long as these conditions were within the competence of their employer and of themselves.

Our principle would allow all bargains as to hours of labour, the amount of wages, the time of their payment, the conditions of dismissal, the penalties enforced in workshops against workmen, the acceptance or refusal of piece-work, the establishment of courts of arbitration, and the time during which any given set of rules, forming part of a contract, shall be binding. No special provision is wanted against murder, theft, intimidation, or violence. All these things are illegal. A provision against threats might be found useful, and is suggested in the proposed Act drawn up by the conference of amalgamated trades.[1]

In addition to the above restrictions, we would forbid all sudden strikes; that is to say, we would require that no contract should be terminated suddenly either by masters or men, but that a notice of from three to six months should always be required. By this we do not mean that a master shall not be at liberty to discharge a workman, or a workman to leave his master, with any notice agreed to under the rules; but that when given rules are accepted by masters and men, neither party shall be at liberty to require a change without a notice of from three to six months. The above restrictions should all apply to associations of masters, or to single masters, treated as the purchasers of labour. Thus they would be prevented from stipulating that union men should not work for other masters who might happen to be obnoxious to the leading employers; and the penalty for any illegal agreement should be equally enforced against master and man, whether proposed in the interest of the former or of the latter.

What then shall be these penalties? We answer without hesitation, Fines levied on both parties to the illegal contract, if this has been completed, and levied on the party proposing the illegal contract, if this has not been completed. To fine a single workman is a farce. To imprison him is a hardship, unless he has committed a crime or misdemeanour, for which, by the law as it stands, he would be personally liable. Nor do the unimprisoned 999 suffer very much from the imprisonment of their herald or representative; they feel very angry, subscribe large sums for him and for his relations, but vicarious suffering touches them little. If unions are to be restrained as a body, they must be punished as a body. The fines may be equal for masters and men, and should be heavy enough to be really felt. It will be said that the unions will never take any collective action in wrong-doing, but will use some scapegoat of a man to commit illegal actions, and that thus they will

[1] This Bill aims at protecting the funds of the societies, and freeing them from liability under the law of conspiracy; it contains a provision as to the selection of juries in cases of offences committed by trade-unions which the men had better abandon forthwith.

escape any joint responsibility. The evidence before the Commissioners, except in cases of outrage, does not show this. It is the union which strikes; it is the union which demands unreasonable and improper conditions. Facts will show whether the union has or has not supported a particular demand on the part of a number of its members. There may be some attempt hereafter at equivocation; but if all members of a union are withdrawn from a given shop, the motive of the strike and the attendant facts will not be easily concealed from a jury. The case of outrage and crime committed by one member of a union, in its interest, will always present greater difficulties, just as the detection of a criminal who has committed murder is always more or less difficult. But even in this case we strongly advocate a punishment for the union whenever complicity of the main body with the criminal can be established to the satisfaction of a jury. We might then obtain informers without indemnification as to the whole union; and we should be spared the degradation of discovering great crimes only on condition of allowing them to pass unpunished. Of course occasionally this would lead to the punishment of some innocent persons along with the guilty; but if innocent persons belonging to an association by their supineness allow the commission of crime or folly by their associates they must suffer, and ought to suffer, precisely as the innocent shareholders of a mismanaged company must suffer, and ought to suffer, by the misconduct of secretaries and directors. If they fear this, they need not join these associations at all. These involuntary accomplices should have their remedy against single branches of the society, secretaries, or others who may have involved them without their consent.

Our recommendations are briefly, Turn trade-unions into legal associations, with power to contract for the sale of the labour of their members; declare what contracts are illegal, and punish the association as a mass for any illegal transaction it promotes, threatens to promote, or sanctions; require publicity, and enforce regular accounts; punish individuals for misconduct as individuals, and punish the body for misconduct as a body.

We have said nothing about arbitration—a pet plan with many well-meaning persons. Compulsory arbitration is a contradiction in terms. Voluntary arbitration is an excellent method for settling small points and avoiding quarrels upon matters of sentiment, which are by no means the least serious quarrels; and courts of arbitration or conciliation will come naturally to be established wherever unions and masters are animated by good feeling; indeed, they have been established, and have worked well. As a means of determining wages, or any of the main conditions of a contract, they are quite useless, except within very narrow limits. Mr. Kettle arbitrated as to wages by the simple plan of finding out what wages were given in the neighbourhood—a very good plan, but hardly applicable on a large scale. Arbitration cannot fix the average price of sugar, land, or labour, though it may decide whether the average price of the day has been offered for any small quantity of these commodities. Until bargains in the market and on 'Change can be replaced by arbitration, arbitration will not replace strikes as a means of determining the market value of labour.

A much more mischievous suggestion has clearly taken deep root in the minds of some of the Commissioners—namely, that trade societies should not be allowed to exist as benefit societies. In the interests of the community, no less than those of workmen, we earnestly trust that the impolicy of this proposal may be seen in time. It has been put forward, as though in the interest of the workmen; but the suggestion came from no working man. No man has complained of not receiving the benefit to which he was entitled. No man has complained that to meet such payments to others he has submitted to vexatious exactions exceeding the subscriptions he undertook to pay. The men are thoroughly satisfied with the mutual assurance system which has grown up. Englishmen of the lower classes find much difficulty in setting by sufficient sums out of their earnings to provide against sickness, accidents, or old age, while retaining command of the capital saved. The recklessness and improvidence of the Englishman is too well known; but in the form of subscriptions to benefit societies they do and can save, being unable to withdraw their deposits. These trade and other benefit societies have induced thousands to save thus, who would never save in other ways; the best unions wholly prevent pauperism among their members. These admirable provisions are to be destroyed! and why? Because, forsooth, the accumulation of funds destined to provide these benefits is supposed to be a temptation to extravagance in striking. In other words, the capitalist is supposed to be more ready to peril his position than the spendthrift or needy man. The evidence is wholly against such reasoning. The societies with large benefit funds are the most reasonably managed. If a large fund is accumulated for trade purposes only, it forms an irresistible temptation to strike. How else can it be employed? Masters would at least have the melancholy satisfaction of being able to foretell when a strike was imminent, by simply watching the accumulation of the trade fund. But a subscriber to a benefit society, who sees the fund applied to trade purposes, knows that he must make good every farthing wasted. The fund that goes is his fund; either he will some day share it, or if it goes he must some day replace it, by extra payments. You say the men are too stupid to understand this; but you are wrong. The men do understand it, and even the dullest are taught when, after a strike, whips and levies come week after week to enable the union to meet its liabilities. They will repudiate, you say; they have not repudiated, and little good is to be got by repudiation when the assurance is mutual. To provide against the conceivable case of all the young men of a trade repudiating a debt mainly owed to old ones, the dissolution of a company or withdrawal of members may properly be subjected to some restriction, though it seems hardly worth while to provide for a contingency which is highly improbable. So strongly do we feel on the subject, that we would rather urge that no trade society should be allowed to exist without certain benefits. No better guarantee could be obtained for a prudent administration of the funds. This is no theory, but a fact. Separate trade and benefit societies involve separate expenses of management, separate governing bodies: if restricted to a given trade, the funds will infallibly be improperly used for trade purposes; if they are unrestricted to one trade, the supervision of each member by all the others, allowing benefits to be cheaply purchased, would be sacrificed. You also sacrifice the *esprit de corps* which brings in the thoughtless lad as well as the sober middle-aged man. In a word, let those who advocate the separation say distinctly in whose nterest they desire it. If for the workman, believe that he knows best what he wishes, and wait for complaints before you force your aid upon him. If you desire the separation in order to weaken unions, say so. It may weaken them, but it will force them to be aggressive, and diminish their responsi bility. A precious plan this to avoid quarrels! you give a man money which he can spend in no other way than in fighting and then prevent him from accumulating other property, so that he can lose nothing in the fray! Of all the folly talked about unions, surely this is the most mischievous, supported though it be by men of real benevolence, who prate of widows and orphans as though hundreds such had been defrauded, as has truly been the case in some of the very friendly societies they so strongly advocate in opposition to trade-unions, which have hitherto everywhere met their engagements.

In conclusion, we have only to urge that before men are condemned for practices which at first sight may seem unreasonable and even unjust, care should be taken to understand the practices, and the arguments should be heard which the men have to urge in their favour. When we speak of the men, we speak of the secretaries or others among them who have the gift of speech. Many English workmen, not dull of understanding, cannot explain themselves, and what is more, they will not do so, in answer to avowedly hostile inquiries. The Press, in notices and articles written for the middle classes, and written by men ignorant of workmen, has so very generally misunderstood and misrepresented the action of unions, as to have raised a feeling of angry contempt, preventing even wise and reasonable advice from being listened to. Above all, let us beware of believing that the men are suffering from hardships, of which masters draw a harrowing picture, but of which no artisan complains. Workmen are wedded to the system of unions from no irrational motives, but because they have by their aid obtained great benefits.

The members find great pleasure in the management of their own affairs, and boast of the kindly feeling and enlarged sympathies which co-operation induces, at least within the pale.

The artisan enrolled in one of the great societies may with some truth speak as follows:—'To unions we owe increased wages, diminished labour, freedom from care; in hard times, and in sickness, from want of work and want of bread, the union protects us; neither by accident nor in old age shall we or ours be paupers; we ask no patronage, receive no charity, fear no oppression; we live as free men, owing our welfare to our own providence, and we shall maintain our power by using it with prudence.' There is indeed a sad reverse to this pleasant picture. The best things may be misused, and trade-unions have been misused; but were we to abolish all institutions misapplied, all rights abused, all customs warped from their true aim, what fragment of society could we retain? Let us neither seek to destroy trade-unions, strong as they are for good and evil, nor yet fear with a firm hand to set a legal limit to their power. With good laws and sound teaching these bodies may yet become the pride of our country, affording one more proof of the great faculty Englishmen possess of self-government. Under bad laws, ignorant dislike, and unsound advice, they may indeed turn to a curse, fostering disloyalty and outrage, fatal to trade, and to the well-being of all classes. God grant that we may be wise in time!

THE GRAPHIC REPRESENTATION OF THE LAWS OF SUPPLY AND DEMAND, AND THEIR APPLICATION TO LABOUR.[1]

I.

Graphic Representation of the Laws of Supply and Demand.

Recent discussions on the laws determining the price of commodities seem to show that these laws are neither so well understood nor so clearly expressed in the writings of economists as is sometimes supposed. Men are too much in the habit of speaking of laws of political economy, without attaching to the word law the same rigid meaning which it bears in physical sciences. There are, however, some truths concerning the subjects treated by the economist which do deserve the name of laws, and admit of being stated as accurately, and defined in the same manner, as any mathematical laws affecting quantities of any description.

The following essay is an attempt to state in this rigorous manner some propositions concerning the market price of commodities, using what is known as the graphic method of curves to illustrate the laws and propositions as they arise.

First, I will consider what determines the price of a commodity at a given time in a given market, and it is unfortunately necessary for this purpose to define the sense in which some words will be used.

The *whole supply* of an article will be taken to mean the whole *quantity* of the article for sale there and then. Supply is in this sense mensurable, and can be expressed in tons,

[1] From *Recess Studies*. Edited by Sir Alexander Grant. Edinburgh, 1870.

quarters, etc. *Supply at a price* denotes the *quantity* which at a given price holders would be, then and there, willing to sell. Supply at a price is also mensurable. *Demand at a price* denotes the *quantity* which, then and there, buyers would purchase at that price.

The *supply at a price* and the *demand at a price* in any given market will probably vary with the price; they may be said to be functions of the price.

Let a curve be drawn, the abscissæ of which represent prices, and the ordinates the supplies at each price. This curve will be called the *supply curve.* A similar curve, constructed with the *demand at* each price as ordinates, will be called the *demand curve.*

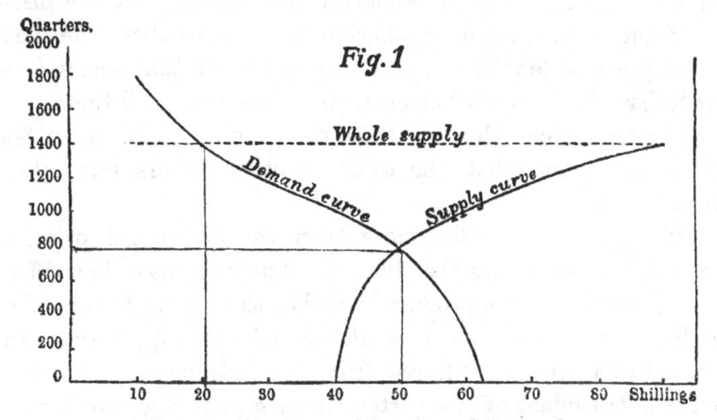

FIG. 1.—First Law of Supply and Demand.

Whole supply for sale at any price, 1,400 quarters of wheat.
Price at which whole supply would be sold, 90*s*.
Price at which whole supply would be bought, 20*s*.
Market price, 50*s*.
Price below which no sale could take place, 40*s*.
Price above which no sale could take place, 62*s*.
Quantity which will be sold, 800 quarters.

Fig. 1 shows a pair of imaginary demand and supply curves for corn. At 60*s*. per quarter the supply is 1,000 quarters; at 55*s*. the supply is 900 quarters; at 50*s*. only 800 quarters; at 45*s*. only 600 quarters; at 40*s*. the supply is *nil.*

At 62*s*. the demand is *nil*; at 55*s*. only 600 quarters; at 50*s*. it is 800 quarters; at 45*s*. 900 quarters; at 40*s*. 1,000 quarters; and continues to rise fast at lower prices.

The first law of demand and supply may now be stated as follows:—

PROP. 1. *In a given market, at a given time, the market price of the commodity will be that at which the supply and demand curves cut.*

This price is the price at which the supply and demand are equal.

Corollary A.—At this price a greater quantity of the commodity will change hands than at any other price.

In practice, the supply and demand curves are unknown, but the price at which supply and demand are equal is ascertained approximately by competition in this wise :—

If some sellers offer their wares below the theoretical price, so many buyers present themselves, that the other sellers, not being afraid of having goods left on their hands, raise the price.

If some sellers begin by offering their wares above the theoretical price, so few buyers present themselves, that others, fearing to have their goods left on their hands, undersell them.

If some buyers offer less than the theoretical price, so few sellers are found that the same or other buyers raise their offers.

If some buyers offer more than the theoretical price, so many sellers are found that the other buyers reduce their offers.

The same effect is produced whether the buyers bid or sellers proffer first. The excess or defect of the supply over the demand at a price is inferred from the briskness of the sales, and is independent of the particular form in which transactions may take place.

Thus the market price, as determined by the first law of supply and demand, is ascertained by competition.

The law thus stated assumes that each man knows his own mind, that is to say, how much of his commodity he will then and there sell or buy at each price, and that the condition of his mind shall not vary.

If this be granted, then the market price will not be changed by the sales. The base line, as each quarter of wheat is sold, will then in our figures gradually ascend, until, when 800 quarters have been sold, it will have reached the position shown in Fig. 2, when no further sales will take place, but the market price will, as before, be 50*s*., any increase finding sellers, any decrease buyers.

If every man were openly to write down beforehand exactly what he would sell or buy at each price, the market price might be computed immediately, and the transactions be then and there closed.

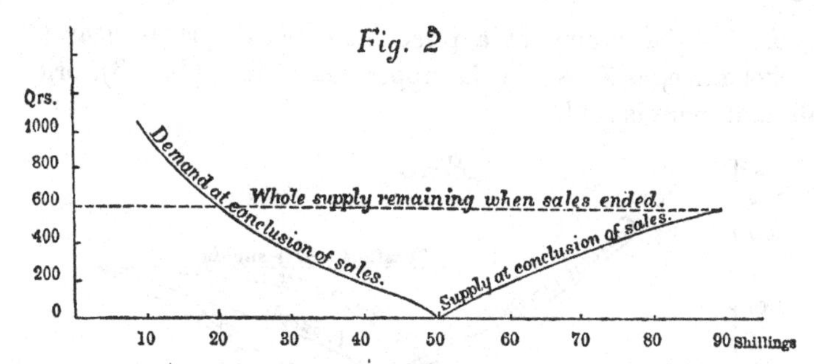

FIG. 2.—Demand and Supply Curves after completion of sales indicated in Fig. 1, without change of market price.

Whole supply at any price, 600 quarters.
Price at which whole supply would be sold, 90*s*.
Price at which whole supply would be bought, 20*s*.
Market price, 50*s*.
Price below which no sale would take place, 50*s*.
Price above which no sale can take place, 50*s*.
Quantity which will be sold—*nil*.

The actual price at which each quarter is sold will be a mere tentative approximation to the theoretical price. Bargainers all day long will be watching the market, to ascertain whether a given price is above or below that at which the quantity to be bought and sold will be equal.

But, in practice, men's minds do not remain constant for five minutes together, and we have found that only one point in the supply curve, viz. the highest, is independent of men's minds. The supply curve can never rise above the *whole supply*,[1] but under this height it may vary almost indefinitely.

Similarly, the demand curve has a limit, which in this case is a limit at each price. The funds available for purchase at any price are limited, and this, which may be called the purchase fund, at each price limits the possible demand; but below this continuous limit the demand curve may in its turn vary indefinitely; and thus, while the whole supply in the market may be constant, and the funds available to support demand may be

[1] Except in those cases where men sell what they have not got, trusting that before the time for delivery they may be able to secure a supply.

constant, yet the market price of the commodity may vary immensely, as men's minds vary.

Assuming the two limits to be fixed and constant, the following additional corollaries flow from the first law :—

B.—If the supply at a price increases at prices near the market price, as shown in the upper dotted line (Fig. 3), prices fall, and more is sold.

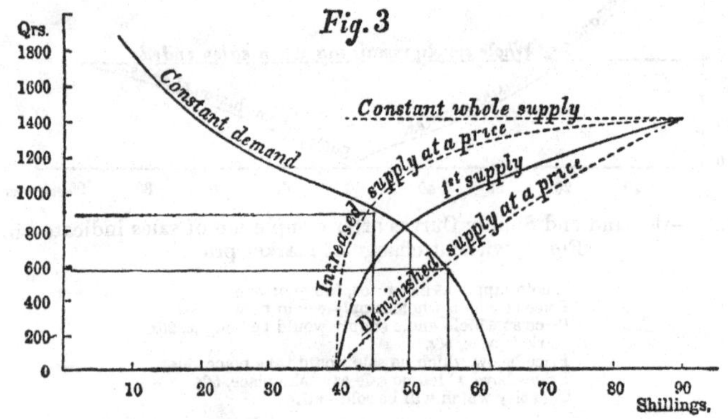

FIG. 3.—Showing Changes in the Supply Curve.

If the supply curve rise to the upper dotted line the market price will fall to 45*s.*, and 900 quarters of wheat will change hands, instead of 800 as in Fig. 1.
If the supply curve fall as shown by the lower dotted line the market price will rise to 55*s.*, but only 600 quarters will be sold.
The whole supply, the price at which all would be sold or none sold, all bought or none bought, may all remain unaltered, as well as the demand curve. In practice some or all of these elements would generally vary when the supply curve varies.

C.—If the demand at a price increases at prices near the market price, as shown in the upper dotted line (Fig. 4), prices rise, and more is sold.

D.—If the supply at a price decreases at prices near the market price, as shown in the lower dotted line (Fig. 3), prices rise, and less is sold.

E.—If the demand at a price decreases at prices near the market price, as shown in the lower dotted curve (Fig. 4), prices fall, and less is sold.

F.—It is possible that both the demand at a price and supply at a price may increase simultaneously, so that the price shall be unaltered, while more of the commodity is bought and sold, or less of the commodity may change hands with an unaltered price demand and supply decreasing simultaneously.

These effects, and combinations of these effects, may occur, while the whole supply and purchase funds are both constant, nevertheless at each moment the first law of demand and supply holds good.

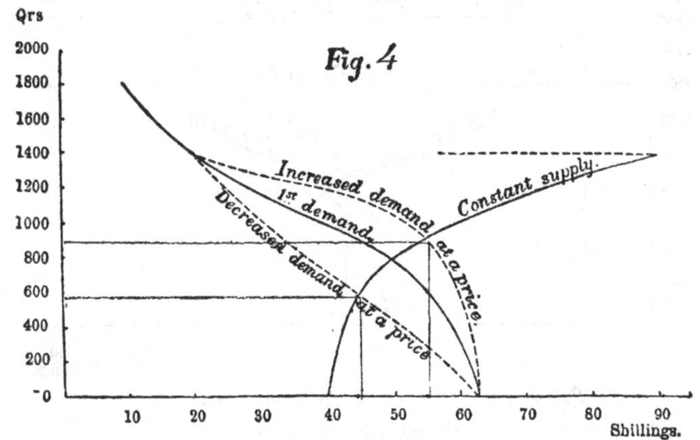

FIG. 4.—Showing Changes in the Demand Curve.

If the demand curve rise, as in the upper dotted line, the market price will be 55*s*., and the quantity sold be 900 quarters.
If the demand curve fall, as in the lower dotted line, the price will fall to 45*s*., and the quantity sold to 600 quarters.
As in Fig. 3, the whole supply, the price at which all would be sold or none sold, all bought or none bought, is left unaltered, as well as the supply curve.
In practice some or all of these elements would generally vary when the demand curve varies.

Let us next consider what will happen if the whole supply or whole purchase fund vary.

PROP. 2.—*If the whole supply be increased, it will most frequently, but not always, happen that the supply at a price will, throughout the whole scale, be increased; prices will then fall, as shown in Fig.* 5.

If the purchase fund be increased it will often happen that the demand at a price will rise throughout the whole scale; prices will then rise, as shown in Fig. 6.

These two statements, which, for convenience, may be spoken of as the second law of demand and supply, are frequently confounded with the first law, from which they are wholly distinct. They are not laws in a strict sense, but express a degree of probability, varying immensely in different cases, and with different materials. I will now apply these laws to some of the current controversies.

It has lately been suggested that the manner of conducting bargains will alter the market price of the commodity; that, in fact, open market, an auction, and a Dutch auction, might

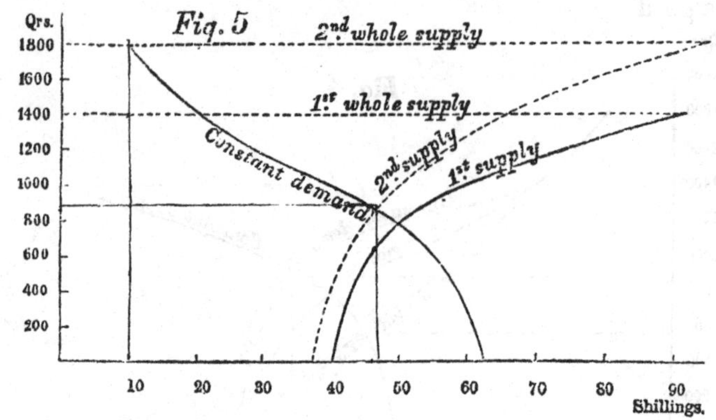

FIG. 5.—Second Law of Supply and Demand.

Fig. 5 shows probable effect of increase in the whole supply.
The dotted line on right shows probable effect of increasing the whole supply to 1,800 quarters.
The price at which whole supply would be sold rises to 95*s.*
The price at which the whole supply would be bought falls to 10*s.*
The market price falls to 47*s.*
The price below which no sale could take place falls to 38*s.*
The price above which no sale could take place may remain unaltered.
The quantity which will be sold rises to 870 quarters.

produce three different market prices. The peculiarity of an auction is that the supply is constant at all prices, down to a

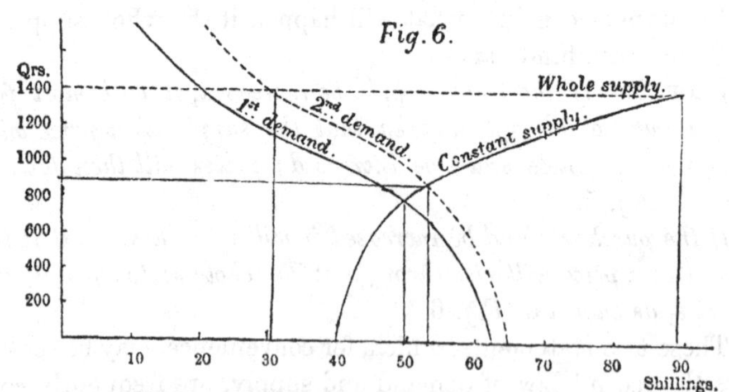

FIG. 6.—Second Law of Supply and Demand.

Fig. 6 shows probable effect of an increase in the purchase fund.
The dotted demand line shows the probable effect of increasing the purchase fund.
The price at which the whole supply would be sold may be unaltered.
The price at which the whole supply would be bought rises to 30*s.*
The market price rises to 53*s.*
The price below which no sale would be effected may remain unaltered.
The price above which no sale would take place rises to 65*s.*
The quantity which will be sold rises to 850.

limit which is the reserved price, and which may be absolutely *nil.* Then the supply at each price is equal at all prices to the whole supply, and it is the competition among buyers alone which

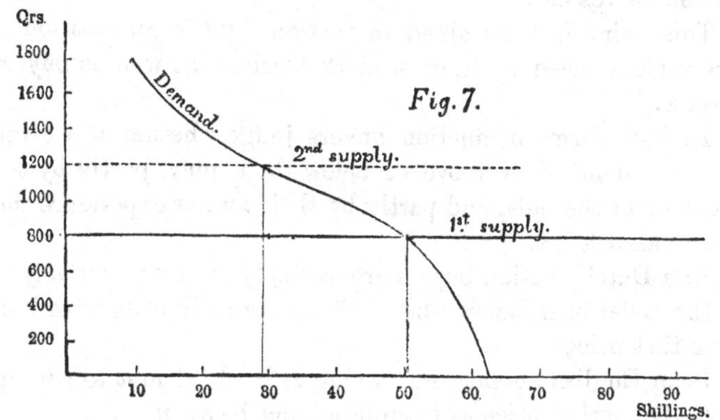

FIG. 7.—Showing effect of selling 800 quarters unreservedly by auction, the Demand Curve being constant, as in Fig. 1.

The supply curve becomes a straight horizontal line.
The market price will be 50*s*.
If 1,200 quarters were to be sold unreservedly, the market price would be 30*s*.
If only 200 quarters, the market price would be 61*s*.

keeps the price above zero. This state of things is shown in Fig. 7, where the supply curve has become a straight line parallel to the base line. Fig. 8 represents the supply and demand curves at an auction, with a reserved price at 25*s*.

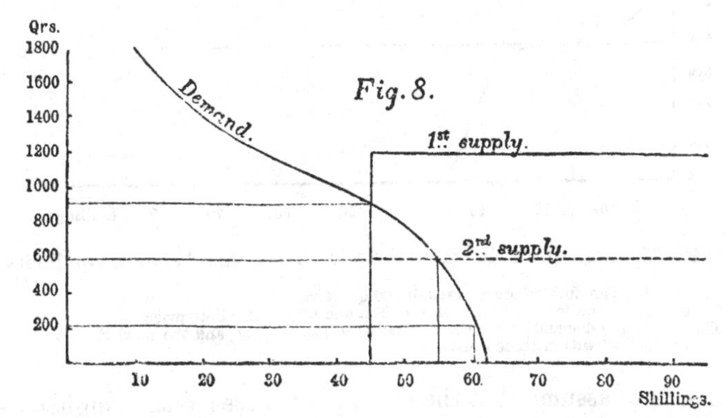

FIG. 8.—Shows the effect of a sale by auction, with a reserved price of 45*s*.

If 1,200 quarters are offered for sale, only 900 will be sold, and the price will be the minimum of 45*s*.
If 600 quarters are offered for sale, they will all be sold, and the price be 55*s*.

Except in the shape of the supply curve, which is here always known, these figures do not differ from the typical Fig. 1. And the market price here, as before, is that at which the supply and demand curves cut.

This point is ascertained in auctions by the competition of buyers alone, whereas in open market sellers as well as buyers compete.

In both forms of auction buyers judge whether at a given price the demand is above or below the supply, partly by the quickness of the bids, and partly by their former experience and general knowledge.

In a Dutch auction buyers are as likely at first tentatively to let the seller offer below the market price as to close with him above that price.

In an English auction, buyers are as likely at first to run up above the market price as to stop bidding below it.

It is only by experience of former markets, and a considerable number of tentative transactions, that the theoretical price is approached.

The device by which Mr. Thornton has made it appear that in a Dutch and English auction there might be two market

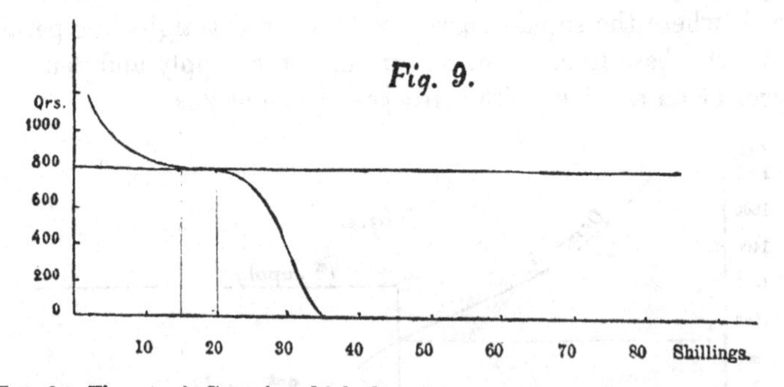

Fig. 9.—Thornton's Case, in which the price is indefinite between two limits.

The supply at an auction, unreserved, is 800 quarters.
There is a demand for exactly 800 at 15*s*., 20*s*., and all intermediate prices.
The supply and demand curves coincide between 15*s*. and 20*s*., and the market price is not defined within those limits.

prices, is to assume that the demand at prices in the neighbourhood of the market price is constant at all prices; that the same number, and no more, fish would be bought at 18*s*. as at 20*s*. In this case the demand curve becomes horizontal near the market price; and as the supply curve is also horizontal, the market price is indeterminate. This case is not peculiar to any form of bargain, but represents an unusual state of mind. It is shown in Fig. 9.

Where only a small number of transactions take place, there can, in the above sense, be no theoretical market price; thus, with one buyer and seller of one thing, the demand and supply curve become two straight lines, ending abruptly, as in Fig. 10.

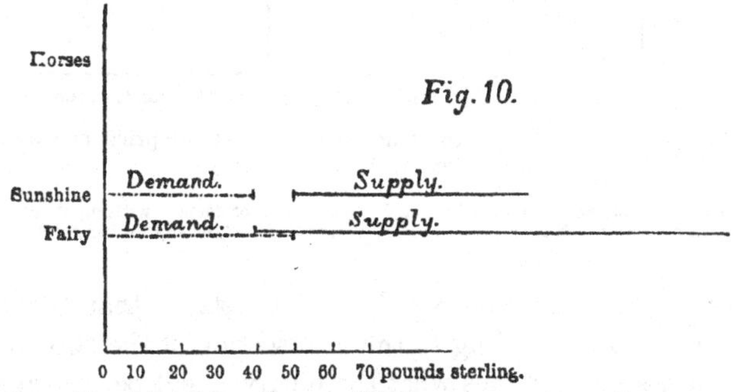

FIG. 10.—Sales of one article by one man to another.

A owns the horse Sunshine; he will sell it at any price above 50*l*. *B* will buy that horse at any price below 40*l*.
The demand and supply curves are two straight lines which do not overlap.
C owns the mare Fairy, which he will sell at any price above 40*l*. *D* will buy the mare Fairy at any price below 50*l*.
The demand and supply lines overlap, and the price is indeterminate, depending on the relative skill of the two persons in bargaining.

If the supply line overlaps the demand line, the sale will take place, and not otherwise; but the price is indeterminate. This is true whether the sale be by private bargain, by Dutch or by English auction.

Fig. 11 shows Mr. Thornton's case of three horses, or any less number, for sale at 50*l*. each, and one purchaser at 50*l*. The curves indicate plainly enough the fact that the sale of one horse is possible. The quantity sold is not equal to the quantity bought, but it is more nearly equal than at any other price.

We are now able to see precisely in what sense and in what cases the first law of demand and supply may be said to fix prices in a market. This law only selects one among many possible prices already determined in the minds of sellers and buyers; it does not in any case determine, or even change, the price at which any one seller chooses to sell or any one buyer

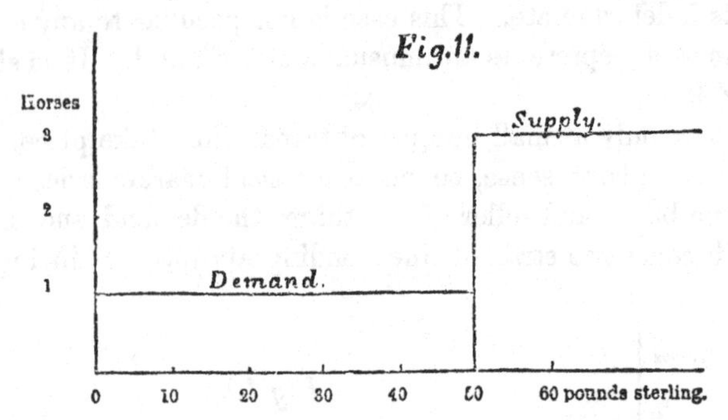

FIG. 11.—Thornton's Case of three horses for sale at one price, one wanted at same price.

Price determinate and also quantity sold, by intersection of demand with supply curve, which indicates 3, *or any smaller number*, for sale at 50*l.* or any higher price.

chooses to buy. In other words, the law states that the price will be that corresponding to the intersection of the two curves, but in no way determines what those curves will be. Moreover, the law only comes into operation where buyers and sellers can approximately estimate whether, at a given price, the quantity wanted or the quantity for sale is the greater. Competition in open market, or at any form of auction, is one mode in which this estimate may be formed.

The first and second laws of demand and supply cannot affect those cases where the buyers and sellers have no means of estimating the relation between demand at a price and supply at a price.

Examples of this occur—

(1) In simple transactions between man and man for one object.

(2) When sealed tenders are sent in for the supply of a given article.

(3) When shares in a new company are applied for before any market is found for the shares.

In the first case, neither demand nor supply curve can be drawn: the price is that at which the buyer can persuade the seller to part with the article.

In the second case, the supply curve can be drawn, but the demand at a price does not exist, and the demand curve cannot be drawn. The buyer waits till the tenders are opened, choosing the lowest.

In the third case, the supply is fixed, being the whole stock of the company: the supply curve becomes a simple point. The demand is unknown until the applications for shares are opened —being all at one price, the demand curve then becomes a point either above or below a supply point. If above, the shares are allotted, but their price to the applicants is unaffected by the excess of the demand over the supply.

As soon as allotment takes place the market price can be determined; and in accordance with Law II., if the whole demand at one price has been in excess of the supply at that price, the demand curve at other points in the neighbourhood of that at which the shares were allotted will probably be above the supply curve, and the shares will reach a premium. Rigging the market consists in the artificial production of false supply and demand curves, with the object of deceiving the public as to the market price.

Even where the first and second laws apply, the price selected by the first law, and the change of price resulting from the second law, depend on the state of mind of the buyers and sellers simply, and not on any material quantity, or on any law hitherto stated. The demand curve and supply curve indicate certain resolutions on the part of buyers and sellers; and these curves are therefore variable within certain limits, and do vary with every cause which can affect the desire of men for the article in question. The limits are set by the purchase fund to the demand curve, by the whole supply to the supply curve; but at what price the limit for the supply curve, and at what quantity and price the limit for the demand curve, shall be touched, is wholly within the competence of the buyers and sellers.

The first and second laws of demand and supply therefore afford little help, or no help, in determining what the price of any object will be in the long run. If the supply be limited, as in the case of pictures by an old master, the desire for the article may rise or fall to any extent both on the part of buyers and sellers. The knowledge that only 200 pictures of a given artist exist helps us in no conceivable way to determine what their value will be, nor whether that value will rise or fall. In a given market, on a given day, the elements of the demand and supply curves already exist in the minds of purchasers and sellers, and both laws will apply, but neither law in the least helps to determine the absolute height of either the supply or the demand curve at any one time.

In this case of a limited supply we are thrown back on the truism that the price of the article depends on the desire felt for that article relatively to others; but it is worth while to observe that the price is not fixed by the buyer, but by the holder. No man, by saying 'I will not give more than 100 guineas for that picture,' can make the price 100 guineas. The price at any moment of a similar article cannot be lower than the lowest price which any holder will accept.

The popular adage that the worth of an article is what it will fetch, supposes the holder forced to sell unreservedly; competition of buyers then fixes the price by the first law of demand and supply.

Leaving on one side the case of a limited supply as unapproachable, let us consider the case of an article, the manufacture of which continues, and of which the quantity made depends ultimately on the price obtainable. It is not necessary to suppose that the possible quantity produced should be unlimited, but only that as the price obtainable increases, the supply manufactured can be increased until the supply at a price equals the demand at a price.

In this category fall most manufactured goods. The average demand curve may vary to an indeterminate extent, but the average supply curve will be found in the long run to depend simply on the cost of production: inasmuch as manufacturers rarely desire the article they make for itself, but make it only with the object of selling it.

If in a given market, or series of markets, they find no demand, or an insufficient demand, for their produce, at cost price (including what they think a fair profit), they will cease to produce, or produce only as much as the demand at that price requires. While, if the demand equals or exceeds the supply at a higher price than cost price as above defined, makers will be tempted to produce more, until by the action of the second law the demand and supply at cost price become equal. From these considerations may be deduced the third law of demand and supply, which really does enable us to estimate the probable price of any article, as well as the probable quantity manufactured, and which may be stated as follows :—

PROP. 3.—*In the long run, the price of the manufactured article is chiefly determined by the cost of its production, and the quantity manufactured is chiefly determined by the demand at that price.*

The average height of the supply curve, over a number of years, depends on the cost of production alone; but at any moment its position above or below that average depends on the estimate formed by producers whether the actual demand curve is above or below the average, and whether the future demand curve is likely to rise above or fall below the actual demand curve. This uncertain estimate varying day by day according to transactions in the market and the dispositions of holders, determines the departures of the supply curve from the average fixed by cost of production. These departures alone (which may be considerable) are the subject-matter of the first and second laws of demand and supply. It is to be observed that this case has one point in common with that where the quantity of an article is limited—viz. the price in the long run is fixed by the holders; the number of transactions or quantity sold, by the buyers. For when we say the average price depends on the cost of production, including a profit to producers, these producers are free to choose what profit they will be content with as a body. This profit may vary, and does vary immensely, in different countries, and at different times. Competition at any one time prevents wide divergence from the average rate of profit expected by manufacturers: but what profit is sufficient to induce a man to produce is none the less a mere matter of opinion. Still, over a long range of years the general opinion on this point in a given country can be ascertained, and thus, by the third law, then and there the probable price of an article can be calculated.

The graphic representation of the third law is very similar

to that of the first. The ordinates of the demand curve represent the average quantity wanted, say in a year, at the several prices. This curve can only be approximately estimated beforehand from past experience, and is subject to no one general law for all materials. For each material experience may, however, show several points of great importance in the curve; as, for instance, whether the curve be nearly horizontal—the total demand being little affected by price—or whether it is sharply inclined, showing that the demand increases rapidly as price is lowered, and *vice versâ*. Statistics collected over several years might also show whether the general character of the curve was convex or concave to the base, and at what rate approximately the average height of the curve increased year by year. Thus, for many materials, approximately accurate average yearly demand curves might be determined by the collection of statistics.

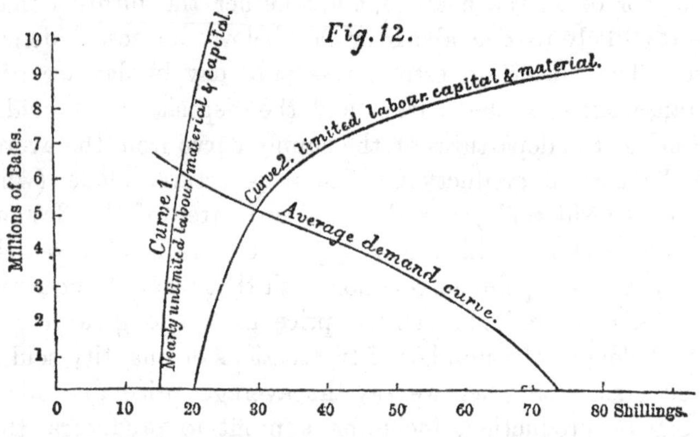

FIG. 12.—Third Law of Supply and Demand.

The ordinates of the demand curve are the average yearly quantities of the material required at each price.
The ordinates of the supply curve are the average quantities which at each price would be manufactured if there were sufficient demand to absorb that quantity in the year.
The price corresponding to each point in the supply curve is the cost of production of the article in that quantity, including in the words cost of production, sufficient profit to labour and capital to induce the production of that quantity.
In other words, the price is the lowest price at which that given quantity will be produced.

The ordinates of the average supply curve represent the quantity which will be produced in a year at each price. This curve will also vary for all materials. In some it will have the character of curve 1 in Fig. 12—terminating almost abruptly at a given price, and shooting very rapidly upwards, so as to be nearly vertical; this characterises the case of articles which, at a given price, can be produced in almost unlimited quantities, there being no material limit to their production in amounts immensely superior to the demand. Toys, for instance, are articles of this kind, requiring little capital, moderate skill, and common raw materials.

Most supply curves will have the character of curve 2 in Fig. 12. In these the cost of production will gradually increase with the quantity produced, owing to the limitation of labour, of capital, and of raw material.

The cost of production, as it must be understood in the third law of demand and supply, is obviously no one fixed cost constant for all quantities, and the supply curve determined by the cost of production may vary just as much as the supply curve in a

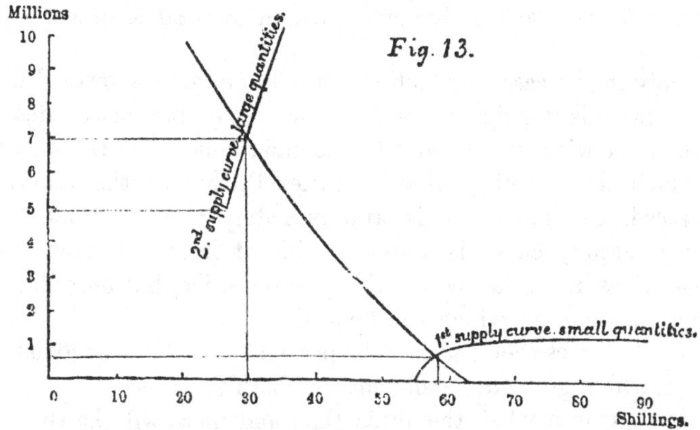

FIG. 13.—Article dearer to produce in small quantities than in large.

Articles produced at the rate of less than 1½ millions per annum, cost so much to produce that the quantities indicated by first curve only are produced at each price. Sale 750,000, price 58*s*.

If made in wholesale quantities exceeding 5 millions, the cost of production is shown by second curve. Seven millions may then be sold at 30*s*.

The demand curve might have fallen below 5 millions at 27*s*., when the price will be determined by the first supply curve only; or, while above 5 millions at 27*s*., the demand curve might fall short of the price of 55*s*., when the lower price would be alone possible.

given market at a given time. The cost of production does not mean the cost at which an article can be produced if unlimited numbers of labourers could be found who would be content with given wages, and if capitalists would apply unlimited capital at some fixed rate of profit. The cost of production varies with the quantity to be produced, because as more is wanted a higher price is generally required to tempt more capital and more labour into the given walk. The ordinates of the supply curve, although therefore truly representing cost of production, ultimately depend greatly on men's will or choice, and may vary and do vary immensely with variations in the quantity required.

Occasionally a commodity may be dearer to produce in small quantities than in large, and its supply curve may have two branches with two possible prices, as in Fig. 13.

The form and character of the average supply curve for many materials might be determined by statistics with fair approximation.

The intersection of the average supply and average demand curve fixes the probable average price.

Average demand curves will in general approach horizontal lines, and average supply curves will in general approach vertical lines.

Only in the case in which the supply curve is a vertical line at one price is it strictly true that the cost of production determines price without reference to demand; but for all those cases in which the cost of production varies little with the quantity produced, the statement is approximately true. When the average supply curve is a vertical line, it is strictly true that demand has no influence on the average price, but only determines the quantity which will be sold.

These curves then regulate the price of manufactured articles, among which gold and labour may be both included.

In a given market, the price then and there will be that at which the quantity wanted and quantity for sale are equal.

Increase in the whole quantity for sale will generally increase the quantity for sale at all prices, and so lower prices, and *vice versâ*; while increase in the funds applicable to purchase will generally lead to an increase in the demand at all prices, and so will raise prices, and *vice versâ*.

In the long run there is no law by which the probable price of an article limited in quantity can be ascertained.

In the long run the price of any article capable of being manufactured depends mainly on the cost of production; the quantity produced depends mainly on demand, which itself is subject to no law.

The cost of production is fixed by no immutable law, but depends on the profit (including in this word all forms of advantage) expected by the producer (including in this word the labourer as well as the capitalist).

Enough has now been said to show how much the value of all things depends on simple mental phenomena, and not on laws having mere quantity of materials for their subject.

The application of these laws to wages requires much care. The difficulty here will be found in determining the meaning of cost of production as applied to human beings, and in ascertaining how far the production of human beings is determined by the same or similar motives to those which regulate the supply of other produce. The first and second laws of demand and supply find their application in limited districts and for short periods, but these are of infinitely small importance compared with the third law, in which we must seek the solution of that great problem: On what does the general level of wages depend? The laws of price are as immutable as the laws of mechanics, but to assume that the rate of wages is not under man's control would be as absurd as to suppose that men cannot improve the construction of machinery. A knowledge of the laws of mechanics allows us to improve machines; a knowledge of the laws of wages may equally lead to improvement in the lot of men.

II.

Application of the Laws of Demand and Supply to the Special Problem of Wages.

The rate of wages paid to a certain class of workmen is simply the market price of their labour, and this price will be determined by the same laws as determine the price of other commodities. This truism is often interpreted as though it meant that no efforts on the part of the masters could lower wages, and no action on the part of men raise wages, the price of labour being, as is supposed, fixed by some natural law depending for its action in no wise on the will of men, but solely on the number of labourers and the amount of capital which is ready to be spent on labour. The natural deduction from this hypothesis is, that whatever wages happen to be paid are the right wages, and that men should thankfully accept them, abandoning all attempt to bargain as a useless waste of time and trouble in contending against the inevitable law of nature.

What is true of wages is true of other commodities, and it would be just as difficult to persuade the holders of any merchandise to desist from bargaining about the price of goods, as to persuade labourers to refrain from bargaining about wages. Nevertheless, it is certainly true that by competition among buyers, a market-price would be settled both for wages and other commodities, as at an auction with unreserved price even if the sellers did not bargain at all; and this is precisely what the more reasonable opponents of trade-unions mean when they assert that wages are fixed by the laws of supply and demand. It is unnecessary now to refute the other absurd form of the doctrine by which it was assumed that, the wages fund being fixed, and the number of labourers fixed, you had only to compute the mean wages by a simple sum in division—it being clearly understood that the wages fund is a fluctuating quantity, altered by every circumstance which affects the minds of capitalists. In an article in the *North British Review*,[1] the writer dwelt at much length on this fallacy, and nearly the same arguments, subsequently used by Mr. Thornton, have been acknowledged as just by Mr. John Stuart Mill. The theory of wages, therefore, presents two questions only:—

1. Does the power of bargaining give the labourer any advantage?

2. What is the cost of production of a labourer in a given trade?

If we can settle these two questions we may feel certain of being able to answer all contingent questions, for we have found that the cost of production ultimately regulates the average of all prices, and therefore the average rate of wages; and in a given market we have found how the fluctuations in the average

[1] *North British Review*, March 1858. 'Trade-Unions: how far legitimate.'

price are determined. The one question which has not been touched being the effect on the market of the power of bargaining on the part of sellers, leaving the competition to buyers only.

I will first discuss this special case, which is that of the labourer who does not bargain as to his wages; it is also the case of a forced sale, as at a bankruptcy, and of any other sale by auction without a reserved price. The curves in Fig. 7 represent this case. The power of bargaining, or, in other words, of reserving some of the goods for sale, may lower the supply curve below the straight line, and thus raise the price, diminishing the quantity bought; but it must be remembered that the demand curve is no fixed line depending on some numerical quantity or material weight, but represents a certain mental state. Now, it is observed that the knowledge that goods must be sold, that in fact there is no reserved price, does tend to diminish the desire for these goods quite independently of the knowledge of the mere material quantity in the market. The knowledge that the bankruptcy of some firm will necessitate the immediate sale of a stock, at once lowers the demand curve while it raises the supply, and by a double action lowers the price. This is often described as the effect of a sudden increase of supply; so it is, but the words, *at a price*, should be added, and whenever we take away the power of bargaining, or of fixing a reserved price, we virtually increase the supply at a price, putting the sellers into the position of bankrupts whose effects must be realised.

Now the legitimate action of trade-unions is to enable the labourer to set a reserved price on his merchandise, precisely as any other dealer in goods can, and to deprive him of this power is to put him to the same disadvantage as the bankrupt salesman. Each labourer who has a reserve fund may bargain for himself, and concerted action is not theoretically necessary to allow of bargaining; practically the individual labourer seldom bargains, but, acting in concert with others, he can and does set a reserved price on his labour, and gets precisely the advantage that any other salesman would. Probably the quantity of labour or any other material bought under these circumstances will be less than when no reserved price is fixed; but this need not be so, and even if it be, it may be more advantageous to the salesman to sell a moderate quantity at a good price than a large quantity at a bad price. It may be more advantageous to labourers that a moderate number should be employed at good wages than a large number at bad wages.

It is said that when men have this power they abuse it, and will set so high a reserved price on their labour, as ultimitely to stop the purchase of it altogether, or at least diminish it so far as to injure both themselves and capitalists. Self-interest is the natural remedy for this. We do not object to fishmongers setting a reserved price on fish, although they may so diminish the consumption as materially to injure both themselves and the community. The only motive by which the demands of a fishmonger are kept within reason is that of self-interest, and precisely the same motive may be trusted to in the case of a combination of labourers, unless they are so foolish as to be unable to perceive the consequences of their own actions —a case which is quite possible, though rarer than is imagined.[1]

Thus we find that the power of bargaining given to the labourer does tend to raise wages; but that it may diminish the number of labourers employed, and often does so.

It will be very rationally asked: How is this statement consistent with the doctrine that price, and consequently wages, depend ultimately on the cost of production? The explanation of the apparent inconsistency is to be found in the fact that the cost of production is itself not a fixed material thing, but depends on men's desires and will in a great measure. The cost of production is the cost at which a man thinks it worth his while to produce the article; this depends on his desires and opinions, and is affected by a vast number of circumstances. If he has a reserve fund, so that he need not perish should he cease to produce, the immediate motive to produce is very much weakened, and he will not think it worth his while to produce except at a higher profit to himself; and when it is said that the cost of production regulates the price, in that cost must be included the profit to the producer. Indeed, if we analyse the cost of production, it will be found to consist solely of the sum of a

[1] In the ordinary method of purchasing at shops we have an example of buyers trusting to the competition of sellers only, and paying whatever is asked, within a limit.

succession of profits reaped by the producers of food, of raw materials, and of tools—each of these profits depending solely on the desires or fancied necessities of each class of labourers. Thus the power of bargaining actually tends to increase the cost of production, and thus the apparent paradox is justified.

While, however, we find that, both in a given market and on an average of years, the power of bargaining will enable a seller to obtain higher prices, this does not assist us in determining what those prices shall be. The first law of demand and supply gives no help, since it requires for its operation that men's minds shall be already determined; the second law, when applied to labour, brings us to the Malthusian doctrine, and when misunderstood, to the erroneous wages-fund theory. It is true that if the number of labourers be increased, without a corresponding increase in the demand at a price, probably wages will fall; it is also true that if the maximum of available capital be not increased, probably the demand at a price will not increase, and therefore it is true that if the population of a country increases faster than at a certain rate, determined by the increase of wealth, probably wages will fall. But all these truths help us little. Certainly, we cannot *à priori* determine the fit proportion between land and population. Many countries are poor, and afford bad wages, when thinly populated or semi-savage; and are rich, giving good wages, when thickly populated. Nor can we fix a ratio between increase of population and increase of wealth. If the population be a productive one, the more it increases the better, until the limit to food production in the world be reached. Only the labourer who, by want of education or natural gifts, is able to produce less than he requires, impoverishes a country; and I see no reason to suppose that as yet any man, born in the most thickly-populated country, need be in this condition. It is often said that to produce is useless if there be no demand, but it is no real production of wealth if the thing made be not wealth, that is, if it is not wanted; the producer, falsely so called, is producing a wrong thing; demand is limited by production as much as production by demand. The demander can only demand in virtue of his possessing or having produced what others want, and hence if all were truly skilled, demand and the power of production would keep pace. Demand falls short of production from two reasons: the thing produced is a thing not wanted, or those who want it can produce nothing in return. If labour were wisely applied, the only limit to the increase of a well-to-do population is the difficulty of producing food. To say that the wealth of a country depends on its powers of production, is as true as to say that its powers of profitable production depend on its wealth; no doubt, the two things are dependent on one another, but there is no *fixed* ratio between them, such that knowing one you may calculate the other. Capital is produced by labour; the relative amounts of these two elements which can be employed in production *at a given moment* is approximately fixed, but fixed only by the customs and desires of that particular country at that particular time, when capital expects such and such profit, and labourers expect such and such comforts. Change these feelings, and the fixed ratio between capital and labour vanishes, so that the Malthusian law, true as it is, gives no help in determining the rate of wages, because it gives no help in determining what profit a capitalist will expect or what comfort a labourer will expect; because, in fact, it gives no help in determining the cost of production either of labour or of anything else.

The action of the increase of population depends on the second law of demand and supply, showing the probable direction of the change in wages under given circumstances, but not helping to determine what wages must be at any time, nor what they will be in future.

To determine ultimately the rate of wages, we are therefore driven, as in all other cases, to the third law, that price is determined by cost of production. But before applying this law to labour, let as run over the mental machinery by which the price of any article is brought to coincide with the cost of production.

The value which buyers and sellers set on any article depends on different motives. The demand in the buyers' minds corresponds to the utility which, in their opinion, attaches to the article; and the causes which help to form that opinion are too numerous for classification. If their desire be slight, they can force no one to produce; if it be strong, they will tempt so many to produce that competition will bring down the price to the cost of production; so that their opinion of the utility of the article does not ultimately affect the price. Applying this to labour, the capitalist, if willing only to pay low wages, can neither force men to enter his workshop, nor to marry and produce children; and if willing to pay high wages he will attract crowds, and produce such welfare as results in many marriages. The competition among labourers will bring down the wages to something which is equivalent to the cost of production. So far the parallels run on all fours.

The value set on any article by the mind of the seller, who only holds it with the object of obtaining money for it, depends on simple motives as compared with those which may actuate the buyer. The seller does not think of the utility of the article, but he will remember the past, guess at the present, and estimate the future demand, and he will compare these with the cost of production which he has borne. He will be loth to sell below the cost price, reckoning in that a given profit to himself. Nevertheless, if he thinks that the future demand will be less than the present he will sell: but if he estimates that the demand will presently rise, he will hold back his ware. The different estimates of the present and future demand determine the differences in the supplies at various prices in any given market; but if the market price be below cost price, he will refrain from producing the article in future; and thus supply will be diminished until the market price rise up to or above cost price, then production will continue at a diminished rate. If the market price is much above cost price, more will be produced, until cost price and market price again agree.

Now, how far do similar motives affect the production of labour? The labourer is precisely in the position of the seller who attaches no value to his goods in themselves. The cost of production of his labour may be defined as the cost of maintaining the labourer from his birth to his death; and if our law is to hold good, his average wages must simply be such as will enable him so to maintain himself; but in what style? Goods will not be produced unless they pay the producer, besides his outlay, such a profit as makes it in his opinion worth while to produce them. Similarly, labourers do not produce and keep a family until their wages equal the cost of maintenance, including such enjoyment or comfort as in their opinion make life tolerable. The producer of other goods, when dissatisfied with his profit, ceases to produce. The labourer, if dissatisfied with his comforts, does not cumber himself with a wife and family—he even emigrates, or, in last resort, goes on the parish, and dies. A man being alive will generally sell himself day by day, even below the rate at which he can be said to care for life, just as in a temporary depression of the market the producer of cotton goods sells below cost price rather than not sell at all; but he will not commence the production or rearing of a family unless satisfied with his wages; and thus the wages of each class depend on the standard of comfort at which they think it worth while to marry and beget children. This action does not, indeed, take place with the view of producing goods for sale; but it does produce goods for sale, and it only occurs when existing goods (i.e. labour) meet with a sale satisfying the producers. The case of labour is here again analogous to that of any other goods produced for profit.

We can now answer the question, Whether men, by any action of theirs, can raise wages? How can they? Not by emigration, unless at the same time the cost of production of labour vary. Such emigration will simply increase the production of human beings, representing, as it does, one form of increased demand for them, and we have seen that, in the long run, demand has nothing to do with price. Not by increasing the capital employed productively. This again will only increase the number of human beings employed, not the wages they receive. Not by limiting the number of children to be born in a given area and time; this will limit trade, but will not permanently raise wages, though the diminution in the number of workmen of a given class at any moment will temporarily raise wages, which will gradually fall, unless the standard of comfort be permanently raised. Not by the limitation of the number of apprentices to a given trade; this will have but the same effect as limiting the number of children. If not all these, what then? Simply by creating an increased standard of comfort among the labourers. Do this, and our young labourer will claim higher wages, and failing to obtain them, will not marry, or will emigrate. Our young apprentice will not enter those trades which do not offer him that profit which he expects. The cost of production of labour will have risen—the price or wages must rise—the quantity supplied will have fallen, unless the demand have risen. A new state of equilibrium between capital and labour will have been reached, and although production of wealth may have been checked for a while, it will afterwards progress as before.

It is futile to preach Malthusian doctrines of self-restraint to a labourer. The upper classes may grasp at those doctrines as an excuse for selfishness, and precisely that class which by education and wealth is fitted to produce valuable members of society may be induced to refrain from so doing; but it is useless to expect that the mass of mankind, having once married, or feeling sufficiently comfortable to think of marrying, will ever refrain from self-indulgence for the sake of humanity at large. It is only by appealing to self-interest that the desired end can be attained, and this end is attained in the middle and upper classes by means of self-interest. Educated men and women will not, as a rule, marry until they see their way to live in such a style as suits them, and the production of the educated class is thus limited probably as much as is desirable; the fear here being, rather that the standard of comfort should be chosen falsely with reference to imaginary luxuries, wholly unnecessary to a really refined and cultivated existence. The learned professions maintain their fees on no other principle than that of maintaining a high standard of comfort. Men fitted to exercise them will not enter them or remain in them unless they see their way to live in the style they have chosen; should this fail them, they change their occupation, they emigrate, or at least remain single, just as in the case of a labourer. The average remuneration of a doctor or of an architect depends on the cost of his production, being mainly the cost of his living in the style which men who have received that degree of education expect. The number of doctors employed at this rate depends on the number of people willing to pay the fees asked. It is frequently supposed to be the limitation produced by the cost of education which enables the professional man to earn higher wages than the mechanic, limiting as it does the supply of educated men. It does this to some extent, but if there were openings for a thousand more doctors to-morrow the funds for their education would be found next day; and again, if the social position of doctors were to fall so that much smaller fees were expected, the profession would be far more numerous than at present.

It appears, then, that our third law, rigid as it is, only brings us to a point at which once more the feelings of men are the true arbiters of price. The cost of production of labour determines wages, but is itself determined by men's expectation of comfort.

Are we then to conclude that by simply asking for higher pay men can always obtain it? By no means.

The alternative presented to each man is, or might be, work at the present rate or starve. However serious his discontent at his present pay, he may well falter at the alternative; but if his dissatisfaction be real, he will, at least, refrain from producing more labour, and thus his discontent will tend to raise the wages as he wished. This is not the old doctrine of simply limiting the number of human beings in order to raise wages, but contains the truth partly taught by the old doctrine. The ultimate effect of the limitation of labour can only be to limit the production of wealth, and the bargain between labour and capital will always, in time, result in driving the labourer back to the lowest standard of comfort at which he will marry; but if we begin by raising the standard of comfort which our labourer desires, we shall both limit the population and raise the wages. The limitation of population is the intermediate machinery by which the desired end is to be effected, but the motive-power must be sought in individual self-interest or discontent.

Every action which tends to make the labourer put up with less comfort, or less security, tends to lower his wages. Teach him nothing, so that he wants no books nor other pleasures of the mind. Abolish bodily amusements, so that he expects no pleasures of the body. Accustom him to live like a beast in a hovel; remove all fear of absolute starvation by a poor-law; use the same machinery to persuade him that he may, with a good conscience, desert helpless children or old parents; give him little doles at every pinch, lest he be tempted to rebel;—do all this, and your dull aimless wretch will plod from day to day, expecting nothing more, and spawning swarms of wretched beings to succeed him in his hopeless contentment with a pittance. All this has been too much done, undo it; teach him, in order that he may desire knowledge; let him know that the world has pleasure in it, so that he may long for pleasure; show him what comfort is, so that he may learn to want, not live as a savage without wants; and when he has learnt how good a thing it is to enjoy life, let him fear poverty and starvation, as that man does fear them who must subdue them or perish; teach him to trust himself; teach him that his wife, his little ones, should trust him and him alone; let him look on miscalled charity as a temptation of the devil; and add to all this the greatest want of all, the longing that he may be that excellent thing, the man who walks in the ways of God; and never fear that this being will grovel out a life of hardship, or fail to learn how to win that which he so much desires.

Discontent with that which is vile is the mainspring by which the world is to be moved to good. Contentment with degradation is a vice—a vice only too difficult to eradicate. The action of these truths is everywhere to be seen. There is hardly a grain of truth in the doctrine that men's wages are in proportion to the pleasantness of their occupations. On the contrary, all loathsome occupations are undertaken by apathetic beings for a miserable hire. The plodding agricultural labourer has small wages; the discontented mechanic lives in comfort; the professional man, who never ceases struggling for more, wins luxuries;—yet the best paid is the most pleasant life. Now the common explanation of this is, that there are few capable of fulfilling the higher and pleasanter functions, and many capable of performing the baser work; but why are there few fit for the higher lot? Simply because of the high standard obtaining among those who rejoice in it. A doctor with half a dozen sons will not make all doctors; he could afford to do this very well. It is not the expense of the education which deters; he knows he could not find openings for all in the one path—he means by this, not in the style to which he is accustomed, and in which he expects his sons to live. It is the standard of comfort held by the present doctors which limits the number entering the profession, not the expense of education: again, a young doctor says that he cannot marry on 300*l.* a year, although he knows that a mechanic on 100*l.* lives in real comfort, and brings up a well-to-do family. Again, it is the young doctor's standard of comfort which limits the number of the well-to-do class, and so keeps up their pay. But, it may be said, if it is not the expense of education which limits the number of doctors, why do not the well-to-do of the lower class make their sons professional men, and so lower the standard of refinement and culture? They actually do so to a limited extent, but no sooner has the lad received the professional education, and learnt the wants of the new and higher class, than he adopts these; and while his father married on 100*l.* a year, he refuses to be contented with less than 500*l.*

Now, there is nothing which so much helps a man to keep up the standard of his desires as the possession of some wealth. A man who has something can hold out. If he cannot enjoy large profits he can enjoy idleness; but the pauper must work or starve. Thus those who have money make the larger incomes, and the poor in a profession make small profits. Money makes money, because money produces a high standard of ease. Trade-unions are one of the most powerful agents for raising wages, because they enable the community of workmen to acquire wealth. They are more powerful than savings-banks or building-societies, by which individuals obtain reserve funds. The individual workman knows that his reserve fund will be nearly useless unless his neighbour has a reserve fund also. If each workman in a strike trusted to his own funds only, the poorer ones must give in first; and these would secure work while the richer, after spending a part of their reserve, would find themselves supplanted by the poorer competitors, and the sacrifice made uselessly. A combined reserve fund gives great power by insuring that all suffer alike. The trade-union, therefore, has a permanent action in raising wages, because it enables men to accumulate a common fund, with which they can sustain their resolution not to work unless they obtain such pay as will give increased comfort. The common reserve fund plays just the same part in raising their wages, as is played by the small patrimony or sustenance given by a doctor to his son, which enables him, without risk of starvation, to refrain from taking unprofessional fees.

Strange, therefore, as the doctrine may sound, it is the wants of men which regulate their wages: and this is a simple deduction from the principle that the cost of production ultimately determines the prices. Where wants are few and simple, there wages are low; where wants are numerous, wages are high; and it is the wants which raise the wages, not the wages which have created the wants: pay a savage more than he is accustomed to, and he simply squanders the money. But while the wants of men determine their pay, it is the demand for men of that class which determines how many shall be employed at that pay. This is the corrective to discontent. If their wants are great, few or no men of the given class may get any pay at all. It is the seller of labour who determines the price, but it is the buyer who determines the number of transactions. Capital settles how many men are wanted at given wages, but labour settles what wages the man shall have.

Surely we may be well satisfied with this conclusion. The laws of demand and supply, rigid as they are, point to no necessary impoverishment and degradation of the labourer as men and means multiply; they require for his improvement no superhuman exercise of prudence on his part; they ask for no bounty from the men of great wealth. Well might men exclaim that the law was a law of death, when they so misread it as to believe that unless an ignorant and brutish population could be taught for the sake of unborn men to restrain their strongest passion, poverty must be perpetuated in an ever-increasing ratio. I do not so read the law; but say, Convince these paupers of their own misery. Teach them what comfort is, and a rational self-interest will lead them to independence, which the more wealthy have reached by the same path. Rational self-interest has done most things in this world, the great duty of the teacher being to distinguish rational from irrational self-interest. Whatever school of religion or philosophy we belong to, we cannot deny that each man, acting rationally for his own advantage, will conduce to the good of all; and if the motive be not the highest, it is one which at least can always be counted on. It were a poor world if higher motives were banished; but in looking to the improvement of the pauper it is well that we may count on a motive which is at least effectual. And, looking to this motive alone, I say that the laws of demand and supply are so far from being antagonistic to improvement, that we need only teach men what they should desire, and excite in their minds a real disgust at their miserable condition, in order, by the very laws of price, to create the Utopia of labour where all men shall get a fair day's wages for a fair day's work, by which vague phrase the workman means that the pay for his labour shall buy him enjoyment as well as bread.

ON THE PRINCIPLES WHICH REGULATE THE INCIDENCE OF TAXES.[1]

It is well known that many taxes do not fall ultimately on the person from whom they are in the first instance levied. The merchant advances the duties imposed on goods, but the tax ultimately falls on the consumer. The problem of discovering the ultimate or true incidence of each tax is one of great importance, and of considerable complexity. The following paper contains an investigation of the methods by which this incidence may in some cases be experimentally determined, and of the principles regulating the incidence in all cases—these principles being stated in a mathematical form.

The author, in a paper published in *Recess Studies*, expressed the law of supply and demand by representing what may be termed the demand and supply functions as curves. The ordinates parallel to the axis O X, Fig. 1, were prices—the co-ordinates parallel to the axis O Y were the supplies at each price, and the demand at each price for the respective curves—the market price is then indicated by the ordinate x of the point at which the curves intersect, this being the only price at which buyers and sellers are agreed as to the quantity to be transferred.

We might write the law algebraically as follows, calling y the quantity of goods in the market, at each price x, we have $y = \phi(x)$; and calling y_1 the quantity of goods demanded at each price, we have $y_1 = \phi_1(x)$; the market price is determined by the equation $y = y_1$. There is, however, little or no advantage in adopting this algebraic form, because we cannot suppose that in any instance $\phi(x)$ or $\phi_1(x)$ will be any tolerably simple algebraic function, whereas the curve for given goods might be determined

[1] From *Proceedings of the Royal Society of Edinburgh*, Session 1871–2.

experimentally by observing from year to year variations of quantities bought or quantities supplied at various prices.

Professor Jevons has since given a much more complex algebraic representation of the same law, which, however, reduces itself to the above simple form.

The graphic method may also be employed to indicate the advantage gained by each party in trade, and to show how it may be estimated in money. Let the two curves indicate the demand and supply at each price for a certain kind of goods. If all sellers were of one mind, and were willing to supply all their goods at a given price x, and were quite determined to sell

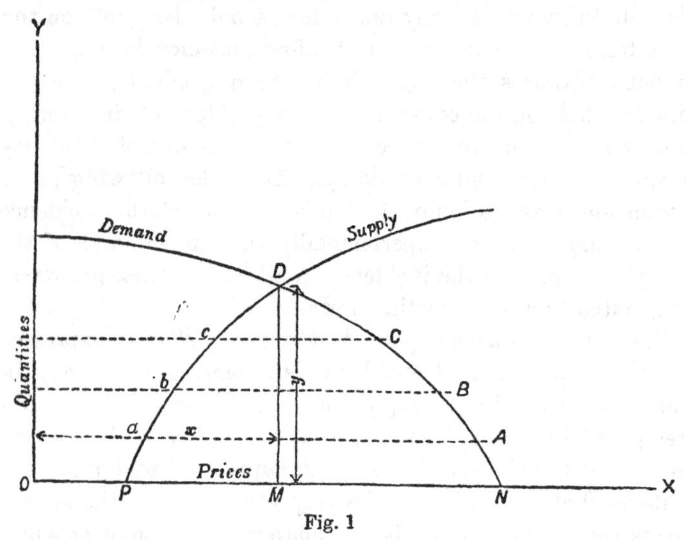

Fig. 1

no goods below that price, the supply curve would be a mere straight line parallel to O X, and ending abruptly at the ordinate raised at x. Similarly, if all buyers were of one mind, and would only buy below a given price x, but were willing to buy all they want at that price, and no more at any lower price, the demand curve would be a line parallel to O X ending abruptly at the ordinate raised at x, and the price would be quite indeterminate. If the two lines overlapped, transactions might take place at any price between that at which the sellers were willing to sell and the buyers willing to buy; there would in this case be no market price. This case does not represent the true state of either buyers' or sellers' minds in any real large market. There are always a few holders who would only sell if the price were much higher than the market price,—these are the people who expect prices to rise; there are some who are just willing to sell at the market price, but who will not sell a penny below; and there are others, weak holders, who expect prices to fall, and these would really, if pushed to extremity, sell below the market price. This condition of things is represented by the supply curve in Fig. 1.

Similarly, there are a few buyers who, if pushed to extremity, would buy some goods above market price; some also will just buy at market price; some will not buy unless the price is below market price. This is represented by the demand curve.

Now, I contend that when the market price is fixed, those traders who are perfectly indifferent whether they buy or sell at that price reap no benefit by the trade; but these will be few in number.

Looking at the demand curve, the ordinate x_A from the axis O Y to A represents the value set on some of the goods by some buyers, but these buyers have got the goods for the sum represented by the ordinate $x = $ O M; the difference between these two ordinates $x_A - x$ is the difference in price between what was given and what might have been given for a certain small quantity Δy of goods. $\Delta y\ (x_A - x)$ is therefore the benefit reaped by buyers from the purchase of the quantity Δy; and integrating the benefits derived from the sale of each successive quantity, we find the area M D C B A N represents the whole gain to buyers by the purchase of the quantity y of goods. Similarly, it is easy to show that the area M D $c\,b\,a$ P represents the gain to sellers by the same transaction; these areas represent the gain in money, each product $\Delta y\ (x_A - x)$ being the product of a quantity by the gain in money per unit of quantity.

Thus the whole benefit to the two leading communities is represented by the sum of the two above-named areas, and the partition of the benefit between the two communities is perfectly definite.

Professor Jevons has used curves to integrate what he terms the utility gained by exchange in a manner analogous to the above; but utility, as he defines it, admits of no practical measurement, and he bases his curve, not on the varying estimates of value set by different individuals each on what he has or what he wants, but on the varying utility to each individual of each increment of goods. The above estimate of the gain due to trade, deduced from the demand and supply curves as originally drawn in my *Recess Studies* article is, I believe, novel, and gives a numerical estimate in money of the value of any given trade, which might be approximately determined by observing the effect of a change of prices on the trade; the curves throughout their whole lengths could certainly not, in most cases, be determined by experiment, but statistics gathered through a few years would show approximately the steepness of each curve near the market price, and this is the most important information.

A steep supply curve and a horizontal demand curve indicate that the buyers reap the chief benefit of the *trade*. The sellers, if producers, may, however, be making important profits as capitalists and labourers.

A steep demand curve and a level supply curve indicate that the suppliers are chiefly benefited by the trade; the community or body which is most ready to abandon the trade if the price increases a little, benefits least by the trade.

When the traders are producers and consumers, the benefits estimated in this way as due to the *trade* are not the only benefits reaped by the community from the manufacture.

In this case, what is termed the supply curve depends on the cost of production of the article, including that interest on capital and that remuneration for skilled superintendence which is necessary to induce the producer to employ his capital and skill in that way. The cost of production increases generally with the quantity of the article produced, otherwise the supply curve would be a straight vertical line; but as a matter of fact, to produce an increase of production a rise of price is necessary, indicating that only a few men with little capital are content with a small rate of interest and small remuneration for their skill, but that to induce many men with much capital to be employed in the particular manufacture, a large rate of interest and considerable remuneration are required, hence the supply curve will be such as shown in Fig. 2, where the price O P is that price or cost of production which is just sufficient to tempt a few producers to produce a little of the article.

Then if O P′ is the actual cost out of pocket required to produce a small quantity of an article, and if O P is the lowest cost at which any manufacturer can afford to produce it, the area P′D′D M represents the whole profit to the producing capitalist when the price is O M. The line D′P′ is not necessarily parallel to D P, nor vertical, the bare cost of production of the article generally increases as the quantity increases; and in that case D′P′ is not vertical. Again, the rate of interest required to tempt additional capital into a particular field is not constant,

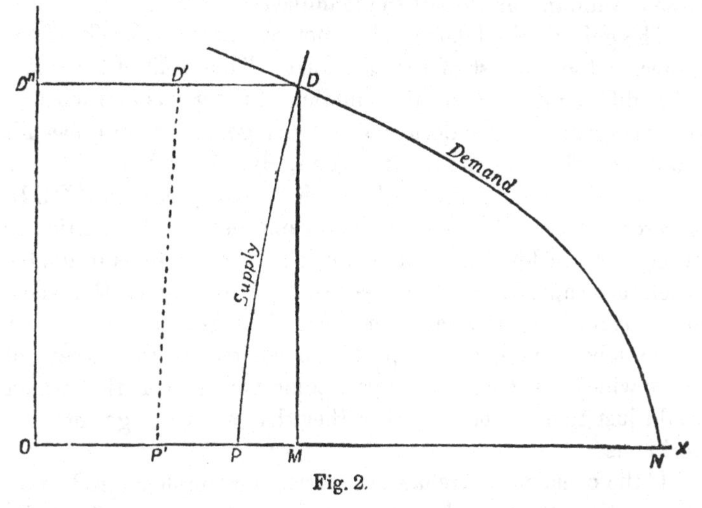

Fig. 2.

but increases, hence P′D′ is steeper than P D. I see at present no means of experimentally ascertaining the gain reaped by producers represented by the area P D D′P′: it can be approximately estimated by considering the prevailing rate of interest in the producing community and the amount of capital required for the production of the unit of the article.

We see that the gain of a manufacturing capitalist may be divided into two parts—the profit as a trader, and the interest as a capitalist.

In safe trades, where there are few fluctuations in price, the former gain may perhaps be the most important; in more speculative trades the latter.

There is yet a third source of gain to the manufacturing community: the labourer who produces the goods earns his wages by the manufacture, and this is an advantage to him. In the diagram, the area O P′D′D″ represents the wages paid for labour alone. The length of the lines between O Y and P′D′ represents the wages of labour per unit of goods, increasing as the quantity of goods required increases. This is lost to the community if the manufacture is stopped. Thus the whole sum paid by the consumer is the area O M D D″; and this is made up of three parts, one of which is the profit to the trader, one the interest to the capitalist, and one the wages of the labourer; all these advantages are lost if the manufacture ceases.

The gain of the labourer does not resemble the profit of the trader, or the interest of the capitalist. The profit of the trader is the difference between his valuation of the goods and what he gets for them. If he does not sell his goods he still has his goods, he only loses the profit. Similarly, if the capitalist does not sell his capital, he still has his capital. Now, the area P′P D D′ represents the profit made by the capitalist on the particular employment of his capital, and this is all that he loses if unable to sell that capital; but the area O P′D′D″ represents the whole sum received by the labourers, not their profit. The profit of the labourer may perhaps be considered as the excess of wages which he earns in a particular trade, over that which would just tempt him to work rather than starve or go into the workhouse.

If the consumer purchases the article for simple unproductive consumption, then the loss to him is only represented by the area D M N. If, however, a community purchases goods, and consumes them productively, then, by the cessation of the trade, they in their turn lose the interest on the capital they employ, and the labourers of the community lose their wages; so that, in that case, the loss to the buyer, who cannot be classed as an immediate consumer, is made up of three parts, similar to those enumerated in the case of the seller.

Taxes on Trade.

Having distinguished between the three distinct advantages given by trade, I will now consider the incidence of a tax on trade, levied as a fixed sum per unit of goods, as one pound per ton, or one shilling per gross.

The effect of such a tax is to produce a constant difference between the price paid by the buyer and the price received by the seller. The market prices are determined in the diagram of the supply and demand curves, by the points between which a line parallel to O X, and equal in length to the tax, can be fitted between the two curves.

Thus, if in Fig. 3, F N be the demand curve, and P E the supply curve, and if the length of the line C C′ be the amount of the tax per unit of goods, then O M is the market price to the

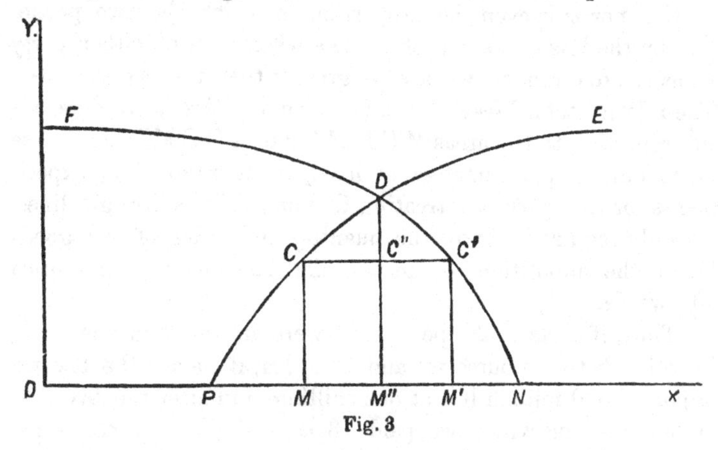

Fig. 3

supplier, O M′ the market price to the buyer, and the difference M M′ is equal to the tax.

The total amount raised by the tax from the transactions represented in the diagram, is measured by the area M C C′M′. The portion paid by the seller is measured by the area CC″M″M. The portion paid by the buyer is measured by the area C″C′M′M″. The whole loss entailed by the tax on the two communities is measured by the area M C D C′M′; the loss to the sellers is measured by the area C D M″M; the loss to the buyers by the area M″D C′M′; both buyers and sellers suffer a loss beyond the tax they pay. This excess of loss is represented by the area C C″D for the sellers, and C′C″D for the buyers.

If the tax be large, the line C C′ will approach the axis O X, the tax will be unproductive, and the area C C′D representing the excess of injury to the buyers and sellers will be large, compared with the produce of the tax. This fact is one justification of free-trade.

There is a certain magnitude of tax which will produce the maximum revenue or value for the area M C C′M′. The ratio in which the tax falls, in one sense, on sellers and buyers is simply the ratio of the diminution of price obtained by the sellers to the increase of price paid by the buyers.

It is absolutely clear that this is the proportion in which the tax is actually *paid* by the two parties, although this may by no means correspond to the relative suffering inflicted on the two parties, nor is it even the proportion in which the two parties lose by the loss of trade profit. The whole loss of either party is, as the diagram shows, always greater than the tax they pay. The relative total losses of the two communities as traders, are in proportion to the areas M C D M″ and M′C′D M″; and these areas might approximately, at least, be ascertained by experiments for this purpose; treating C D and C′D as straight lines, we only require to know the quantity and price of the goods before the imposition of the tax, and the quantity and price afterwards.

Thus, if a tax of 2*d.* per pound were imposed on the trade in cotton between ourselves and America, if before the tax we imported 500 million lbs. at one shilling, and after the tax 300 million lbs. for which we paid $13\frac{1}{2}d.$, and the Americans received $11\frac{1}{2}d.$, the total loss to the two communities as traders would be $600+200=800$ million pennies, the produce of the tax 600 million pennies.

England would pay of the tax 450 million pennies. England's total loss would be 600 million pennies. America would pay of the tax 150 million pennies. America's total loss would be 200 million pennies. The incidence would be the same whichever government levied the tax.

It follows from the above principles, that if a holder sells unreservedly, trusting to the competition between the buyers to produce the market, the whole tax falls on the seller; the supply curve becomes a vertical straight line. If a buyer buys unreservedly, the whole tax falls on him; in this case the demand curve becomes a vertical straight line.

Thus, if sales by auction were subject to a tax *ad valorem* or otherwise, and if sales were quite unreserved, the number of transactions not being altered, the prices would be unaltered, but the sellers would only get the prices minus the tax.

This case does not practically arise, because, if auctions were really so taxed, although in each auction that occurred the sale might be unreserved, auctions would, as a whole, be checked; fewer people would put up their goods for sale in that way,—the prices would rise, the number of transactions would be diminished, and the tax would really be borne in part by the buyers and part by the sellers.

If the trade between two countries really consists in the exchange of goods, effected by the agency of money as a unit for expressing value, but not involving the actual transfer of coin, the above principles show the whole gain by the exchange to be the sum of two gains which each party would make by each trade if it alone existed.

If by duties one portion of the trade be extinguished or much diminished, both parties lose, but if the other portion of the trade remain uninjured, then, although there may be no exchange of commodities other than of goods for actual money, nevertheless the full gain on that which is untaxed remains intact. Thus, although the French may tax our goods, and so inflict a loss on themselves and on us, this is no reason for our inflicting an additional loss on the two communities by taxing the import of their goods.

House Rent.

I will next consider the effect of a tax on house rent.

Landlords are here the sellers, and tenants the buyers of what may be termed a commodity; not the house, but the loan of a house for a term of years—the tenant buys what might be called, by the extension of a suggestion of Professor Jevons, a *house-year* from his landlord.

The difference between the house and other commodities such as food or dress is, that the house remains, whereas they are consumed. The house-year is consumed year by year, but it is reproduced year by year without material fresh expenditure on the part of the landlord. This permanency alters the incidence of taxation.

If the demand falls off the landlord cannot remove his house—he cannot cease to produce his house-year, which therefore he must dispose of. Hence, in a stationary or declining community, where no new houses are being built, but where year after year a sensible proportion remains unoccupied, the landlord must sell his house-year unreservedly, and any tax imposed on house rent would fall on him alone; that is to say, he would receive a rent diminished by the full amount of the tax, and the tenant would pay no more rent for a house of a given class than if no tax were imposed. The supply curve becomes a straight horizontal line, and is unaffected by the tax; the demand curve is equally unaffected by the tax; the number of houses let is unaltered by the tax, but the landlords lose as rent the whole amount raised by taxation.

This reasoning is based on the assumption that the supply curve has become a straight horizontal line unaffected by the tax. This condition is altered in any prosperous or growing community. There new houses must be built, and a considerable number of houses are always unlet, not because they are not required by the community, but because the speculative builders are holding out for higher terms. This produces a supply curve of the kind common to all other kinds of goods. At higher prices more goods are forthcoming. A newly-imposed tax will then be distributed between sellers and buyers, landlords and tenants in a manner depending on the form of these curves. A sensible check will be given to the letting of houses, tenants will be content with somewhat less good houses, and landlords with rather smaller rents. This is the immediate effect of the tax—the greater portion would probably fall on the landlords at first, at least in the new houses where fresh contracts are being made. But after a few years the conditions would have altered. New houses are only built because the builders obtain the usual trade profit and interest on their capital—the check to letting consequent on the imposition of the tax will therefore diminish the supply of new houses until, owing to diminution in supply, rents have risen to their old average. Then builders resume their operations. The whole tax by that time will be borne by the tenants; that is to say, if there were no tax they would get their houses cheaper by the precise amount of the tax, because rents so diminished would suffice to induce speculative builders to supply them. The rents through the whole town are ruled by those of the new districts. There is a certain relative value between every house in the town, and if the rents of new houses are dearer, the rents of the old houses are increased in due proportion. In fact, when new houses need to be supplied year by year, houses are commodities which are being produced, and the tax falls on the consumers.

The above principles determine the incidence of a tax; whether nominally levied on the landlord or tenant, but in their application account must be taken of the mental inertia of both landlords and tenants, as well as of the fact that many contracts for houses are not immediately terminable. These two conditions will for the first few years after the imposition of any new tax cause it to fall on the party from whom it is nominally levied.

Precisely as a tax on trade not only falls on the traders, but injures capitalists and labourers, a tax on house rents injures the capitalists who build houses and the labourers they employ —not that the capitalist pays the tax, but he is prevented from finding a useful investment for his money owing to the diminution in the number or quality of houses required.

Taxes on Land.

The question of the incidence of taxes on land is peculiarly interesting. Land differs from all other commodities, inasmuch as the quantity of it does not depend on the will of any producer. The number of houses in a flourishing community does depend on the will of speculative builders; but land can only be increased in quantity by such processes as enclosing commons, or breaking up private pleasure grounds. We will neglect these small disturbing influences, and assume that all the land in a country is available for cultivation, where such cultivation is profitable; and that the absence of profit is the only reason for neglecting to cultivate any portion of it.

It is well known that the rent of each acre of

the excess of annual value of that acre over the annual value of the poorest land which tenants think it worth while to cultivate. We may classify all land according to the total return which it will yield per acre upon capital invested in its cultivation; and we may draw a supply curve of land such that the ordinates will be the total quantities of land which will return each successive percentage on the capital required to cultivate it. The supply diminishes as the rate of percentage increases, that is to say, there is less land which will return ten per cent. on the capital than will return five per cent., and still less land which will return twenty or thirty per cent.

If, therefore, tenants as a body, considered as capitalists, will not cultivate any land which does not yield twenty per cent., there will be far less land in the market than if they will be just satisfied with ten per cent.

Again, all tenants are not of one mind, and we may construct a demand curve in which the ordinates are the total quantities of land which would be let, if the land paying no rent be fixed at each successive percentage. The actual quantity of land let will be determined by the intersection of the two curves, and is represented by the height M D, Fig. 4.

If we now build a solid on the base O D′D N, such that its height all along each ordinate x is the number of hundreds of pounds of capital per acre required to give the percentage corresponding to the length x, then we shall have a volume standing on O D′D N, the contents of which will measure the total annual returns from all the land cultivated.[1] The rent is the volume standing on M D N, the profit received by the farmers is the volume standing on O D′ D M, and this is in excess of what would have just tempted them to cultivate by the volume on M D P. We may, therefore, considering the farmer as a capitalist and a trader, call the volume on M D P his trade profit, and the volume on O D′ D the interest on his capital.

The effect of any tax on the land is to reduce the interest which each class of land is capable of returning on the capital

[1] If 150*l.* per acre are required to give the percentage x of any one class of goods, the height of the ordinate perpendicular to the plane of O D′ D N will be 1·5.

employed. This it will do in very different ways according to the manner in which the tax is levied.

If the tax be an *ad valorem* duty on rent, it will modify the supply curve only between D and N. There will remain just as much land as before capable of paying rates of interest less than O M, but the quantity of land capable of paying the higher rates will be diminished; in other words, the rate of interest which the poorest land worth cultivating pays will not be affected, for this land pays no rent and remains untaxed—hence no land will be thrown out of cultivation, but the supply curve will be altered from D N to D N′, diminishing the volume representing rent, but leaving the other quantities untouched; hence any tax assessed on rent

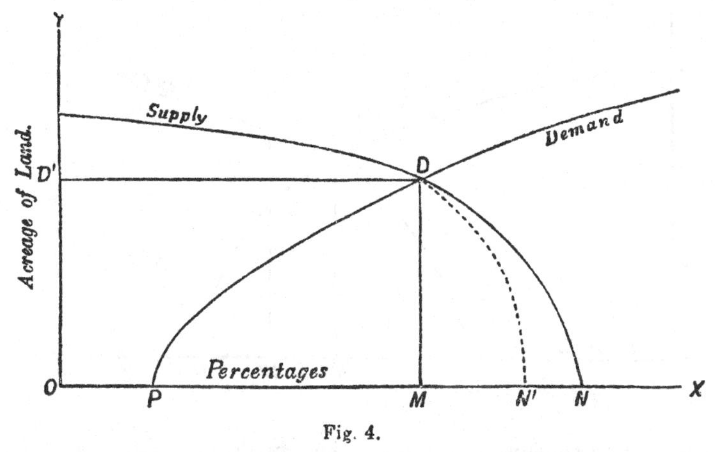

Fig. 4.

is paid wholly by the landlord. The amount of the tax is the volume standing on D N N′. It is curious to remark that this tax in no way falls on the consumer. The tax on rent simply diminishes the excess of value which some land has over others; no land is thrown out of cultivation, and no less capital employed in production than before; no one suffers but the landlord. If, instead of being assessed on the rent, the tax is assessed on the produce of the cultivation, the incidence of the tax will be greatly modified. The cultivation of land will no longer be so profitable; *i.e.* the returns from capital employed on the land will be less; in other words, the whole supply curve of the land will be modified, falling everywhere if the produce taxed be that which is produced on all qualities of land. Some land will fall out of cultivation, and only part of the tax will be borne by the landlord; part will fall in the first instance on the tenant, but he, like any other manufacturer, will recover almost the whole of his portion from the consumer. Tenants will be injured by the limitation of the number of transactions, and labourers by the diminution in the amount of work required. This is the effect of an octroi duty.

Sometimes a tax is assessed not on the rent, but on an assumed value per acre. Such a tax can never be raised on land which pays no rent, for the owner would rather abandon possession of the land than pay the tax. It might, however, lead to the abandonment of the cultivation of poorer soils; it would then injure

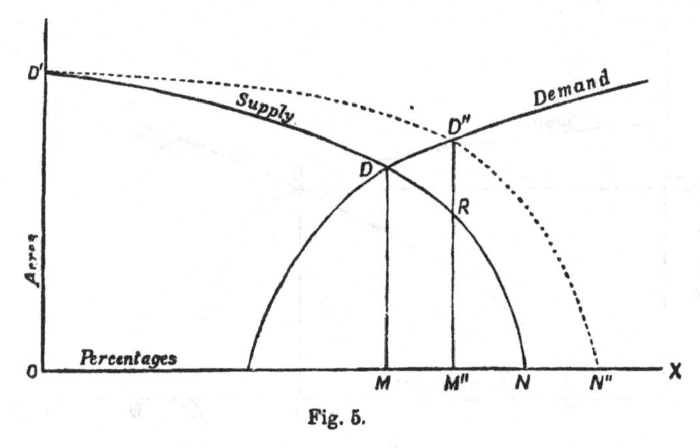

Fig. 5.

tenants and consumers, although they would not pay one penny of the tax; for taxes cannot be paid out of lands which lie waste; assuming that the tax is always less than the rent, as it certainly always should be, it will be paid wholly by the landlords. The tax in this case does not diminish the supply of land.

A cognate question of great interest is, Who reaps the benefit of any improvements in agriculture, making land return more than it previously did? This improvement may require, and probably will require, increased investment of capital. The whole supply curve will be raised; assuming the demand to remain the same, Fig. 5, M″ D″ will be the new increased number of acres in cultivation, but land will be left uncultivated which would have returned the interest O M on capital. The volume standing on D′ D″ N″ will be much greater than that on D′ D N, for the third dimension will also have increased; the average rate of interest and the trade profit of the tenant will have increased, and it is highly probable that the volume standing on D″ M″ N″ may be greater than that which stood on D N M; but this is by no means certain. It might at first be actually smaller. In all probability, however, the demand curve is very nearly vertical, a small increase of profit tempting a largely increased investment of capital in farming. If this be so, then the landlord also reaps considerable benefit from the improvement, for if the farmers were contented with nearly the same rate of interest as before, the solid standing on D R N N″ D″ which he gains would be larger than the solid on D R M″ M which he loses; moreover, the volume on R N M″, which he retains, is increased. Labourers and consumers also gain.

THE TIME-LABOUR SYSTEM:

OR HOW TO AVOID THE EVILS CAUSED BY STRIKES.[1]

MANY people think that strikes are due either to wickedness or to ignorance; that men on strike are endeavouring by illegitimate means to obtain higher wages than they are entitled to by the law of demand and supply; others consider that strikes are the only means by which the men can ascertain what wages this much-quoted law entitles them to claim. All, however, will admit that the process is singularly barbarous and clumsy, entailing much loss on all concerned. The first step in looking for a remedy is to ascertain as clearly as possible the cause of the evil. Why should a strike be required to ascertain the state of the labour market when no analogous action is required to fix the price of other goods? Some deny that the strike is necessary. Let us therefore examine in what way the law of demand and supply does fix the price of goods in general, and then we shall easily see whether the process be or be not applicable to the determination of the price of labour.

We all know that when the demand exceeds the supply prices rise and *vice versâ*, but let anyone who thinks that this statement closes the question consider how, if he were appointed arbitrator, he would decide what rise of wages should follow a given increase of demand; nay, how he would measure the increased demand. This is the difficulty. The common answer would be that no arbitrator can decide any point of this kind, which must be settled by the operations of nature. This is done so far as it goes; but nature does not act by supernatural methods. Her methods can be examined, and in some cases new and better methods may be set in operation than those which have crept in unwatched, as when we substitute emigration and sanitary science for famine and pestilence. Now a strike

[1] From unpublished MS., 1879–81.

is the famine and pestilence of trade. Even famine and pestilence serve useful ends, and so does a strike; but instead of submissively accepting strikes as the only means by which, in the present state of the body politic, wages can be determined, let us examine carefully how prices of other commodities are determined by nature, with the hope that when the special difficulty in the case of labour is clearly understood, we may find the means of evading or overcoming that difficulty.

The price of any one article, say eggs, in a given market is that at which the number of eggs required by the buyers is equal to the number offered by the sellers; there is no such thing as a definite demand or a definite supply irrespective of price. We can only think of demand as definite when we name a given price: at that price buyers would be willing to take, say 100,000 eggs; but perhaps at this price sellers would only supply 90,000; if so, the given price is below the true market price. At some higher price 100,000 eggs may be ready for sale, but at this price buyers decline to take more than 90,000; this price then is above the market price.

There is some intermediate price at which perhaps 95,000 could be disposed of by sellers and 95,000 also bought by the buyers. This price would be the market price and this number the number sold on that market day—provided—and this is a very material point, buyers and sellers could accurately ascertain one another's minds. But there is no mysterious law of nature bringing this price to light without visible agency—an approximation to this theoretical market price is actually formed by a tentative process. Some one offers eggs at what he guesses or wishes the market price to be. Then one of two things happens; either this offer is snapped up at once, and buyers crowd round him, or buyers shake their heads, and say that they will wait. If sales at this price are numerous the market is said to be brisk and the sellers raise their prices; if there are few transactions the sellers lower their prices. By this power of trial they find out roughly a price at which the number of eggs for sale and the number required are about equal. This process of determining the price of an article will hereafter be called the *market process*, and the price thus ascertained, the *market price*.

Price in a market varies in a manner so resistless as to

suggest that it must really be determined by some process which is independent of the *will* whether of purchasers or sellers: but this is not so. Each transaction shows that one man is *willing* to part with some goods at a given price and that another man is *willing* to purchase goods at that price. The briskness or slackness of sales simply brings into evidence the condition of the *wills* of the traders. In a market there is never any serious discontent, you are free to offer your eggs to a hundred buyers at what you think a fair price—you find that not one will buy—you see eighty or ninety neighbours willing to sell for a lower price than yours—you go home with your eggs unsold, disappointed no doubt, but feeling no suspicion that injustice has been done you, believing on the contrary that next day in another town you will get a still better price, when you will rejoice at your foresight; your state of mind would be wholly different if you saw that all the buyers had agreed that they would not compete with one another but would simply give sixpence a dozen. They would no doubt get a supply even at this price, perhaps 20,000 eggs instead of 95,000, the number which would have changed hands if the price had been settled by the market process, but every seller, even those who sold their eggs, would feel sore. Men see that when buyer competes with buyer, or seller with seller, a natural tentative operation settles the price; sellers are revolted by any attempt on the part of buyers to fix a price arbitrarily, just because they have money. Buyers feel the same thing if sellers combine to say they won't sell their eggs for less than a shilling a dozen. They must have some eggs; they buy perhaps 20,000, and see 80,000 rotting taken home. Monstrous of the sellers to try to fix an arbitrary price just because they have got eggs and we have none!

Nor are matters a bit improved when buyers and sellers both combine—the sellers will not sell under a shilling a dozen, the buyers will not give more than sixpence a dozen; each party abuses the other and all the eggs rot. This is a strike. After some dozens of market days spent in this way one of the two parties finds out that it is more inconvenienced than the other; perhaps that a substitute for eggs is in the market; that eggs are coming from Timbuctoo, and they come to an agreement fixing some intermediate price. But what a wretched way of finding out the market price as compared with the usual one where every egg that changes hands helps to determine what the true price ought to be, the price which all can give and receive without rancour.

How does it come then that the price of labour cannot be settled by the simple natural process which fixes the price of eggs? In some cases it can be and is. Domestic servants' wages are settled by the market method. There is a continual market open. Employers give different wages; not only do various degrees of merit command various rates of pay, but the same degrees of merit are paid for at different rates; above all, there is no combination among masters not to give more than a certain rate for a housemaid; nor among housemaids not to work for less than a certain rate of pay. Competition is free and incessant. The condition of the market is patent to all: either there is or is not a difficulty in procuring the labour at the old rate of wages and the price rises or falls accordingly. The rise or fall stops when the supply and demand are equal. The reason why the price of labour in factories cannot be settled in a similar way is obvious enough.

When a hundred men are working side by side at weekly wages, doing work of the same quality, they will never be contented unless they are paid at equal rates. When B. gives 18*l.* to a housemaid this does not entail a rise in the wages paid by A. to his housemaid of equal merit, engaged six months before at 16*l.* Not only are the housemaids engaged for longer periods, but there is a tacit understanding that the agreement, though nominally for a quarter, really is intended to apply for much longer. But this state of things is only possible under the condition that the housemaids are not in the same house or hotel, working for the same master. Now this is at the root of the whole matter. A master wanting more workmen than he can get at the actual rate of wages, cannot have recourse to the natural plan of offering a slight advance in the wages of new hands, unless he is prepared to raise the wages of the thousand workmen he is already employing—a much more serious affair. It is as if a buyer having bought a thousand eggs at sixpence a dozen and wanting a hundred more, could not offer sevenpence a dozen for the extra lot, without going back on his old bargains and paying sevenpence for all.

Now this peculiarity of the labour market does not arise from any *defect* either in men or masters. There has been far too much uncalled-for asperity on this subject, which the Press has rather fomented than allayed. The difficulties of the case are inherent not accidental. No set of men can be expected when doing work of equal merit to work side by side at different *weekly* wages. If they had bargained to stick to certain wages for a year, they would hold to their bargains, but to consent to work for 30*s*. a week when another man is getting 31*s*., is to confess your inferiority, and if you are not inferior you will not do it and I have no right to ask you.

Let no one here go off on the side issue that unions insist that good and bad shall be paid alike; that is a separate and important question, but has no relation to what is now being spoken of. Grant as a postulate a hundred men of average merit working at 30*s*., the master wants a few more of the same sort; he cannot offer these 31*s*. without raising the wages of all the others. He could not do it whether unions existed or not, and this essential difficulty distinguishes the labour market from other markets. We thus find ourselves precluded from applying the tentative process by which the rate of wages could be ascertained at which demand and supply are equal. To make the labour market like the egg market, we must imagine week after week all labour of a particular kind standing ready to be hired; no combination among the sellers; all buyers equally uncombined. An employer says, 'I want a thousand men at 30*s*.;' only 900 come forward. They are booked; then the employer says, 'I will take another hundred at 31*s*.' Ninety come forward, and he says, 'Well, I don't mind giving 32*s*. for ten more.' The market price of labour would obviously be rising.

This ideal state of things can, of course, never obtain. The workman to be useful must not be changed week by week, he must work on year after year in the same shop; we cannot fling all labour on to the market every week; an approach to the true market method would be made, if week by week the master could say 'I want ten hands more, if I can get them at such a price, or I must discharge you ten men unless you will work for less.' The men, as sellers of labour, could also employ the tentative process, saying 'We ten will leave, unless you raise our wages,' or 'We ten unemployed will come at lower wages.' In theory this would give a tentative method by which the market price of labour could be ascertained, but neither method is conceivable because of the natural and laudable self-respect of the men. There is no disgrace in selling eggs at a lower price than your neighbour got; the market is falling and you may regard it as a misfortune, but there is no question of shame. You are, however, disgraced if you sell your own labour for less than that of your neighbour, knowing it to be as valuable and as good as his. In consequence of this feeling of pride, any change in men's weekly wages must occur unnaturally at a bound instead of gradually. No system at present exists by which the change can occur gradually, but one may perhaps be found. Just as there is no competition among the men and can be none among men of equal merit, so there is and can be very little competition among masters. One master will now and then offer a slight advance of wages beyond his fellows, if he finds a difficulty in getting hands, but there is not and never has been such a brisk competition with a continual fluctuation in prices as will enable a market price to be tentatively determined.

A very large number of writers, especially in the Press, frequently insist on the fact that if men would compete with one another this would settle a market price. If men of equal merit would consent to work at different wages, this would be so, as explained above, but the writers alluded to often speak as if the competition were to be between men of different degrees of merit, which is a totally different thing and beside the question. The market price of a certain quality of goods is in the main settled by competition between the buyers and between sellers of that particular quality; not by the competition of one quality of goods as against another quality. It is certainly desirable that the best men should be paid more than the worse, but supposing masters and men were to agree in ranking the men in categories, each of which was to be paid five per cent. more than the one below it, this would do nothing to settle the rate of wages of the whole mass. Even if there were a keen competition as to the category in which a man was to be classed, this would not ascertain the rate of wages of any one category. You cannot tell how tall a child is by learning that he grew an inch last year. The fact that in consequence of competition Tom is to get five per cent. more than Jack, does not settle the wages of either. The fallacy may perhaps be seen by considering whether it is the least necessary in an egg market that in order to determine the market price there should be many qualities of eggs. It may be, and is desirable that good and bad eggs should be sorted and sold separately, but if they were not, but were simply all mixed, the average merit would be quite well known, and a market price fixed quite as easily as if the eggs were sorted in lots each containing eggs of one quality. It is not competition between different qualities that is required, but competition in the case of each quality.

The fallacy on this subject is so wide-spread that space will not be thrown away in endeavouring to explain it. The fallacy lies in using the word 'competition' in two different senses, in some such way as the following. Without free competition between individual men, the market price for labour cannot be ascertained; there is no competition when men belong to a trade-union, because all its members receive equal wages. Therefore trade-unions prevent the market rate of wages from being ascertained by the natural tentative process. The first proposition may be accepted as true, or at least true enough for our present purpose. The first part of the second proposition is also true, but the reason given, 'because they receive equal wages,' is false, inasmuch as it treats competition as a competition in respect of excellence of work, whereas 'competition' in the first clause is used to denote readiness on the part of men to undersell one another. Competition in excellence and competition in price are two different ideas, which have got entangled together because buyers and sellers who are competing in price also compete in excellence, urging that two apparently similar qualities are really different. Moreover, the tentative process by which the price of each quality of a thing is fixed serves to arrange the goods in categories of excellence, but the mental process of judging the relative merits of two classes is a different one from that of fixing how much you will give for either; you may fix in your mind that one article is worth ten shillings more than another and yet be quite undecided what price you will give for either.

Again, this question may be looked at from a purely practical point of view. Let us try to imagine a competition in excellence used to settle wages in a case where men of equal merit will not undersell one another. A man shall come forward and say, 'I am worth a shilling a week more than you are giving the hands generally; give it me or I leave.' Let us suppose that the master admits it and grants his advance, then comes the effect on the mind of his mates. If the man is really, obviously better, and the men are just, they acquiesce; but then he will have done absolutely nothing towards changing the rate of wages for men of other degrees of merit. If a hundred or a thousand men so improve themselves as to get this advance, they would not have changed the wages of the workmen who possess the former degree of merit. If in sufficient numbers, they might diminish the demand for the other class of labour, and so, indirectly, lower the wages of men of the old average quality, but their demands have done nothing to ascertain by any tentative process whether the market is rising or falling. It may indeed be said that masters would be more willing to allow the claim when trade was brisk and less willing when trade was slack, so that by watching how many men got their advance and how many were allowed to leave, the necessary information could be got. When trade was brisk numerous individuals would make their claim and get it allowed and so a market price would be obtained tentatively, but this would be no competition in excellence, this would really be a competition between equals, some being willing to remain at the old wages, others not, and this would be the true underselling competition. Moreover, the merit of artisans differs little from a general average, except in a few exceptional cases, and employers know well enough that when the form of payment by wages is adhered to very few men can claim more than the average pay, and very few are employed who are not worth their wages. The better workman makes a better average throughout his life by being employed more constantly, the inferior workman is continually thrown out of work, but this is quite beside our present point.

The fact that one article at 35*s.* may really be underselling an inferior article which is offered at 30*s.* is true, and in this sense competition between different qualities helps to determine prices, but not otherwise. It is the underselling which is the point, and workmen think this inadmissible.

Much might be written as to the propriety of the feeling which is finding an expression in the law restricting the immigration of Chinese labour to Queensland, but the consideration of this would lead us far afield into the question of what ultimately determines wages and prices in general over a long period, whereas, in this article, the one question for consideration is how the market price of goods at any one time and place is actually discovered. We see that underselling and outbidding are essential to the process, that these methods are practised with material goods, because there is no idea of pride or disgrace attached to the labour obtained, but that with labour these processes are opposed by very strong human feelings—men's pride will not let them work along with an equal but at lower wages; men's generosity will not let them try to turn an equal out of his place by offering to work for less than he is getting: outbidding by masters is prevented by the feeling of common interest amongst employers and by the fact that it cannot be done gradually and tentatively for a few hands, but must be done suddenly for all the men they employ. If these facts were once understood there would be some hope that a remedy might be found.

The remedies hitherto proposed do not meet the case. Courts of arbitration are merely a device, useful enough in its way, for allowing disputing parties to give way with a better grace, but would be as ridiculous when applied to settle wages permanently, as a committee would be, employed weekly to decide the price of corn after hearing arguments from counsel employed by importers and purchasers. Sliding scales by which an agreement is made that wages shall bear some relation to the price obtained for the manufactured article, may in some cases be convenient and serve to render men contented when prices are rising; but they afford no solution of the general problem. The workmen if discontented can as easily ask that the whole scale should be raised, as that a single rate should be altered.

Employers can as easily insist that, owing to changed circumstances, the whole scale must be lowered. Payment by a share in profits, besides being extremely inconvenient, is liable to the same objection as the sliding scale. The workman remains at liberty to demand a larger share—the employer to offer him a smaller share. The question for solution is not what is the best *form* of wages, but what is the *amount* which a given state of the market will entitle the workman to claim. Payments by sliding scale or by shares in profits are changes in the form of payment, convenient or inconvenient, but leaving the main question 'how much' absolutely untouched. Co-operation need not be spoken of here. The cases in which it is possible are unfortunately rare. Piece-work is a good enough mode of payment, but is not very generally applicable. Employers like it because men will compete under this form of contract, inasmuch as the underseller can frequently say with justice that it is because of his excellence as a workman that he can afford to undersell his competitor; so that the low price at which he tenders may actually be a subject of boasting, if his gains at the end of the week are large. Trade-unions do not like it, because upon this system no general rate of wages can be fixed or known, and because it tends to introduce uncomfortably hard work. Whether piece-work be in the main good or not, need not be argued here, for it is evident that piece-work does not settle the rate of weekly wages, the problem with which we are now occupied. There remains so far the one solution of strikes, which do settle wages, but at a great loss to workmen, employer and consumer.

The object of the present article is to suggest that under a new system of contract between employer and workman, a true labour market would be established: a market in which day by day the demand and supply could be watched and in which wages might tentatively rise and fall so as at all times to secure an approximate equality between supply and demand. Moreover, the system suggested is one which calls for no sacrifice of self-respect either by man or master. Recurring to an original description of the process by which prices are determined in a market, we see that underselling only occurs when buyers buy few goods at the price first quoted: outbidding occurs only when sellers part with few goods at the original price. It is the briskness or slackness of the market which precedes the underselling or outbidding. The underselling and outbidding are tentative processes by which the extent of the change necessary to equalise supply and demand is determined; the necessity for the change is apparent beforehand. Similarly in the labour market the fact that more hands are wanted, or that hands are being discharged, is, even now, notorious enough; what is required is the tentative process by which we can ascertain what rise or fall of wages will equalise the supply and demand. Now suppose that men, instead of being engaged at weekly wages, were engaged say by the year; not from term-day to term-day, but from every day or any day at which they happened to be wanted, and suppose that, trade being good, more men were wanted; it being seen that extra hands were being engaged, it would be quite open to applicants to say 'We will not engage ourselves except at an increase of pay.' Masters might very easily be led to engage a few men, if they wanted them much, at this higher rate, involving a very trifling increase on their total expenditure, but they would not engage so many as they would have taken if no increase had been asked. Moreover, as the old engagements gradually came to an end, of course the men would not re-engage except at the new rate. When the masters found it no longer their interest to engage fresh hands at this advance they would decline to do more than simply re-engage old hands at the new rate. There would then be an equilibrium between supply and demand. If the new rate were too high to allow all the old hands to be re-engaged with profit, the masters would decline to receive some of the old hands, the numbers employed would fall off, the labour market would be seen to be falling; when the refusal of the masters had been ascertained to be genuine by its permanence, the men might tentatively lower their wages and re-engage until a fresh equilibrium was found between supply and demand; that is to say, until there was neither increase nor decrease in the number of hands employed. There would be no disgrace in working at different rates in the same factory when the difference was clearly seen to be due to the state of the market when the engagement was entered into, and thus the plan suggested would entirely save the self-respect of the men. The plan is quite compatible with the action of trade-unions: they could by using this method ascertain tentatively exactly the wages which kept the average number of their members employed. Of course they would resist every decrease in wages until they saw by the gradual, but persistent decrease in the re-engagements, that trade was really against them. Of course masters would never offer an advance till they failed to get men enough at the old rate, and if trade was obviously good, men, whose engagements expired, would refuse to re-engage at the old rate. Masters again would give way with perfect self-respect, if, after trying to lower the rate of wages, they found they could not get all the men they wanted. They could at any moment return to the old rates, saying 'We used to employ a thousand men at the old wages; we cannot employ that number and so discharged those unwilling to work at a lower price, but we can find profitable work for eight hundred at the old rates, so we will return to them: only you will see that we shall permanently employ fewer hands than before.' The plan would allow the course of the labour market to be watched with just the same precision, though not with the same speed, as any other market, and provides for continual fluctuation, so that the rates might even differ week by week. Every order received would produce its effect on the market, as it ought; every depression in trade would be instantly felt in the labour market, as it ought. All violent and sudden changes would be avoided, and it is these violent, unforeseen and often unjustifiable changes that do all the mischief. The method suggested does not introduce competition between workmen waiting to be engaged, but when we examine the action of the sellers of any kind of goods, we see that competition is not wanted, because of the struggle it involves between those who compete, but simply because it allows different prices to be tried tentatively until the price is found at which demand and supply are equal. The new method does this without competition since, week by week, or day by day, a new rate might be tried, gradually raising or lowering the rate until the balance was reached. The action would be slower than that of a common market with competition, but there is no need of extreme rapidity.

Of course it may be said that men engaging at different rates in different months are competing, but if they are, this would be competition with the sting taken out. The plan does give different rates of wages for the same degree of merit, which is the one essential thing required to determine a market value.

It is not in the least necessary that all engagements should be for equal times. If an employer takes a contract which he thinks will require twenty men for six months, let him engage them for that time. Even weekly engagements are compatible with the method, but the bulk of engagements must be for much longer. Engagements for different periods might be at different rates; sometimes short engagements and sometimes long ones would be most highly paid: there would be a true market with all its sensitiveness and all its convenience. Of course by coalitions attempts would from time to time be made to force the market one way or the other, but the futility of these attempts would soon be experimentally proved. Men are not monsters, and when they have recourse to such rude methods as strikes or lock-outs, it is because no other action is open to them.

It has now been shown that engagements for long periods, starting at all periods of the year, would provide a real market determining wages by the same process as the price of other commodities is fixed; but the objection will inevitably occur that even this benefit may be purchased at too dear a rate. Masters would have no control over their workmen if they were unable to dismiss them at a week's notice. Workmen might be subjected to annoyance, inconvenience, even oppression if they were bound to work in a particular shop for a whole year. This brings me to the second feature of the new proposal.

Each engagement ought not to be a contract with a man to keep him *individually* for a whole year or other period, but should provide on the one hand that the master shall employ one competent hand for the given time at the given rate, and on the other hand that the workman shall not leave without providing a competent substitute who will work until the end of the given time at the given rate. This system might be called 'the time-labour system' of engagement as contrasted with the 'personal system.' An employer would on the new system buy a unit of labour for a definite time, instead of, as at present, buying the labour of one *individual* for an indefinite time as is really the case now, for although wages are paid weekly it is well understood that they are not to change weekly.

Let us now consider what disadvantages might be alleged as inherent to the novel system of purchasing labour. The first objection would probably be that men are not units and that an engagement must necessarily be personal. This objection will probably be made by those who are not familiar with large factories. A personal servant could not be engaged on the plan suggested, but hands at a spinning frame or fitters at a vice are really very nearly of equal merit and for practical purposes are already treated as of exactly equal merit. A substitute coming forward to replace a man wishing to leave would be bound as at present to satisfy the foreman that he was of the same class as the workman leaving. Disputes would of course arise under this as under every other kind of contract, but large employers of labour will be the first to admit that, even now, they engage labour units rather than men whose personal peculiarities are important.

In some trades the number of men required fluctuates so greatly that it may really be necessary to limit the term of engagement to a very short period such as a week, but these hands are few in number. Most employers would only be too glad if they could feel certain that the labour they required was secured for the next six months or year. Even if they did not care to enter into so long an agreement with the whole of their workmen they might safely do so with the majority. Long engagements are shunned by employers only because they prevent dismissal in case of incompetence or negligence, and by the workman because they limit his freedom of leaving a shop when for any reason he may desire to do so. When the engagement is no longer personal as regards the workman, but is merely a contract to employ a labour unit, both these objections fall away. The employer, if dissatisfied with A, may dismiss him with no cause assigned; he is simply bound to take another workman, B, C, D or Z at the same pay for the unexpired period of the contract, even should wages meanwhile have fallen. If they have risen, he will have to pay the extra rate for a new hand. On the other hand a workman would be at liberty to quit a shop without reason assigned, on the simple condition that he found a substitute who would work for the same pay during the unexpired period of the contract, even should wages meanwhile have risen. The plan could be worked with or without trade-unions, but in all probability better with their assistance than without it, as they could always readily find a substitute for a man who was discharged, or who was anxious to leave his employment. If the man did not belong to a trade-union he might inform the gatekeeper of his shop that he desired a substitute, and under the new system applicants for employment would make it their first business to inquire whether any time-labour berths were open, for what periods, and at what rates. If the market-rate happened to be falling, the man who wished to leave would have no difficulty. If the market was rising he would have to supplement the payment made by the master, paying a kind of smart money or premium to his substitute.

Long engagements dating from a *term-day* would be objectionable in most employments. Employers do not as a rule know on any one day of the year how many men they will require for that year, and moreover if all hands were free to leave on one and the same day, the whole merit of the proposed system would vanish. The labour-market could then no longer be watched and each term-day would involve something very much like a strike or a lock-out. If however men were engaged, as they now are, from every day in the year, there would be no dangerous period of the year and no pressure put on any employer to make up his mind on any one day as to the number of men he might require.

The greatest objection is perhaps the novelty of the plan, but on consideration it will perhaps be found to be less novel than it seems and the novelty, such as it is, might be adopted gradually and tentatively.

After strikes, agreements are not uncommonly made, that a given rate of wages shall last for a year or even two years. This is a much more violent interference with general custom than is now proposed. It implies that whatever number of people are employed, wages are, for the given time, to remain constant. This may be most inconvenient if the state of trade alters. If trade becomes brisk, the employer is not sure to find his old, or average number of men at the covenanted rate, for his workmen may emigrate to other towns. Again if trade be dull the men are not sure of finding work; they may be dismissed one by one although their wages can not be lowered, and thus by the attempt to enforce a constant rate when the labour-market would give a varying rate, both parties suffer. If to this it be answered that there is a tacit understanding in such cases, that workmen are not to emigrate and that masters are to employ the full number of hands, then I say this is the time-labour system imperfectly developed; imperfect in so far as it does not allow the labour market to be watched, since the engagements of all the hands end on one day. The compromise, by which wages are maintained constant for all during a long period, flies in the face of natural laws and may make bad worse, but the complete time-labour system enforces no unreal constancy; the labour-market under that system might fluctuate day by day and although each unit might be engaged for a year, it would be bought at its true value on the day of sale.

Two seemingly contradictory objections may be made. 1st, that the plan diminishes the power of the workman by mitigating the pressure which a sudden strike inflicts: 2nd, that it diminishes the power of the masters by mitigating the pressure which a lock-out inflicts. Both objections are well founded; the new plan would mitigate evils which neither side has any right to inflict on the other, and which can not be justly inflicted on the community by traders in any commodity; but while the suffering inflicted by strikes and lock-outs would be mitigated and indeed perhaps abolished, the legitimate objects of these operations would be attained more perfectly. Wages would rise rapidly and certainly when trade was good; they would fall gradually but certainly when trade was bad, and in all states of trade the maximum number of men would be employed that could be properly employed profitably both to masters and men. This is not the case under the present convulsive system. Wages often remain lower than they ought to be, because no action short of a convulsion can raise them. They often remain higher than they should be for the same reason; but in this latter case employers retain a much smaller number than they could usefully employ, if wages followed the market. Both employers and workmen would benefit by the proposed plan, which is nothing more than a contrivance for letting them see the course of the market and ascertain tentatively the true market rate. It affords new and better machinery for ascertaining the market rate than exists now and it gives time for the process. Neither party would, by adopting the time-labour system, gain the smallest advantage over the other; a fact which may, for some time, prevent it from finding an advocate.

The writer is however not without hope that it may ultimately find its way into use, inasmuch as it may be adopted gradually and tentatively, being worked by the side of the existing plan. A single master in any trade might put up a poster stating that he was willing to engage men on the terms suggested.

Here is such a notice.

'A. B. is prepared to engage competent fitters for a term of six months from the date of engagement at the rate of 35*s.* per week of 54 hours. Each man engaged will be liable to dismissal as if engaged as hitherto week by week, but A. B. undertakes to receive as substitute any competent workmen at the same wages for the unexpired period of the engagement. Each man engaged will be at liberty to throw up his employment as if engaged week by week, but only on condition of finding a competent substitute willing to work during the unexpired period of the engagement at the same wages.'

Similarly any trade-union might offer to adopt the plan as a mode of terminating a strike. Their offer might run as follows: 'The strike shall be at an end if 500 hands are given work at 35*s.* per week of 54 hours. Of these 500 hands, not less than 100 shall be engaged for one year, not more than 100 shall be engaged for six months and the rest shall be engaged for intermediate periods to suit the convenience of both parties; the employer shall be at liberty to dismiss any hand so engaged on condition of receiving an efficient substitute from the union at the same wages for the unexpired period of his engagement. Similarly any hand shall be at liberty to quit his employment on finding a substitute.'

There are numerous questions connected with wages which have not been alluded to in the course of this article. The most interesting of them is the question what 'cost of production' means, when the article to be produced is a skilled workman. Cost of production determines the cost of all commodities in the long run. The law of demand and supply is limited in its action to each successive market and each successive day. The scope of the present proposal is neither to raise wages nor to lower them. Neither this, nor any other system of ascertaining a market price will in the long run determine the average rate of wages or the average price of any article. The higgling of the market ascertaining the result of the relative demand and supply in that market does not in the long run determine the price of either eggs or tea; it simply finds out the price which has really been determined by quite different means. To state what these are would require long explanations not wanted here. This article aims at no more than showing that no higgling of the market exists for labour nor can exist with the present form of contract between employers and workmen. Higgling determines the market price by showing the briskness or slackness of sales at a particular price. This information could be obtained in a true labour market by changing the form of contract in the mode proposed. Wages could then be settled with as little difficulty as the price of any other commodity.

IS ONE MAN'S GAIN ANOTHER MAN'S LOSS?[1]

TRUTH can take good care of herself in most sciences, for in these no human longing for the impossible disturbs our vision. But in political economy the most natural and universal desire is always combating the truth. Not one man or woman or child is free from the wish to obtain what he wants without paying for it. Political economy proclaims that this shall not be. Its sayings are intended to make clear how we can best obtain what we want by paying for it. Other modes of acquisition may have more artistic charm and be morally preferable, but of these we will not speak to-day. We only propose to uphold the thesis that under some circumstances a man may become rich without incurring any moral stain. We have been moved to this daring attempt by some words which Mrs. Oliphant has put into the mouth of a clever lad who has been reading political economy.

We fancy that a large number of people share Jack's opinion and think that what is one man's gain must be another man's loss, but in this matter political economy for once sends a gracious message of peace. Buying and selling are simply modes of barter; coin or cheques are merely pieces of machinery facilitating the exchange of what we have for what we want. After each exchange freely and honestly conducted, each party is richer than before. Both have made a profit, no man a loss. If I have a dog and you have a cat, but I want the cat and you want the dog—we shall both be richer for swopping. It may be horribly selfish to swop, though even that seems doubtful since I please the other man as well as myself, but we are clearly both richer after the bargain is concluded. I prefer the cat to the dog which I had, therefore I would not sell her for the sum

[1] From unpublished MS., 1884.

which would have bought the dog, she went up in value the moment she became mine; so did the dog the moment he became my neighbour's. So that if the live stock of the country were valued before and after the transaction, it would be clear that the aggregate wealth of Great Britain had been increased by the transaction. To believe this you must of course admit that there is nothing valuable but thinking makes it so. Some people try to draw distinctions between a necessary and a luxury—sheer nonsense. Necessaries are only necessary when we want them and luxuries have no value when we don't. Circumstances however may force a sale upon you which shall leave you poorer: for instance—you may be ten miles from a house and very hungry—you may meet a man carrying a good luncheon—he may decline to give you a share unless you give him your signet ring, and being very hungry you may do it. This is a forced sale. You know very well that in three hours' time when you have reached town, you will consider the ring of much more value than the food. Your preference of the food was temporary and you knew it. This transaction enriched one man to the detriment of the other, being destitute of that element of permanent preference which confers additional value on goods freely bartered. Esau's was a case in point, and men who say that the wages of labour are not determined by a free sale of that commodity, deserve an answer. But one proposition is enough at a time, so let wages wait awhile and let us hold to the question whether a free exchange of goods can ever make anyone poorer.

If at every sale every one grows richer, where does poverty hail from? Poverty is no child of Commerce. She is the unwelcome offspring of that Physical Necessity which Sir William Thomson calls the Dissipation of Energy. We may paraphrase this law by saying that everything wears out. Everything we want requires continual renewal, continual production as we term it. Everything we have is for ever losing value by decay. We call the use we get of things Consumption, and this consumption ends them so far as their use is concerned. Consumption makes us poor—Production makes us rich; that is clear enough, but barter adds to the wealth which production creates and it does this without taking anything from any one; simply by letting each man get that which he wants most by giving that which he wants least. This is merely a paraphrase of the free-trade maxim often quoted now as if it were the ne plus ultra of cynical egotism—buy in the cheapest and sell in the dearest market.

If commerce were confined to direct simple barter between producer and consumer the word 'profit' would never be applicable. A man cannot become rich in the ordinary sense by direct barter. He reaps a benefit. He does not make a profit. This profit is a word derived from money transactions with the help of middlemen who neither produce nor consume the articles in which they trade. Let us not think of them to-day but examine how men might be rich and become rich if all men were simply producers and consumers dealing with one another by exchange. Let us see whether under these very simple conditions wealth would of necessity be gained only by the loss of others.

It is not possible to show in a short article any clear picture of the vastly complex system of trade which has taken the place of simple barter, but it may be possible to show how by simple barter in a limited community every one is better off than he would be without it, and that nevertheless even in this primitive state of things we might have rich and poor though no man ever gave anything without receiving that which he deliberately preferred.

Let us conceive a little elementary community of say five families, isolated in a little fertile island, all honest people and each taking up a distinct branch of production—farmers—fishers—hunters—weavers—and carpenters. They can read and write and keep accounts, using bits of slate for the purpose. They are in easy circumstances because each family can with moderate labour turn out considerably more of its own produce than is required by the whole community. One sixpence is the whole coin of that realm. They reckon prices in pence and in early days conducted their business in rather a clumsy fashion. The weaver bought six-pennyworth of grain, the farmer six-pennyworth of fish, the fisherman six-pennyworth of cloth, and round went the sixpence, starting from the weaver and coming back to him. But the weaver wanted more grain and had to begin again at once, buying another six-pennyworth. The farmer this time bought six-pennyworth of venison, and the hunter six-pennyworth of cloth, the circuit was again complete and the sixpence flew round it in one direction while the goods went round the other way. By maintaining a sufficiently rapid circulation the sixpence sufficed for very large transactions, but there could be no hoarding of money and no money profit. They soon learned better. They invented credit—locked up the sixpence in the national bank—kept accounts and had a sort of clearing-house where from time to time balances were struck and carried to the new account. So long as the transactions were all conducted on the cash down principle, it was abundantly clear that the whole process of buying and selling must come to a standstill if the sixpence did not continually return to each person who paid it. A closed circuit (as we might say, borrowing a metaphor from electrical science,) was necessary, round which the sixpence travelled from consumer to producer while the goods went the other way from producer to consumer. The simplest circuit was that of two people, as when grain was exchanged directly against cloth, but not unfrequently all five producers were in the circuit at once. The weaver wanted grain, the farmer fish, the fisher venison, the hunter wood, the carpenter cloth. The order in which they took the goods could be varied as much as the laws of permutation would allow, but ultimately each family got for the purposes of consumption an amount of goods of exactly the same value as that which they produced for sale. The laws of supply and demand were in full force and fixed the prices which rose and fell in open market. Property was divided on principles of the strictest egotism, and yet no one could make a profit of one half-penny though all reaped great benefit from the mutual interchange of produce.

When the sixpence was locked up, these closed circuits of trade became much less obvious, nevertheless a little thought will show that they were still a necessary feature of the trade. A made what B used, B could buy it, because he made what C used, and he could buy it because he made what A used, and so each got the worth of his work. Even in this state of things there were rich and poor. The rich man was he who could produce most, that most corresponding to a perfectly definite idea, being measurable in fact in pennies, though no pennies ever passed. If we imagine all our closed barter circuits drawn as lines from man to man with arrow-heads to show which way the goods travelled, every man would have as many lines coming into him as went out, but the rich man would have many lines running through him, and the poor man few. Not only were there rich and poor, but whole classes might become poor by no fault of theirs and by no visitation of Heaven. Even a so-called improved mode of production would do the harm.

As a matter of fact in course of time the Hunters became wretchedly poor. It happened this way. The farmers bred their cattle judiciously and improved their modes of cultivation, and with very little trouble to themselves they produced much meat and corn. They consequently let the weavers, fishers, and carpenters have beef very cheap. Simultaneously game went out of fashion. The hunters killed as many deer as ever, but, alas, there was no demand for venison. So they could not buy the beef, cheap as it was, nor indeed anything else. The circuits round which the credited sixpence travelled, never included the poor hunters. They could produce as much as ever in mere bulk of meat, but nobody wanted their produce. Nothing was taken away from them, the whole wealth of the community was apparently increased, for they could produce as much as ever and the farmers could produce more, yet the hunters were ruined, and had to live on the charity of the four other families, who were sorry for them and good to them, but, in the beaten way of barter, were unanimous in declining game of any sort. Every now and then they went through a pretence of buying a little antelope, which they threw to the cat, who, comically enough, also refused to touch it, but these purchases were only too obviously charitable—a mere pretence and not the real thing.

The island was in a great puzzle. They all loathed game: even the hunters themselves felt that they were degraded by eating it. What was to be done? After much discussion it was settled that as the farmers appeared to be richest of the families, the hunters should take to farming. This did not answer well. The two families could now produce twice as much meat and grain as was wanted. But no matter how much they produced they could not induce the other families to give them more fish, clothes and shelter than the single family of farmers got before. Indeed, owing to the large production the price of farm-produce fell, so that the hunters and farmers were now both very badly off. The hunters had gained something at the expense of the farmers; but the farmers lost more than the hunters gained. The community was poorer now than it was before and yet no national wealth had been taken from it. It had lost nothing but a want. The game was there and the hunters were there, but the consumers had vanished and the whole community was the poorer because it had lost its relish for wild fowl. Redistribution of the produce helped some and harmed others. Moreover, there was much idleness among the farmers and hunter-farmers, and this led them into mischief.

One day time was hanging very heavily on young George Hunter-Farmer's hands. He was a lad with a taste for pretty things and he took to executing a sort of embroidery with shells and feathers on his hat. Charlie Weaver, passing, said 'Awfully jolly hat, that.' 'Glad you like it,' said poor George. 'I say, I'll buy that hat from you,' quoth Charlie, and this was a turning-point in the history of the island.

The Hunter-Farmers proved to be possessed of very superior taste. They gave up farming and, taking to the production of ornamental clothes, rapidly became the richest people in the island. The farmers again got as much fish, cloth and carpentering as they ever had, and the spirits of the whole island rose materially as a result of their new and tasteful costume.

Now let us get back to Jack's question and see what answer is given by the facts in our little community. In the ideal island one class or one man could only be richer than another in virtue of producing goods of more value: the value being measured in that commonplace but definite measure of which one penny is the unit. The question of one man in a given class differs from that of one class relatively to another. Let us take the single man first. The strong clever farmer would be richer than the weak stupid one with equal opportunities. Represented in our diagram of barter circuits, he would have a large number of lines flowing out to the consumers of his produce and the same number coming back from those whose produce he consumes. His productions benefit all those who receive them and the goods he consumes are enhanced in value by his demand for them. He benefits those who buy from him and those who sell to him, but the goods which he receives might have gone to his competitors if he had been less strong and clever. He does not get rich at the expense of the weavers or carpenters; to them he is a pure benefit, but he is a nuisance to the weak and stupid farmers. They receive less clothes and carpentry than they would if he were no better than they. If the productions of those with whom he trades be a fixed quantity, then the whole excess of wealth which one producer of a given class obtains beyond that reaped by his competitors must be at their expense.

It is the knowledge of this which makes workmen of one class agree to work at one rate of wages. The clever strong man takes his advantage in ease of work and certainty of employment, not in extra wealth. This is commonly thought very wicked—by those who are not workmen.

If we now pass from the individual to the class, we may easily suppose one class, say that of the farmers, to be richer than the others at the outset of our story. Were they rich at the expense of their neighbours? No. Their riches benefited all their neighbours: as a class they would have more barter lines going out and coming back than any other class; that is to say, they could have plenty of the productions of all other classes, whereas the fishers might have to do without game and to be content with poor clothes. Nevertheless the fishermen would be the richer for being able to barter their fish for plenty of meat and corn. It seems pretty clear that the rich might exist as a useful class on two conditions; that they shall only be rich in proportion to the value of what they produce; and that their productions shall not compete with those of any other class. A stickler for equality might even under these circumstances insist that the poorer members of the community ought to take to the business of the rich until these are brought down from their eminence, but we found that the plan did not answer for the hunters and it would not have answered for the poor weavers either. This plan only benefits the poor at the expense of the rich, and injures the whole community besides.

We found that an improvement in production did not necessarily act well in an otherwise stationary community. It ruined one class of competitors in the food market. This is the same case as that of the strong and weak individual competing in the same class. If the improvement in production had occurred in a class which could not compete with any other, say the weavers, the improvement would have been pure gain. All the community would have become sensibly richer. But if the rest of the community had been stationary the weavers would not have become relatively richer than before—or if they did—must have done so at the expense of others, giving the fishermen for instance better clothes, but depriving them of some corn or carpentry. Remark, moreover, that the fishermen might not prefer the better clothes to the corn and meat which they might lose. They might be out-bid by the prosperous weavers.

How is it then that in one community, one class may *be* rich and benefit all the others, and yet can not *get* rich otherwise than at the expense of others? The explanation is that the gain in wealth of certain classes can take place without injuring the others during any period of general advance. Some may advance farther than others and so come to be relatively rich, but when all are becoming richer, it would be an abuse of terms to say that those who are advancing fastest are gaining wealth at the expense of their neighbours. The general advance was shown in its simplest form when, simultaneously with the improvement of agricultural methods, we had a new general want discovered and supplied by those who were thrown out of work in consequence of the diminution in the number of hands required to supply food. Under these circumstances every one might have as much cloth, carpentry or fish as before, but the farmers might have more of their corn and of the new produce than any one else. The hunters and all the others might have food which they preferred to the old food and also handsome clothes. So every one was richer and no one poorer; but remark that the two improving classes, farmers and hunters, got no more from the other classes than before. Yet they benefited those other classes.

The extension is quite easy. In a rapidly advancing community those who make improvements may become rapidly rich relatively to those who make none, but they nevertheless benefit the stationary classes. Their own benefit is wholly derived from their own productions and from the productions of those who, like themselves, make improvements. The relative gain of various classes may be very different, but it is clearly possible that one class may gain much more than others if its goods are much desired. That class may then become rich and remain rich and injure no one, but benefit all. If then any youthful Jack asks us whether a man who gets rich does so at the expense of others, we must answer he may; and again he may not. We must ask him to read this article and then to remember that the existing state of society with limited land, with capitalists and classes who are commonly called non-producing, is far more complex than that of our ideal island. Nevertheless, if he follows us so far, we may venture to tackle the complications introduced by these additions to the problem. The closed barter circuit, as a graphic conception, is a useful device for disentangling this complexity.

Poverty, take it as you will, must be hard to bear, but much harder if we think we have been robbed;—that the rich fellows have got our share of the world's goods by some hocus-pocus which is morally theft. Now this conception of poverty would be impossible in our ideal island. No hunter was robbed. There were no monopolies. There was land enough. And yet the hunter was poor till that lucky chance of the embroidered cap turned up. Vice, idleness, sickness and infirmity, all cause poverty. Bad laws, bad government, an over-population may cause poverty, and have caused poverty, but the main cause of this great evil is the same in our Great Britain as in the little fancied island.

The bulk of our poor are poor because they produce the wrong things. What then are the right ones? Trade is always working at this problem, solving it ill or well as our traders are wise or foolish.

The question whether one man really could under a system of pure barter be rich at the expense of his neighbour is altogether of secondary importance and greatly a matter of words. The man who produced much would be rich, but his produce would be enjoyed by all who gave him what they valued less. Nevertheless if by producing much he threw others out of work, he might be said to be rich at their expense; but this would be a very unfair way to word it, since when they found out new right things to make, they would all be rich and be richer than before.

It is absolutely false to suppose that the well-being of one man must entail the ill-being of another, but it may produce it.

What does all this mean? It is an attempt to show that poverty might be found where all men laboured, where the fertile soil was no monopoly, where no capitalist claimed a share of produce and where no profit was made by trade. These things may or may not cause poverty, but let us put them on one side for the moment. When they are all removed, we can still picture to ourselves a state where a whole class of men might be poor with no injustice, no moral blame attaching either to the poorer or richer classes.

In order to show this two assumptions were necessary: 1st. Our labourers were not self-sufficing; a division of labour was treated as inevitable: 2nd. Some men produced the wrong thing by their labour. Why was the thing wrong? Both when our hunters still hunted and when they gave up hunting for farming, they produced food, which at first seems not only a right thing, but the most right of all things, being a universal need. Their produce of venison was a wrong thing, because others wanted that special food either not at all, or so little that they did not care to purchase it by giving what cost them continual labour. To avoid poverty a man must not only produce what another may want, he must produce what is wanted by a second man who is both able and willing to give in return something which the first wants more than his own produce. There is no remarkable selfishness involved in this law. No socialistic reformer would propose that all men should be condemned to eat a food they did not like in order that the producers of that food should be richer. They would at once admit that the hunters, when venison became unpopular, should be set to produce a right thing instead of a wrong one; and the right thing is something which others would be willing and able to buy. Even food may be a wrong thing to produce when there is plenty of other preferable food for every one in the market. We found that matters were mended very little by setting the hunters to produce preferable food. A right thing was produced, but too many men were employed in its production. This is a second cause of poverty wholly independent of all the stock bogies nursed by social agitators. If ten men can grow all the wheat really needed by a community, and twenty are employed to do it, they will be poor. They could afford to be idle no doubt, but you can have too much even of idleness, especially when you are poor. And what was the remedy against these two modes of poverty? Not an improvement in agriculture; that would make the twenty poorer than ever, five men then could grow wheat enough for all, and putting twenty men to do it would be worse than before. The one conceivable remedy is to put the surplus hands not really required for the food production to produce a new right thing. That is to say, to produce something satisfying a new want felt by those who can give back that which the producers need; or that which will get them what they need by exchange. When this new want was found and the equilibrium between production and consumption re-established, then indeed the improvement in agriculture was found to be a real gain. The community had more wants satisfied than before.

Direct barter was not in the least necessary, provided the barter circuit was closed in the way described. The easiest way

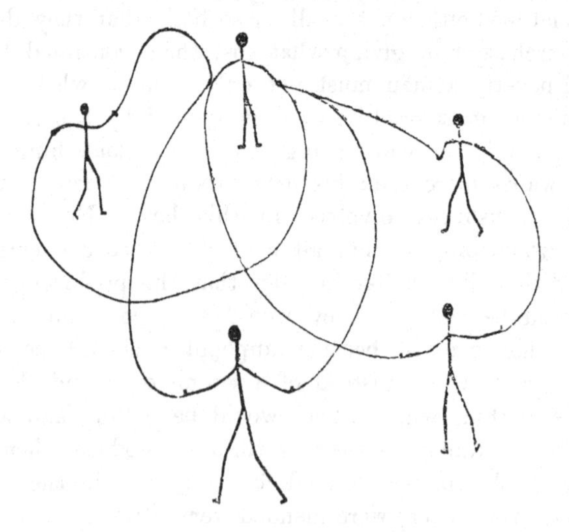

of representing this condition to the mind is by a diagram. Let the reader sketch five little rudimentary people standing upright: let lines going out of the strokes to the right indicate produce, which each has to sell—one line representing the unit of produce per annum. Similarly let a line arriving at his left hand represent the goods which each man consumes, one line representing one unit of consumption per annum. Now if any two join hands by means of a line, this line forms a closed circuit showing that produce of equal value has been interchanged. All the lines that go out from a man's right hand must be followed home to his left without using the same line twice. This indicates that all the produce has been sold. On a system of pure barter he never could get rid of any produce until one of these imaginary circuits was closed. But the circuit might be a very long one. By the introduction of the system of credit under which the coin of the realm was all in the national bank, a man might sell his produce and get nothing for it. One of the lines would not come home, he would never be paid his due. But excluding this in our ideal community we see that riches and poverty may exist where there are no capitalists—no bad people—no money. Further, that it does not arise by any process analogous to robbery or even profit. If a man can make nothing wanted by other persons able to give him back an equivalent, he is thereby reduced to his own resources. He must in that case make his own food, clothes and shelter. That will always mean that he is poor. We may further see that the way to make him rich is not to teach him to produce that which is already produced in quantity sufficient for the community, but to teach him how he may produce that which will be consumed by those who already produce what he wants in greater quantities than they require for their own use. If the weavers, fishers, farmers and carpenters had only just been able to make what they required, the hunters' lucky cap would never have started a new industry. Now the actual condition of the world with capital and wages is not so very different from that of our ideal island. The poor people are those who make the wrong things, and the right things it will pay them to make are the things which the rich want. We have shown that poverty and riches may co-exist where the conditions of existence are very primitive and where no man makes a profit either in money or in kind: that even under these primitive conditions an increase of production by one class, as by the farmers, may benefit some classes and impoverish others, and that on the other hand the introduction of a new industry may, by supplying a new want, restore wealth, making every one richer all round. On the other hand it might be the bread out of the mouths of one or more classes. There is no general rule.

In this primitive community with no capital, no wages, no money in circulation, consisting wholly of producers and consumers, we nevertheless see that many of the difficulties arise which we meet with in our complex body commercial. There were rich and poor; the rich were those who could produce much that was wanted by others, who in their turn produced what the rich desired. The poor were those who could produce little for any one able to repay them with what they considered necessary. It is not enough to say that in our imaginary island the rich are those who produce most. It is not enough even to say that the rich are those who produce most of those commodities that are wanted. We have to go farther and remember that want felt by a man who can not pay is useless to the producer, and that the only real payment is the return of something which he can consume. Nevertheless, when we so arrange the conditions as to make the bartering obvious, we still find that a change in taste or in the method of production may result in a completely new distribution of wealth. Moreover, it can not be affirmed as an absolute truth that an improvement in manufacture must benefit the whole community. When the farmers improved their agriculture they ruined the hunters. They took no goods from them, but they deprived them of that demand which gave their goods value. This truth is extremely unpalatable to the political economist, who dearly loves to think that every increase in output must be an increase of wealth benefiting all. The elementary case chosen shows how an increased output of an absolutely necessary production may nevertheless ruin another class of producers.

This result followed the improvement because the two classes really supplied the same want, that of food. The increased wealth of the farmers and of the other producers was therefore gained to some extent at the expense of the hunters. These were thrown out of work and could only get their living by learning a fresh business. They naturally made the mistake of taking up what seemed a profitable trade, but more people were still employed in producing food than were really required for that purpose. Hence they simply shared their poverty with the farmers, and gained at their expense. In the end the whole community did benefit, but only after a new want had been discovered, and the surplus labour was employed to supply this new want. Then the improvement in farming and the supply of a new enjoyment made them all better off than before. Thus we see that a benefit resulting from barter may or may not be obtained at the expense of others. An improvement in production may or may not be followed by increased comfort all round. It would not be difficult to distinguish between the cases, but these distinctions when written out would make tedious reading. Our present contention is merely this. In a community of honest people, all producers, with no capitalist and even with no money, there would still be rich and poor classes, and improvements in production might result in suffering to some classes; but this suffering is not necessarily the result of any profit made out of the process of buying and selling; it was due to one of two causes. Either some producer was making a wrong thing, something which no one able to pay for it wanted; or else too many producers were engaged in the production of a right thing. The consumers were only willing to give so much of their own produce for meat as would support a limited number of meat-producers. When however a new want is discovered, those who supply it become rich and enrich others.

No one suffers if the new supply competes with nothing else. The old things are wanted in the same quantity as before, and if, as we assume, they can be supplied in sufficient quantity for all by the number of hands actually employed, the new want can be supplied and enjoyed with absolute benefit to every one. The danger is that with complex wants, men may be tempted to work too hard. In civilised communities our wants are innumerable: they are satisfied in a very remarkable way. Even the very poor consume more produce than the very rich in uncivilised communities, where the wants are few. But the savages have most leisure. Each savage works moderately and supplies his own wants, which are few. The increase of civilisation enables comparatively few men to supply the wants of the whole community as regards any one article. This sets large numbers free to supply new wants. This process goes on gradually, for reasons. In the first place, the improvements in manufacture which increase the output per man in a given industry come gradually. And in the second place, the new wants to which the surplus labour can minister are only gradually developed and gradually discovered. In the third place, the channels of distribution by which what we have termed the closed circuits of barter are established can only be discovered tentatively and by slow degrees. But progress in wealth and its distribution by barter are both compatible with increased comfort all round. In trade one man's gain is not of necessity another man's loss.

SCIENTIFIC AND TECHNICAL EDUCATION

TECHNICAL EDUCATION.[1]

THE resolutions arrived at by the Conference on Technical Education, and the able address of your President, prove that this Society is fully alive to the necessity for improved scientific instruction throughout the country. I will not, therefore, detain you with any arguments upon general principles which have been affirmed by every class of society, but taking it for granted that we are all of one mind in believing that the education of our artisans, our manufacturers, and our engineers should be improved by the improved and extended teaching of science, I purpose to-night to confine myself chiefly to practical suggestions of steps which I think might be immediately taken towards the object which we have all in view.

I well know the difficulty and danger of making practical suggestions. The man who confines himself to general principles, and the critic who assails existing abuses, is sure to carry a large portion of his audience with him. The abuses are often palpable, and the general truths soon become popular truisms; but the man who brings forward new proposals for definite action, cannot and ought not to expect an equally ready adhesion to his schemes. They must in their turn run the gauntlet of criticism, and be subject to many successive amendments, before any large body will consent to put them in practice; but a man who will thus subject himself to criticism, with the object of attaining an avowedly good end, may at least ask for an indulgent hearing before the criticism begins, and this indulgence I ask from you to-night.

Probably the general improvement in the scientific education of the community at large can only be effected by the adoption of some such scheme as that proposed by the Schools Inquiry Commission, having for its result a complete system of schools of

[1] An address read before the Royal Scottish Society of Arts in Edinburgh, Jan. 11, 1869.

different grades under local management, but subject to general regulations laid down by a department of the government. The Commission have shown how the necessary funds for such a system can be obtained, and have proposed a definite scheme for the administration of these funds. If schools, of several grades systematically arranged, were once established, it would be an easy matter to insist on the introduction into a given number of each grade such distinctly scientific training as would warrant the appellation of science schools, as distinguished from classical schools, and results would soon show whether science does or does not afford an excellent material for mental culture, besides that merely useful information which it is not the chief object of education to impart. If a system of graded schools existed, we could apply towards their improvement the masses of information which have been collected as to foreign courses of study. Now, even when we know our own wishes, what bodies are we to attack? It is hopeless to expect that a successful experiment on any great scale can be made so long as the schools of the country are under an infinite number of different trustees, corporations, and committees wholly incapable of combined action. I do not purpose to-night to examine the proposals of the Commission, but will direct your attention, in connection with this part of the subject, to the report of the Sub-Committee on Technical Education appointed by the London Society of Arts.

I fear, however, that we shall have to wait some years before a complete system of graded schools can be established; and it is most desirable that even if a complete system of scientific instruction cannot be created offhand, something should be done to remedy our deficiencies at once. I will therefore make a few suggestions as to means by which the scientific education of artisans and foremen, on the one hand, and manufacturers and engineers, on the other hand, might be improved under existing institutions.

Workmen and teachers of workmen now receive some scientific instruction from teachers in schools, assisted by the Science and Art Department of the Committee of Council on Education; and in the Report for 1866, on Science Schools and Classes, Captain Donnelly says, that 'the various modifications and enlargements made from time to time in the system of aid to science instruction have now rendered it a system which may fairly be said to meet the requirements of the country, as far as elementary science is concerned.' From that assertion I most thoroughly dissent, but I am not prepared to join the ranks of those who think that no good thing can come out of South Kensington. South Kensington has done something for elementary scientific instruction; it has established a system of examinations and rewards, which have stimulated the study of the elements of science. Although I do not approve of all that has been done, although I think that much remains to be done, I will not therefore refuse such assistance as is given, but will rather endeavour to state how I think the value of that assistance can be greatly increased and the sphere of its action greatly enlarged. If therefore, in what I am about to say, I criticise freely the acts of the Science and Art Department, I beg you will consider that the criticism is not hostile, but is made with a view to the improvement, not to the injury, of the department.

Now, I will draw your attention to what has been done by the department in support of mechanical drawing, and you will thus understand better why I dissent from Captain Donnelly's opinion that the system meets the requirements of the country. I choose mechanical drawing, because I think that a knowledge of the elements of this art is the most useful of all to the workman. Why do we think reading, writing, and arithmetic so useful and essential? Because, without a knowledge of these, little accurate knowledge of any kind can be acquired. These are the tools we put into all men's hands with which to work at the mine of learning, and the knowledge of mechanical drawing is almost as essential a tool as the three others for the working man. Without a knowledge of this art he cannot represent his own mechanical ideas; worse still, he cannot understand the ideas of others, which he is called upon in his business, as a mason, mechanic, carpenter, joiner, fitter, erecter, and so forth, to carry out; perhaps, worst of all, he is debarred from learning any of the principles of his special business from books or journals, because he cannot understand the diagrams. So true is this that most skilled artisans and all foremen do contrive, after infinite labour, somehow to pick up such a knowledge of drawing as allows them to understand the simpler representations of machinery and construction. Not a few teach themselves how to draw a little; but they have to teach themselves, or at least to learn in little classes, where the appliances and assistance received is almost pitiable. The men requiring this knowledge are the whole class of skilled workmen; and so important is this knowledge to them that the French Government Commission on Technical Education came to the conclusion, that for the working classes technical education meant *instruction in mechanical drawing*; and I entirely agree with this conclusion because, as I have said before, it is a tool or key with which the able and energetic can for themselves open fresh paths to knowledge, and because experience has shown that it can be taught, and taught successfully, to men who have only learnt how to write, read, and cipher.

Which is really the most important to the working classes in Great Britain, such a power of representing objects by free-hand drawing as can be given to workmen, coupled with such an appreciation of the fine arts as they can be expected to acquire; or the power of representing the elements of machinery and constructive details, coupled with the power of understanding the representation of mechanism and structures?

The first branch of knowledge will be useful to all those classes engaged in the design and execution of artistic produce. The second branch of knowledge will be useful to all engaged with those productions which require the use of machinery or structures. Can we doubt which is the most important class in England or in any country? The whole agricultural population of this country, and the whole manufacturing population except the artists, using this word in its largest sense, require a knowledge of mechanical drawing, and the artists would be the better for having it.

Now, let us compare what Government does for mechanical drawing, as compared with artistic drawing—for art as compared with science.

There are 99 Government schools of art in which free-hand drawing is taught, and aid is given to 560 other schools. 17,210 students learn free-hand drawing in the Government schools;

and altogether, in 1866, 105,695 were taught drawing through the agency of the department—an admirable result, and one which probably does meet the requirements of the people. 2,306 teachers were examined, and 1,583 of these obtained certificates of competency. Turning to mechanical drawing, the contrast is very great. 16 teachers were examined, and 13 passed. 13 as compared with 1,583 for free-hand drawing. 20 schools receive some assistance, because they give some instruction in mechanical drawing. 20 as compared with nearly 700 for free-hand drawing. The total number of persons receiving instruction in mechanical drawing in Great Britain and Ireland is 1,207. This is the number we must contrast with 106,000 taught free-hand drawing. Lastly, turning to the payments for results, we find 34,851*l*. paid to encourage free-hand drawing, and about 340*l*. paid for the encouragement of mechanical drawing. I ask, Does this proportion of one hundred to one correspond to the relative importance of artistic and scientific drawing? Is this sum of 340*l*. such a sum as can fairly be said to encourage mechanical drawing in such a way as to meet the requirements of the country? Why, the total payments on results for the 23 branches of scientific instruction under the patronage of this department is 5,000*l*., about one-seventh of the sum spent in encouraging free-hand drawing! About 7,000 students learn these 23 branches of science, as compared with 105,000 learning the elements of free-hand drawing. The department has not yet, I think, earned the right to sit down complacently, satisfied that the requirements of the country are met.

But you may fairly ask me, how it is possible, if mechanical drawing be really so important to us as a manufacturing nation, that it is so little taught? Does not the number of art students prove that there is a demand for artistic training, and the small number of mechanical schools show that there is no demand for the scientific branch? The answers to these questions are, I think, simple. Our countrymen have so great a natural aptitude for all matters connected with mechanics, that all who need this special knowledge do, by hard labour under sheer necessity, acquire it. Only those plants and trees which are not indigenous to the soil require costly apparatus and careful watching. Art is not indigenous here, but science is, and above all, the science of mechanics; therefore, while under the fostering care of governments and academies, our fine art languishes, though pampered with wealth, mechanical arts flourish wherever two or three men in fustian combine to work together. But what should we think of the farmer who neglected the produce for which his soil was fit in favour of exotic plants, which could only be coaxed into feeble life with infinite care and ruinous expenditure?

Another reason for the neglect of mechanical drawing in favour of free-hand drawing is to be found in the dilettante interest which the more cultivated classes take in art. The patrons of schools, squires, clergymen and their families, can sketch a little, or at least know what a free-hand drawing looks like, while they are generally utterly ignorant of the meaning of a cross section or an elevation. But when the schools and classes are initiated by workmen, then we see a class of mechanical drawing invariably instituted as of the first necessity; and if any one will question workmen and foremen as to their desire for a knowledge of artistic drawing as compared with mechanical drawing, they will soon cease to doubt the existence of a real demand, although in all Scotland there were only 61 persons in 1866 who received this kind of instruction through schools assisted by Government through Captain Donnelly's department.

I hope you are convinced of the importance of this branch of elementary scientific education, and will now proceed to discuss remedies which ought at once to be adopted.

It is quite insufficient to pay teachers on results, for the following reasons:—

1st. The persons really competent to teach mechanical drawing cannot be professional teachers, but must be practical draughtsmen. The knowledge which the workman wants cannot be given him at a school where the master teaches reading, writing, arithmetic, geography, and has just managed to get a certificate that he understands geometrical projection. The knowledge the workman wants is the knowledge acquired by the draughtsman from long familiarity with practical constructions. To this man plans, cross sections, and elevations of machinery are all as real and as natural as the things themselves. He does not look upon them as results of geometrical propositions—he generally does not know the meaning of an orthogonal projection, and is wholly ignorant of the scientific part of descriptive geometry; but he is familiar with the methods of representing the most complete structures in such a way that the most accomplished mathematician shall have no fault to find with his work, while it will be intelligible to the workman of average experience. These draughtsmen cannot be had in parish schools, and to offer those in towns a few shillings a head for such of their students as may pass an examination is a farce. These small payments are of great use to the professional teacher, whose business it is to teach, and who can send up men for examination in many classes, but payment by results will hardly ever induce draughtsmen to form classes. Then, for mechanical draughting, a large space is required for each pupil, a good light, at least two hours' continuous work at each lesson, a collection of somewhat expensive drawings and much more expensive models, and all these things cannot be paid for out of the 340*l*. allotted to this branch of science in Great Britain, nor out of any increased sum earned on the same plan. Lastly, the instruments used are expensive, and such as young mechanics can rarely afford to buy. On all these grounds, which distinguish this branch from most of the other branches of mechanical science, it must be separated from these, and encouraged on a wholly different plan.

I therefore propose that every Government school of art or design, of which there are about 100, *should be required to open classes of mechanical drawing, and that these classes should be taught by professional draughtsmen*—who should receive a salary such as would make it well worth while for the leading men in each town to compete for the appointment—in addition, the teacher should receive a portion of the fees paid by all classes of students, for I am no believer in gratuitous instruction.

Some classes are even now open for this branch of drawing, and the students in these do not seem to be counted in Captain Donnelly's Report, nor do I find any special mention of them in the general report of the department. I will refrain from criticising the work done in these classes, taught by men who are really good masters of artistic drawing, and who might, perhaps, teach elementary geometrical projection, having obtained certificates of competency in this respect, but they are not professional mechanical draughtsmen, and cannot teach things with which they are unacquainted. The contrast between the artistic work done in our schools of design, and the mechanical work, is as glaring as the discrepancy between the 340*l.* and 35,000*l.* paid for the encouragement of the two branches.

Instead of the one little heterogeneous class taught by a certificated teacher, there should be in each school four distinct departments—one for mechanical drawing proper, one for the drawing of buildings, one for plans and surveys, and one for geometrical projection. In each department there should be an elementary and an advanced class, and each class should be taught by men acquainted with the special work.

There should be prizes given upon the plan adopted for artistic drawing, and these prizes should be equal in importance to those given for excellence in the fine arts.

In each school there would therefore be prizes—

1st, For highly-finished shaded and coloured drawings.

2nd, For linear drawings, neither coloured nor shaded.

3rd, For large scale detail drawings, boldly coloured in the style required for use in the machine shop or in the field.

4th, For plans and surveys.

5th, For drawings of the constructive details of buildings.

6th, For writing and printing.

7th, For geometrical projections.

In the larger schools I would add prizes for perspective drawings, accurately corresponding to given plans and elevations.

There should be an annual exhibition of the work executed by the students, and there should be a national exhibition of the prize drawings and national prizes for the whole of Great Britain. This would be encouragement on a very different scale from the prizes now offered for little linear drawings, five of which are to be executed in four hours, and which test the power of the draughtsman to enlarge little lithographed sketches. I am convinced that every one of the above classes would fill in every considerable town; but if, instead of teaching the practical work, geometrical projection and the French form of descriptive geometry be taught, hardly a student will be found.

Another objection may be, that mechanical draughting ought to be taught in connection with scientific establishments rather than through schools of art, inasmuch as it is more closely related to mechanics and mathematics than to anything æsthetic. My answer is, that the arrangements in schools for fine arts are adapted for drawing both as to space and light; that Government establishments for the fine arts exist already all over the country, to which the new classes can be affiliated, whereas the schools of science do not exist; and lastly, that the scientific part of mechanical drawing is wonderfully small. There are excellent draughtsmen who are wholly ignorant of Euclid and guiltless of algebra. The very fact that our workmen do come to understand the drawings of the work they have to execute is a proof that no scientific training is required to render these intelligible. I have known a smith who, taught in a mechanics' institute by a locomotive draughtsman, learned in a very little while to make an admirable set of drawings of the smith's shop in which he worked, with the building, roof, forges, blowing-machinery, etc., in full detail. I have seen a whole class of lads of thirteen and fourteen executing excellent mechanical drawings—so good that I could have afforded to pay them to work in my office; and I am therefore certain that no considerable scientific training is required to enable lads and men thoroughly to understand mechanical drawing, and to represent accurately any ordinary structure. So far am I from thinking that it requires previous scientific training, that I believe mechanical drawing to be the one form of elementary science which practically can be taught in all schools. If we had in our Government schools of art such classes as I have suggested, we should soon find that native draughtsmen would replace the French, Germans, and Swiss, who, to our shame, now fill our drawing offices.

I will now for a time take leave of the Science and Art Department, and consider what should be done in primary schools to carry out the second resolution of the Conference, that elementary science should be taught in all schools. I often see proposals that drawing should be taught in all the primary schools of the country, and the proposition is supported by the argument, that drawing educates the faculties of observation. If this be true to some extent of ordinary perspective drawing, how much more is it true of what I call mechanical drawing? The greater or less accuracy of a perspective sketch of any object is a matter of appreciation. The accuracy of a plan, elevation, and section are matters of certainty. I hold that a child who has acquired the power of measuring the simplest forms, and representing these by a plan and elevations on which the dimensions can be written, has acquired a faculty which will be of use to him in every calling of life, and will have had mental powers awakened which no mere copying would ever awaken, and which will lie dormant even if he learn to represent actual objects by free-hand drawing, a degree of proficiency which cannot be hoped for in primary schools.

It may seem to many that the plans and elevations of natural objects are more difficult of execution than the perspective sketch. This I wholly deny, in Great Britain at least. A child of seven or eight years old can be taught to understand and make the plan of a room, to measure the dimensions, and write them correctly on the plan. The very fact that children understand maps shows the readiness with which the mind receives this simple geometrical conception.

A plan and elevations of any object are only the maps of the top and two sides. The apparatus absolutely necessary to teach the elements of this branch of drawing, is a small collection of simple geometrical wooden models. The figured hand sketch of these on a slate would be all that would be expected from the youngest or poorest scholars; add to this some paper scales and set squares for the older boys, and the parish school would be completely equipped with all that is necessary for the elements of this branch of education—except the teacher. This, then, is the elementary science which, in accordance with the resolution of the Conference of this Society, might, I think, be introduced into every school. When elementary science is mentioned, some think of botany, physiology, the laws of health, political economy, geology, mineralogy, astronomy, and so forth. I hold the unpopular opinion that not one of these things should be taught in any primary school. The sciences may be divided into two classes—the mother, or fundamental sciences, and the derived sciences. Mathematics, physics, and chemistry as a branch of physics, may be called the mother or fundamental sciences. Without a knowledge of these it is useless to attack the derived sciences, except in the most superficial manner as mere sciences of classification. Mathematics and physics, including chemistry, should therefore be the basis of a scientific education, as Latin and Greek have formed the basis of a literary education. Again, the simpler branches of mathematics must precede the study of chemistry or the other branches of physics. Arithmetic is the first branch of mathematics; and the representation by plan, section, and elevation of simple structures may conveniently form the second branch of applied mathematics. What algebra is to arithmetic, geometry is to mechanical drawing, and it is not only possible to give a fair knowledge of practical arithmetic and drawing in primary schools without algebra or geometry, but it is the only way in which we can hope to give that knowledge. Physics and chemistry require comparatively costly apparatus and highly trained teachers; very little that is worth knowing could possibly be learnt in any primary school on these subjects, and, unpopular as the opinion is, I would not make the attempt. Much that will interest the intelligent and awaken new ideas, may be learnt in after life from lectures such as are now being delivered in connection with our admirable museum. I heartily hope that similar lectures will be given in all large towns, but they will rather be a healthy form of intellectual recreation than a means of giving scientific education; and in our primary schools for the working classes I see only one possible branch of science open for our adoption, namely, the scientific representation of mechanical and other simple structures. In making the attempt, we shall start with the support of the whole artisan class, which would be delighted to see its children acquiring precisely those elements of knowledge which as skilled workmen they will have to apply; and we shall start with a strong national bias or talent for the work, whereas in the cultivation of artistic representation the work is impeded by frequent natural incapacity.

The second suggestion which I have, therefore, to press upon your attention is, *that the form of elementary science which is best adapted for introduction into primary schools, is the representation of bodies of simple form by plans, sections and elevations drawn to scale, or with figured dimensions.* To carry out these views in England, it would be necessary that the Committee of Council on Education should add clauses to the Revised Code, under which they should, out of the annual grant made by Parliament, undertake to aid inspected schools giving instruction in elementary mechanical drawing, by annual capitation grants conditional on the attainment of a certain standard of proficiency as attested by the inspector; in other words, this first and most elementary branch of science should be put on the same footing as reading, writing, and arithmetic. It would also be necessary that the teachers trained in normal schools should be required to take up this special study. By these simple regulations the new study would gradually but surely be introduced throughout the primary schools of the country.

I regret that I am not yet sufficiently familiar with the machinery by which the primary schools in Scotland are managed, to be able to make any practical suggestion as to how and where the experiment should here be first tried, but one application will suggest itself to all our minds—I allude to the great charitable hospitals with which Edinburgh is surrounded. Some, at least, of these are devoted to the education of artisans, and all are intended for the instruction of the productive classes—the classes which use machinery if they do not make it. I believe there is no attempt to teach mechanical drawing in any one of these, and I assert that it should be taught in all.

But when a suggestion of this kind is made, it is continually met with the question, How are we to find time? the boys are already fully occupied, and what can we give up? As an answer to this, I will read what was the programme of a well-managed Edinburgh hospital and that of a Prussian industrial school.

Scotland.—Latin, French, English, writing, drawing, singing, dancing, drilling.

Prussia—1st Course.—Mathematics, chemistry, physics, German, arithmetic, drawing, architectural drawing, machine drawing, modelling.

The Latin, French, dancing, singing, and drilling have disappeared, none but the mother or fundamental sciences are taught, and the three forms of drawing have one-third of the whole school time allotted to them. Architectural drawing is no doubt a misnomer; by this is meant the drawing of ordinary buildings.

In the higher course lessons are added in the knowledge of machinery and in mechanics—two very different things—in chemical technology, physical technology, the construction of buildings, the knowledge of materials, and chemical analogy; but drawing has still one-third of the whole school time given to it.

In the Edinburgh hospital only the least useful form of drawing was taught; it got two hours a week, and all the inspector could say of it was, that the course was well adapted to develop any power in this direction which a boy might possess.

Can any Englishman doubt which of these two courses of study is best adapted to train either foremen or manufacturers? who does not know that the little Latin will vanish without leaving even a tincture of literary training? that the French will never be learned in any form which can be practically used, and will never be required by nineteen out of twenty pupils. What foreman has any need of French? Who really thinks that dancing and singing are to be compared in value with any one of the subjects in the Prussian course? Why, then, do not our great industrial schools boldly throw off these traditional bonds, and follow the well-tried course of the German schools? At one time Latin was the one thing which could be accurately taught, and was therefore the one means of mental culture. Long after this ceased to be true it remained a necessary passport to a rise in social position; but except for the Church, who now can suppose that a knowledge of Latin is likely to aid an ambitious youth to rise in life as a knowledge of physics and chemistry would aid him?

I am the last man to decry literary cultivation, but Latin, as taught in middle-class schools, gives no literary culture; French gives still less; and I call, therefore, on all interested in middle-class education to give literary culture through the mother tongue, and mental culture through natural science, which would be learned with an avidity only to be equalled by the distaste with which scraps of dead Latin and imperfect French are rejected by a lad who knows that not a grown man of his class retains any knowledge of either, or cares to acquire it.

Let me not be mistaken. For the highest class of education I look on the dead and foreign languages as essential, but they are useless unless pursued until the student has them well under command, and can forget the instrument through which he communes with the thoughts of men of other nations, and of all ages. I am no fanatic, who would destroy a great instrument of culture by which many generations have been trained, but I say, that for workmen, foremen, and the smaller class of manufacturers and tradesmen, such mental provender as is afforded by a little Latin, French, music, dancing, and drilling, is garbage as compared with the healthy food in the Prussian school.

I will now draw your attention to what can be done in a charitable institution in Paris, worthy of imitation in many respects, although managed by a Catholic religious body called the Frères Chrétiens, who are by no means universally popular. I allude to what is called the school of St. Nicolas in the Rue Vaugirard. This is a boarding-school for the poorer classes, administered and taught by the religious body, the members of which live in the school, and receive their food, and about 10*l.* annually out of the profits of the school. About 1,700 workmen's sons are boarded and clothed in this establishment, and they pay 1*l.* 4*s.* per month during ten months in the year; the school is slightly assisted by charitable bequests, and the average cost of each pupil is 17*l.* per annum, the total revenue being 28,000*l.* The boys are received between the ages of seven and ten, and some do not leave till about eighteen or nineteen. The mechanical drawing executed by boys of thirteen or fourteen in this school is wonderfully good. There were plans of the kitchens to a large scale, showing the details of the cooking apparatus, steam-boiler, and so forth, executed from dimensions taken by the pupils themselves. I saw the drawings in all stages of progress, and each drawing differing in colour and arrangement according to the disposition of the draughtsman, showing that the work was really that of the pupil, and not merely a dead transcript from the teachers' work.

The average cost of a pupil in a Scottish hospital is 41*l*. 10*s*. per annum. Exclusive of bursaries, each pupil in Donaldson's Hospital costs 24*l*. 15*s*. per annum. The cost of each pupil in the French establishment being only 17*l*., may be considered wonderfully small; perhaps the economy will be best understood when I remark, that the Scottish hospitals, for 44,000*l*., educate, board, and clothe 1,064 scholars, while the French school, for 28,000*l*., boards, clothes, and educates 1,700 boys and youths. The dormitories and arrangements for cooking appear excellent, and the youths contrast very favourably in appearance with the pupils in the great French Lycées. The French establishment is also remarkable for a curious and excellent device by which the Frères Chrétiens teach their pupils trades while retaining them within their walls. The plan is devised with the object of retaining moral control over the youths up to a late period in life, and is probably not worthy of imitation in our hospitals, since I believe that a youth of ordinary morality is better fitted to do his duty by mixing in the world than by semi-monastic training; the plan might, however, perhaps be applied to our industrial and reformatory schools. I will, therefore, so far digress as to sketch the system.

The school provides a series of workshops, fitted for the production of numerous articles made by various trades in the town. Thus, in Paris there are workshops for mathematical instruments, levels, lenses, musical instruments, bronze statuettes, packing-boxes, the design of shawls, etc. The school then allows a manufacturer, say of optical instruments, to use one of these workshops with the tools it contains. The manufacturer employs as apprentices, or rather as workmen, as many pupils of the school as the workshop will accommodate, introducing one skilled workman into the shop to teach and direct the lads, who remain for four years as apprentices, receiving nothing for their labour. The master makes a profit by these apprentices, but is bound to accept any lad the authorities choose to give him. The school receives each year from the apprentice the monthly payment of 1*l*. 4*s*., and continues to board and clothe him. The apprentice works nine hours per diem in the shop, has classes in the evening, and learns to make exactly the articles required in the trade, so that on leaving the workshop he is able to command the full wages of a workman, and, indeed, in some departments, as in shawl-designing, a pupil in the fourth year may be worth 70*l.* or 80*l.* to the master. I saw 140 apprentices thus occupied, and there is no doubt that the plan is thoroughly successful. It might be applied to reformatories, so as to give lads a real trade education, and interest them by allowing them to make things which they know are of real use; but my chief object in drawing your attention to the École St. Nicolas was neither to insist on its economy nor its successful industrial education, but to the admirable mechanical drawing which is executed by lads at a very early age. I fear that I must have wearied you by ringing continual changes on those two words, and I will endeavour to justify the high value which I have attached to this branch of elementary education by quoting the following passage from the report of the French Commission on Technical Education:—

'*Drawing, with all its applications to the different industrial arts, should be considered as the principal means to be employed in technical education.*'

Granting the great importance of drawing, you may fairly ask me what analogous proposals I have to make as to mathematics, mechanics, chemistry, and physics? I may remind you that I have already expressed the opinion that these great studies will not be successfully prosecuted without a great reform in all our middle-class schools, effected by parliamentary legislation of a very difficult nature. I fear the elementary schools, inspected and aided by the Council on Education, can do nothing to forward these studies, except by improving the foundation on which they ought to be built; and I am certain that the grants administered by the Science and Art Department can do very little to meet the requirements of the country as to these great studies. Indeed, I am not prepared to hand over the scientific training of the country to that department. The subject of the introduction of public graded schools, with classical and scientific courses fit for this great country, is too vast a subject to be treated of as part of an address. I am not propounding a complete scheme for the scientific regeneration of the country, but only endeavouring to make some practical suggestions, which might be carried into effect without extensive parliamentary legislation, and, omitting any reference to what may be called the secondary schools, I will now make some remarks on professional and scientific education of the higher grades.

I am opposed to the creation of special colleges for the education of special professions. In my introductory lecture delivered at the University I explained at some length my reasons for preferring that our universities should be developed so as to train men for the new learned professions. It is said, and said truly, that the universities will never be able to train so fully for any one profession as a special school would, and this with me is a reason for preferring the university. A special college will attempt to teach a man to be an engineer, and I hold that it will necessarily fail in doing this, for that practice is the only training which can ever give a man that knowledge which is essential before he can be called an engineer. I apprehend that what is true of engineering, civil and mechanical, is also true of the cognate professions of architecture, building, and of the management of large factories. These things cannot really be taught in classes; and a personal inspection of the very excellent special colleges abroad only convinced me that they were (so far as a great portion of professional training is concerned) merely passable makeshifts, replacing the English plan of apprenticeship by an inferior system.

But while the business or profession can only be taught by practice, the preparation for that practice can and ought to be given in schools. This preparation is admirably given in the foreign colleges, but here again the special college labours under a special disadvantage. The preparation for the learned professions should consist in the acquirement of the fundamental sciences of mathematics, chemistry, and physics, with some derived or secondary sciences, such as mechanics, geology, etc. Now all these things are already taught at universities, and the universities command the very best men as professors. It is better that all the architects, engineers, and manufacturers should learn their mathematics, chemistry, and physics from one man in one class-room, for they will then learn from the very best man who can be found; and, abroad, those special colleges are the best which act on this principle, becoming really universities in all but the name. Another disadvantage of the special school is, that they are led to carry this preparatory training much too far. A glaring instance of this is to be found in the Polytechnic School in Paris, where every Government engineer receives a mathematical training such as would fit him to be a wrangler at Cambridge, and very generally unfits him to be an engineer. Moreover, I desire for every engineering student a liberal education. It is probable that wide differences of opinion would be found among us if we were called upon to define a truly liberal education; and the liberal education of to-day will not be the liberal education of to-morrow. But we may all agree that the general course of education given in a university, where men of all professions and every turn of mind mix together, is more likely to be liberal and wide in its scope than the education given in a college devoted to any one profession, be that profession law, physic, or divinity itself. I am quite aware that, as subordinates, comparatively uneducated men are often as useful and trustworthy as their more showy competitors; but in the higher walks of the profession, a good general education is of great use in dealing with all forms of business, and it is of incomparable value to its possessor, by directing his intellect and his tastes to the purest and noblest food.

I hope, therefore, that our professional men will continue to pass through our universities. But in order that this may be the case, the universities must be developed, so as to meet the requirements of old professions as they extend and new professions as they arise.

Dr. Lyon Playfair drew my attention to the fact that existing universities almost all arose as professional schools of law, medicine, and divinity, for a long time the only learned professions. Now that there are new learned professions (I may surely claim that title for engineering at least), the universities should recognise the fact, and provide the necessary curriculum, and the necessary degree or attestation that this curriculum has been profitably followed. We boast in Great Britain that our institutions grow; whereas foreign institutions are too often shackled by such bonds, that, unable to develop, they grow old and die. This fate has actually befallen the scientific faculty of the University of Paris. The class-rooms are deserted, the professorships are despised sinecures, and the whole scientific training of France is given in the new special schools, such as the Polytechnic School and École Centrale. I hope we may never see our ancient universities wither in like manner; but to avoid a similar fate, they must avoid similar conduct.

Edinburgh at least has incurred no reproach as yet. The foundation of the chair which I have the honour to fill is a proof of the munificence of the patrons of its University. The reception I have met with proves, not my merit, but the interest with which this new development is regarded in the city; and therefore I am emboldened to urge those measures upon you which I think the University should be encouraged to adopt.

The system of pupilage for an engineer must be maintained, and pupilage should begin at the age of eighteen or nineteen. The student has, therefore, no time to acquire the higher mathematics, or to follow any large number of courses on special engineering subjects. If our young engineers could enter the offices and workshops as pupils possessing a competent knowledge of geometry, the elements of algebra, trigonometry, physics, chemistry, mechanics, and drawing, they would be able during their pupilage to make a really good use of their time, instead of, as at present, too often employing these years in learning, in a very rude way, projection, mensuration, tracing, and such other elementary branches as they should have mastered before entering the office.

For engineers, at least, no other courses of lectures are required than are now open. What is really necessary is, that parents and intending students should be induced to take advantage of the facilities which already exist. No pass or competitive examination bars the entrance to our profession, and a good fee and some personal knowledge of the candidate are the inducements which lead engineers, in the south at least, to accept pupils. Now I cannot urge too strongly on the profession that the improvement of the education of the younger members lies in their hands. If they will require a real preparation from their pupils, if they will show a real preference to the well-prepared pupil, the cause of scientific education will be won. How can they practically show this preference? So long as there is no recognised curriculum leading to no recognised examination, they can at most ask if the candidate has attended certain classes, such as my own, and with what success. The profession cannot institute a sort of matriculation examination, and a mere general inquiry as to previous training will lead in future to no better results than hitherto; but if the University institutes a recognised examination, conferring a recognised degree or diploma, the profession can then select their pupils by asking a perfectly definite question, Has he or has he not passed this degree? and if so, with what credit? This is one reason why I think it is incumbent on all universities professing to prepare engineers that they should institute definite engineering degrees. The particular examination which should be employed to test the fitness of a pupil to enter an engineer's office or workshop should embrace only the elements of those subjects which I have named above; but these would ensure that the pupils in engineers' offices who had passed that examination should be very differently prepared from the pupils with which I have come in contact (in the south).

I call, therefore, on my professional brethren to select their pupils with reference to their previous training, and I call upon the University to organise such a curriculum, and to institute such an examination, as will enable pupils to be thus selected.

In course of time, should the engineering instruction prove successful, it will be necessary to institute some new chairs, such as that of architecture, and possibly to divide engineering into two branches, with lectureships on special subjects; but every step should be justified by the success of each previous advance.

The University can, however, at once do more than simply test the fit preparation of engineering pupils by the institution of a degree corresponding to that of Bachelor of Arts. It might also offer to test how far, at the end of his pupilage, each pupil has benefited by his practical training, and then give a real diploma attesting his capacity in the particular branch which he has studied. This was the proposal contained in the University Calendar, but abandoned for the present, owing to legal informalities attending its adoption.

Some disapprobation has been expressed at the proposal to

institute new degrees, and the opposition has come from men of different classes—for instance, from Conservatives, who object to new degrees because they are new, and who fear a degradation of academical distinctions. To these opponents I would point out that engineers now form a real new learned profession, as much entitled to academical recognition as the older professions were when the old degrees were instituted. To institute new degrees in the old universities is to carry on the old traditions unbroken; and if this be not done, then the new professions will create schools and distinctions of their own, and the old institutions will fall as old trees fall when their growth is ended. The strictest watch should, however, be kept, to ensure that the new degrees are, if anything, harder to attain than the old ones.

The second class of opponents are those who say that no examinations can test the proficiency of a professional man. I have very little doubt that this was the feeling of some of the older medical men when degrees in medicine were first instituted; but if engineers could only be examined after the fashion of some recent examinations the papers of which I have read, I should cordially agree with the opponents of the measure I now support.

As an illustration of my meaning, I will take the liberty of criticising one point in the examination which is to be held for the Whitworth scholarships, under the management of the Science and Art Department. This examination might be expected to correspond with that which all should take before entering an engineer's office or workshop. Its object is to select those youths who, by two or three years of additional study, are likely to become useful engineers. Consequently, we find a list of subjects closely agreeing with that which I proposed for our University examination; and among these subjects is 'Applied Mechanics.' The French and Professor Rankine mean by Applied Mechanics the application of the purely theoretical mechanics to the problems which occur when dealing with material bodies. Thus Professor Rankine's book on Applied Mechanics treats of the principles of statics, the theory of structures, the strength of materials, kinematics, the theory of mechanism, and the principles of dynamics—all things which the pupil entering on practical work should know. But the examiner of the Science and Art Department has understood Applied Mechanics to mean a knowledge of machinery—a knowledge which no man can acquire except by long practice, and which a pupil entering a workshop can by no means have acquired; indeed, no man in his whole lifetime ever acquires it thoroughly except in certain branches of machinery.

One question asked is—

15. In the older vertical saw frames a ratchet wheel was employed to urge the timber forward for the cut; in modern frames a similar arrangement is employed, but with this difference, that the teeth of the ratchet wheel are dispensed with, and a frictional grip, adjustable to any rate of feed motion, is substituted. Make a pen-and-ink sketch of any such contrivance by which the object can be effected.

Another is as follows:—

16. Lathes for boring guns are now being constructed, in which the power is to be accumulated by worm gear instead of spur or bevel gear. The accumulation of power is 150 to 1; a worm wheel is fixed on the main spindle, and the worm works in oil. Make a sketch of the head stock only, showing the driving pulleys and the gear.

Now, unless a man be told beforehand, 'You will be examined in sawing machinery, and in machinery for the manufacture of ordnance,' I hold that his knowledge or state of preparation cannot be tested by questions like these. If I am to be at liberty to pick any details from any class of machinery, and ask men to make a sketch of it, the merest accident will determine who would give me the best answer. Probably, if I were to ask the examiner to sketch the form of break employed in retarding submarine cables during submersion, and describe its principles, he would be as much puzzled as I should be to sketch a detail in gun-boring machinery. The questions require no particular knowledge of the principles of mechanics, but an absolutely unlimited knowledge of the details of machinery; and to my mind it is absurd to ask from pupils about to enter on a course of practical study any but the most elementary knowledge of the elementary parts of machines.

This, then, I take to be an example of an examination, devised by a very competent engineer, who has not sufficiently considered what examinations can be expected to do. He might defend his paper by observing that the Science and Art examinations were meant to test the proficiency of teachers, and for this purpose the examination is only faulty in requiring superhuman knowledge. I sincerely hope that no similar paper will be put before young students.

On the other hand, neither in theoretical nor in Applied Mechanics is there a single problem as to the strength of materials and the stability of structures, the proportions of which ought to exist between firebar surface, heating surface, the dimensions of cylinders, etc., in steam-engines. Nothing is said of the geometrical methods which engineers find so convenient for solving the problems of strains on framework, nor are there any questions testing the capacity of the student to understand the graphic representation of results by diagrams; whereas I hold that to those points the student's attention should be directed before he enters the workshop, and that he cannot possibly be expected to make a pen-and-ink sketch of a ten-ton hydraulic crane, of a steam hammer, of a turbine in equilibrium adapted to secure the best results from a fall of sixteen feet, or even of a hydraulic accumulator.

I wholly disbelieve in any attempt to ascertain a man's mechanical knowledge by shutting him up for four or five hours without any books of reference, and asking him to answer eight such questions as these, ranging over every variety of machine. The best men in the country could not pass such an examination; and a man who, in after life, attempted to design machines of which he had only a general knowledge, without referring to all the drawings he could find of similar contrivances, and without giving the subject a day or two's consideration, would be an exceedingly bad engineer.

Then, you may ask, to what kind of examination would you subject your candidate for an engineering degree? First, I would subject him to an examination in some one or two branches of theoretical knowledge, allowing him to choose the branch, but then making the examination one which would entitle the successful candidate to honours in that branch.

Next, in the practice of his profession, I would also allow the candidate to choose his department. Thus, he might elect to be examined in railway machinery, or in marine engineering, or in tools, or in telegraphy. Then I would set him a real piece of work to do, giving him perhaps a month or two months to complete the work. At the end of that time he would bring up his designs, calculations, estimates, precisely as if I were his chief engineer and he were a resident or head draughtsman. He should have free reference to all books and persons whatever; but when the designs were laid on the table he should be subjected to a searching cross-examination as to his reasons for adopting the given dimensions, materials, and forms; he should be called upon to justify his estimates and explain the motives which guided him in drawing each clause of his specification; and I venture to say that a man who passed such an examination as this, before a board composed in part of professors and partly of practical engineers specially acquainted with the branch of engineering in question, would prove a really competent engineer in that branch; and I also think that there are hundreds of men in England who would be willing to purchase a diploma by undergoing such a real test as this, who would laugh at cramming themselves with undigested pen-and-ink sketches for a three hours' examination in machinery at large.

I have drawn no fancy sketch. The examination I recommend is simply the form of examination actually held in Germany with great success. It is usually taken by men who have been some years in practice, and is, I believe, a sure passport to advancement.

A degree given in this way would be a degree worth having in England also; nay, it would be especially valuable to the young engineer, who at present has no way of gaining any distinction except in the very limited circle of the shop or office in which he works. The public credit due to all works goes to the chief engineer or the responsible manager of the firm; nor is this essentially unjust, though young men sometimes rebel at it. The responsible man must reap the shame or glory, but owing to this very fact young engineers would be peculiarly glad of an opportunity of gaining real distinction.

Another motive for giving degrees in engineering is, that the want of such an official diploma subjects Englishmen abroad to a serious disadvantage. The French or German candidate for work pulls out his official diploma for inspection, while our Englishman has nothing to show but an informal note from somebody saying he is an excellent man in a general way. What wonder if the duly accredited engineer be preferred, so that we find the French and Germans boasting that foreign Powers now always send to them for engineers, and no longer to England! I believe this is a fact, and that the absence of degrees to a great extent explains it. I hope, therefore, that this Society will give its hearty support to the institution of these new and much-wanted academical distinctions.

I will now very briefly review the several recommendations which I have ventured to make.

After approving the report of the Schools Inquiry Commission, by which real public schools of all grades would be established throughout Great Britain, I drew especial attention to mechanical drawing as the branch of elementary scientific education which I thought most important to our workmen and foremen; I suggested that the Government schools of fine art should be extended so as in all cases to include a series of classes for instruction in all the branches of draughtsmanship; I recommended that these classes should be taught by practical draughtsmen, not by men whose profession it was to teach; that prizes should be offered for each branch of drawing, with national exhibitions and national prizes; and that the more costly and permanent articles required in the class should be provided for the student.

I further suggested that this branch of elementary scientific knowledge is peculiarly well adapted for introduction into primary schools, because it stimulates accurate observation, giving a boy the means of expressing his ideas accurately, and of acquiring information from books and periodicals, and because it is suited to the genius of this country; I endeavoured to show that this important step could be made at once, by putting this kind of drawing on a par with reading, writing, and arithmetic, in the payment for results.

Passing to the higher professional training, I deprecated the establishment of special colleges, but, on the one hand, called upon the universities to institute new chairs and new degrees, to meet the requirements of new professions; and, on the other hand, I called on these professions to support the universities, by according a real preference to the men who were well prepared over those who declined to take advantage of the facilities given them.

I have now the great pleasure of informing you that I have to-day received a proof of the preference for educated pupils felt by one of the leading firms of mechanical engineers in Edinburgh. I cannot better explain my meaning than by reading the following extract from minutes of a meeting of the directors of T. M. Tennant & Co., Limited, held on Tuesday, January 5, 1869:—

'Mr. R. W. Thomson moved, and it was unanimously agreed, that to encourage the new mechanical engineering class in the Edinburgh University, this company will grant a free pupilage, once in three years, to the most meritorious student of the class.'

This valuable endowment has, I think, been given in a form peculiarly well suited to stimulate the scientific education of our young engineers, by showing that the professional men consider that the road to practice should lie through the college gates.

ON SCIENCE TEACHING IN LABORATORIES.[1]

THE thesis which in the present paper I propose to develop will not, I trust, arouse much opposition, inasmuch as my intention is simply to record that which I believe to be the best existing practice. I have no striking novelty to propose for discussion. I hope, however, that I may be able to explain certain facts in connection with laboratory teaching in such a manner as may be of some use to those who, not being themselves teachers of science, nevertheless endow or control the teaching in laboratories. The manufacturer or commercial man who desires to benefit technical education is almost invariably anxious that the teaching which he is willing to promote by giving time or money, or both, should be of a practical character, and, for my own part, I believe this desire to be perfectly justified. What is called pure science has, and should have, devoted followers, but it is also desirable that applied science should be fostered, and, as an engineer, I have a natural sympathy with those branches of science-teaching which tend to be more immediately fertile. Hence I am disposed to enforce this doctrine, that practical teaching in the experimental laboratory is that which, of all others, deserves and requires endowment; but in asking for practical or technical teaching, a disposition is sometimes shown to expect that our universities and colleges should teach matters which shall be not only ultimately, but immediately, useful in a given trade or manufacture. It is even thought that the teacher ought to lead the way in improving the practical methods used in the factory; that the college may, in fact, become a model factory, or contain many model factories. On the other hand, the teacher of pure science is sometimes tempted wholly to

[1] Read at a Conference at the International Health Exhibition, London, 1884.

disregard practical teaching, trusting that if the principles of science are correctly understood sound application will follow as a matter of course. Placed between these two extremes, I would side rather with the man of science than with the man of trade. The principles of each science are few, positive, permanent, and such as can be learnt from lectures and books. The applications are innumerable; their results cannot be made the subject of absolute calculation. The methods vary from year to year and from month to month; they can neither be learnt nor taught in the lecture-room. I say these things dogmatically, having no fear that any teacher of experience will contradict me. Are we to conclude, then, that no practical teaching of importance can be given in our universities? Far from it. There is no teaching more practical, more immediately fertile in results than that which can be given by the man of science in his laboratory. This teaching is of two kinds—each valuable. That which is most popularly known and most appreciated is practical instruction in research. The teacher is himself engaged in the research of some scientific truth, and he finds in the best of his students a willing band of workers, ready to devote their whole energies and time to the prosecution of minute and prolonged inquiry; the young men are inspired by the teacher with his own ardour; they imitate his methods, sympathise with his aims, and emulate his success. In a few years these generous and unknown assistants will themselves be leaders. The process is natural, healthy, and successful, but it is incomplete. It reaches only those who are born with a great natural aptitude for scientific inquiry. The rank and file of the students cannot be employed in this manner by the teacher; they would waste their time, spoil an indefinite amount of apparatus, hinder the advanced student, occupy the attention of the teacher unworthily, and perhaps try his temper; and yet the rank and file—the ordinary well-meaning student who will never become a leading light in science—is worthy of our attention. If he is well educated he may become a successful manufacturer, contractor, engineer, or farmer, and sensibly increase the power and wealth of our country. It seems to me that this student is not so well provided for in our scientific teaching as is desirable. And the main question I propose for discussion is, how we are to improve the education of this second-best young man. My own answer put briefly, is that we can teach him systematically the art of measurement. We cannot give him the hunger for knowledge, the acute logical discrimination, nor the imaginative faculty required for research; but we can teach him how to ascertain and record facts accurately; we can bring home to him the truth that no scientific knowledge is definite except that based on the numerical comparison which we call measurement; we can teach him the best modes of making that comparison in respect of a vast number of magnitudes, and in teaching this we shall teach him to use his hands and eyes. This practical teaching gives clear conceptions to the minds of many who receive a verbal definition as a mere string of dead words. I should be glad if it were generally proclaimed that the elementary training in all our science laboratories should be a training in the art of measurement. I wish that the classes were called measurement classes. Then a student of ordinary intelligence would know that by entering a given class he would learn how to measure those magnitudes with which he will have to deal in after life. The attempt to measure them will lead him to consider their nature, and he will approach scientific study in the class-room with a faith in the reality of science which no verbal exhortation will ever give him. You may define the absolute unit of electrical resistance as accurately as you will, and your definition shall affect the average brain to no perceptible extent; but a young man of very ordinary education and intelligence can learn to measure resistances in ohms, and having learnt this, an ohm becomes a reality to him. Not only does the knowledge he has acquired make him a more valuable assistant to the engineer and contractor, but having acquired a working faith in the existence of ohms, he is prepared to take some trouble to understand the scientific definition.

Let me again repeat that I am here urging no new thing. I am merely, as I believe, stating the practice of all well-arranged laboratories—they are schools of measurement—a fact long since recognised by the chemist, but less explicitly recognised in other branches of physical science. The student of heat or light may come to the laboratory thinking vaguely that he is to make experiments—and to him an experiment does not imply a measurement. I have heard a young man describe as a very interesting experiment, performed by his teacher, the blowing-up of a horse pond by an imitation torpedo. Now if in that college the elementary practical class of physics had been called a class for measurement, this so-called experiment would not have been shown, and the young man would not have been wholly misled as to what physical science meant. The teacher would not have thought of blowing up the pond until his pupils were capable of measuring the resistance of the leads, their insulation, the electromotive force of the battery, and other magnitudes.

The use of the word 'measurement' in naming a class would be in itself a safeguard against the peepshow style of teaching which at one time was far more common than is now the case. Moreover, it would allow examinations to be held of a practical kind, in which students in the same or different colleges might ascertain their relative skill. It is possible definitely to group students in the order of their merit by comparing the measurements which they make. The range and accuracy of their knowledge as to what instruments they should employ can also be tested by examination, and although an abuse of competitive examinations is certainly an evil, nevertheless one test of what can and what cannot be taught is to be found in the consideration whether an examination paper can or cannot be set. In quantitative analysis this mode of examination is universally adopted, but practical examinations in electrical and thermal measurements are not so common in our universities as they might be, and practical examinations in the measurement of velocity, force, or work are even rarer.[1] Is it expecting too much to ask that, wherever physical science is taught, the students should have an opportunity of systematically learning how to measure every magnitude which can be expressed in numbers? The distinct recognition of measurement as a thing to be taught would serve as a guide in the purchase of apparatus—it would serve to distinguish the toy from the scientific instrument.

Let me not be misunderstood. There are innumerable experiments of the highest interest to the trained physicist which are mere magic-lantern slides to the student, and even the magic-lantern slide has its place. A lecturer may with propriety use a mere spectacle to give his audience a more concrete view

[1] Examinations precisely such as are here recommended are held in the Cavendish and many other laboratories.

of the subject on which he discourses; he may even sometimes employ a mere spectacle to afford relief from overstrained attention. I will go further, and say that measurement classes are not, in my opinion, suitable for secondary schools. They require in the student an interest in accuracy, and a belief in the importance of detail rarely found in boys or girls. It is sufficient for the boy to learn that a magnetised needle may move to the right or left under the influence of an electric current. The commonplaces of science are at one time of life interesting novelties; but there comes an age when the young man feels that he knows nothing of electricity unless he can predict the force which will be exerted on a given magnet under given circumstances, which are themselves capable of being defined accurately by the aid of numbers; and he can only learn this knowledge by the aid of practical classes in the laboratory.

Another advantage of the measurement class is this; it brings the teacher of science into direct contact with the practical man. It even enables the practical man to some extent to control the teaching of the man of science.

If a practical engineer comes into a scientific laboratory, he can tell whether lengths, areas, angles, forces, and so forth, are being well measured, or whether the class is being taught in an antiquated or perfunctory manner. It will be obvious to the least educated of our practical men, that measurement is required; and they can judge what measurements are required. Hence, we may expect that measurement classes, boldly so called, will readily find endowments; and, as an incidental advantage, they may help to extinguish the popular fallacy of college workshops. Having worked for three years at the bench in a Manchester locomotive shop, I have always protested against the endeavour to set up in colleges or universities workshops, with the object of giving students any considerable practical knowledge of any art. In the secondary school I believe a workshop may be useful as an adjunct, providing certain boys with the means of acquiring a little skill in a pleasant way. There is much pleasure, and some profit, to be got from tinkering among models when we are boys, but when a young man has chosen a trade or art, he can only learn that trade or art by working at it, and by working under the actual conditions of the trade or art: little girls may pleasantly and usefully dress their dolls, but no woman could in two or even three college terms learn to be a successful milliner by cutting out dolls' clothes for an hour three times a week; and yet I sometimes hear what is no better than this advocated as a necessary adjunct to engineering teaching at a university. The young professional engineer does not simply learn in the works how to file and chip. He learns the time required for all manner of jobs, the finish required in each class of work, the way the various parts are handled, the forms which are convenient, the routine of the shop, the character of the men —the system of storage, the materials and sizes to be bought in the market, and hundreds of other facts, which can only be made his own after contact with manufacture on a full scale. We cannot imitate this in college.

But the workshop, in connection with the measuring class, is a legitimate and almost necessary complement. The work done in this workshop is not the same as that of any trading concern, although it bears some similarity to that of the practical optician. In such a workshop, the student may be usefully occupied in adjusting, repairing, and modifying the apparatus he requires; he may thus learn to use both hand and eye, and he may gain some practical knowledge of materials; he can, in fact, acquire such skill in a number of the minor arts as will be of much use to him in experimental work; used in this way, the laboratory workshop may teach him much which he cannot easily learn in large engineering or manufacturing works.

Scientific research for the most advanced and best endowed students; measurement classes open to all in all branches of exact science, and a common laboratory where apparatus of all kinds can be repaired, adjusted, modified, with the help of highly skilled workmen. This is the general picture which I have endeavoured to draw of a college fully equipped for practical scientific teaching. I have not touched on the study of theory, which must precede or accompany the practical training. This lies outside my subject. I have laid most stress on measurement classes, because it has seemed to me that while the importance of this teaching is patent both to men of science and men of practice, the organisation of these classes admits of considerable improvement and great extension.

In order to bring out more definitely what I mean by a measurement class, and to emphasise the fact, that these measurements are not so fully or systematically taught as is to be desired, I will conclude by giving a list of some of the measurements which might be usefully taught at college to a student who looked forward to becoming an engineer :—

1. *Measurements of Length.*—These would range from micrometrical measurements for standard gauges up to the modes employed in measuring the base lines of surveys. They would include rough workshop methods and the practical methods used in ordinary surveying and navigation, so that the student might learn not only the maximum accuracy attainable, but the degrees of accuracy required in practice. The methods would include indirect measurements by optical apparatus as well as direct methods—lineal measurement for valuation would be included.

2. *Measurements of Surface.*—These would range from the smallest plane area to be perceived in the microscope up to the areas measured in geodetical operations. The various drawing-office methods of computing plane areas, or measuring these by special instruments, would be practised. Curved surfaces of all degrees would be measured, and various classes of integrators applied. Superficial measurement for valuation would be included.

3. *Cubic Contents.*—These measurements would range from determinations of great accuracy, such as are required in scientific research, up to the measurement of earthworks, the contents of barrels, tanks, timber.

I may here remark that these three kinds of measurements alone would require a very large collection of apparatus, and that this collection would require to be extended year by year. I also venture to think that all this information can be far better given in college than during a practical apprenticeship. No single workshop or engineering office contains nearly enough apparatus, nor is it the duty of any one to teach the use of such instruments as may be found there.

In the following list of heads I abstain from pointing out the large range of measurement required in each. I simply give the subject-matter of that measurement :

4. Angles.
5. Time.
6. Velocity, including angular velocity.
7. Acceleration, including angular acceleration.
8. Mass: under this head I would teach measurement of weight and density.
9. Force.
10. Intensity of force, including the pressure of gases and fluids.
11. Work and energy.
12. Power.
13. Friction. Solid on solid; fluid on solid, and fluids in themselves.
14. Strength of materials in various forms, including their elasticity and distortion.
15. The efficiency of gearing.
16. The efficiency of motors.
17. The flow of fluids.

In addition to the above subjects for measurement, the engineer requires to know how physical measurements are made in heat, optics, electricity, and magnetism. The measurement classes in each of these subjects would embrace a range even exceeding that sketched out above for applied mechanics.

There are, I am glad to say, laboratories in this country where the student can learn many of the measurements of which I have spoken. It almost seems to me as if, of all the subjects spoken of, the fundamental measurements in engineering had been most neglected. The object of this paper will have been attained if it in any degree leads to an increase in the opportunities given to students of studying that which is surely the basis of all exact science as well as all practice, namely, measurement.

APPLIED SCIENCE

PREFATORY NOTE.

By Professor J. A. Ewing, F.R.S.

Three papers, each in its way representative, have been chosen for republication from amongst Fleeming Jenkin's technical writings. The first gives, in popular form, an account of the scientific development of submarine telegraphy, a work in which he took a large share during the earlier part of his professional life. Apart from its historical and personal interest, this paper is still valuable as a description of present-day practice in cable making and cable laying. In the second paper, on 'Telpherage,' Jenkin introduced to the public the invention on which his later years were spent: at the date to which the paper belongs he had satisfied himself, by trial on a large scale, that his ideas were practicable, and he was then busy in elaborating the details. The third paper is a somewhat abstruse but most important contribution to engineering theory in a department already occupied by Jenkin in an earlier work. Though of no popular interest it is second in value to none of his writings, and only needs wider publicity to be more generally appreciated by engineers. The papers are reprinted without substantial change; only a few obvious clerical errors have been corrected.

Following the three reprints is an abstract of Fleeming Jenkin's other scientific writings, which has been prepared with the purpose of showing, in some detail, the character and place of his contributions to science, and of facilitating reference to the original papers.

SUBMARINE TELEGRAPHY.[1]

At a time when the first successful submarine cable has been laid across the Atlantic, and a second has been recovered from depths once thought unfathomable, many persons will probably be led to consider how far these great achievements, following on failures almost as great, have been due to mere good fortune, or to a real progress in knowledge. The object of this article is shortly to explain the advances which have lately been made in theory and practice by those who carry out the manufacture and submersion of telegraph cables. To make this explanation intelligible to the general reader, it will be well first to describe what a submarine cable is, and what are the functions it has to perform, although probably few who read this article will be so entirely ignorant of the subject as to suppose, with an ingenious correspondent of the 'English Mechanic and Mirror of Science' that the copper conductor is a long rope which slips backwards and forwards inside a gutta-percha tube, so as to ring a bell in America when pulled by the clerk in England.

The electrical conductor in a cable really is a copper rope in almost all cables now made, though a single wire is still sometimes used; when small, three wires generally form the strand; when larger, seven wires are used. Single wires were first employed, but they sometimes broke at a brittle part, and when large were inconveniently stiff, tending to force their way out through the insulating sheath of gutta-percha. The seven wires of the strand never break all at one point, and the fracture of any one produces no sensible effect on the conductor as a whole; for although the strength of a chain is limited by that of its weakest link, the conducting power of a wire or strand is in no way limited by that of its smallest section. The large Atlantic

[1] From the *North British Review*, December 1866.

strand might be cut in two and joined by a short fine wire barely visible to the eye, without any difference being felt in the rapidity with which signals could be transmitted, or in the magnitude of the currents observed in the cable. The thin wire would produce no sensible effect, unless the length over which it formed the exclusive conductor bore some sensible proportion to that of the whole cable. Six, therefore, of the seven wires of a conductor may be broken in a thousand places without any injury to the cable, provided any one wire at each spot remains not wholly broken; nor is it, of course, necessary that this one wire should always be the same. Of course the seven wires forming the strand act as one conductor, and transmit only one message at a time.

The interstices between the several wires are filled with an insulating varnish known as Chatterton's Compound. The object of this varnish is to prevent the percolation of water along the strand, should any water ever reach it, and also to produce a more perfect adhesion between the strand and the gutta envelope, so that it becomes very difficult to strip off the insulator, even should it be cut or abraded. In older cables it was by no means difficult to pull the insulator off the copper in the form of a gutta-percha tube, and in great depths water was very generally found to have penetrated to the copper throughout its entire length. This was not necessarily fatal to the cable, for the water inside might be quite well insulated from the water outside, owing to the extreme minuteness of the pores by which it had gained access to the interior; but this water was the cause of serious difficulty and danger in joining a fresh piece of cable to an old one during repairs, and it was also probably dangerous by its tendency to produce an oxidation of the copper conductor. In cables as now made, there is no space for the water to lodge, and no water is ever found between the insulator and the copper.

The insulator employed in every cable of importance hitherto laid has been gutta-percha. The copper strand is passed into a vat of semi-fluid percha, and is drawn through a die of such size as to allow a convenient thickness of insulator to be pressed out round it. This first layer of gutta-percha receives a coat of Chatterton's Compound, and the process is repeated until the copper is covered to the specified thickness by a succession of alternate layers of gutta-percha and compound. Three or four coats of each material are generally used; the largest wires with their insulating cover are nearly half-an-inch in diameter, the smallest in practical use for cables are about a quarter of an inch in diameter; but it is quite possible to cover in this way copper wire no thicker than a hair. The dangers encountered in this part of the manufacture are, impurities in the gutta-percha; eccentricity of the conductor in the insulator, leaving a dangerously thin coating of the latter; and, lastly, air-bubbles which may lodge in the insulator unperceived, and do serious injury. In time, water is certain to penetrate to these air-bubbles; it becomes partly decomposed, the gas generated bursts the bubble, and exposes the copper to the water. The slight leak thus formed is, by the action of the battery used in signalling, easily developed into a very serious fault. Fortunately, the manufacturers have been able almost, if not wholly, to prevent the occurrence of these dangerous cavities.

If the cable is to have only one conductor, as is the case in most long lines, the insulated wire is served or wrapped with hemp or jute, which acts as a padding between the gutta-percha and the outer iron wires used to give strength. This serving used to be tarred, but Mr. W. Smith pointed out that the tar was occasionally squeezed into small faults, and was a sufficiently good insulator to prevent their detection during manufacture, though not sufficiently good to prevent these flaws, under the action of the battery, from developing into serious faults. Since then, wet tanned hemp has been generally used. Outside the hemp serving come the iron wires, laid round and round the core, so as to give the whole the appearance of a simple wire rope.

These iron wires are very generally galvanised to prevent rust. In many cases they are further covered by a double serving of hemp, and a bituminous compound of mineral pitch, Stockholm tar, and powdered silica, patented by Messrs. Bright and Clark. This compound is used in the Persian Gulf Cable, the Lowestoft-Norderney (Hanover) Cable, and several less important lines, and seems to answer well. In other cases, as in the present Atlantic Cables, each iron wire is separately covered with a hempen serving, and the served wires are then laid round the core as before: the cable in this case looks like a hemp instead of an iron rope. Many other forms have been proposed and a few adopted, but before these can be discussed, the duties which the cable has to perform, as a rope, must be understood; and before entering on this subject, which is purely mechanical, it will probably be better to return to the insulated conductor and its electrical properties. Its form and materials have nominally undergone hardly any change since the manufacture of the first cable laid from Dover to Calais in 1851. The copper strand was substituted for the single wire in the Newfoundland and Cape Breton Cable, laid in 1856. Chatterton's Compound was used in the cable between England and Holland, laid in 1858. The interstices in the copper strand were filled with compound in the Malta-Alexandria Cable, laid in 1861; and since that time absolutely no change has nominally been effected either in the form or materials used. Now, inasmuch as an overwhelming proportion of the cables laid in deep seas have failed, have we any right whatever to expect that cables will be permanently successful, of which the vital portion is nominally identical with that of the old Atlantic, the Red Sea, the Sardinia-Malta and Corfu, Sardinia-Africa, the Toulon-Corsica, the Toulon-Algiers Cables, which, in the aggregate, represent about 8,000 statute miles of wire, which, after a more or less brief period of working, became wholly useless, as may be supposed chiefly from electrical defects? Did it not seem almost madness to attempt to cross 2,000 miles, in depths exceeding 2,000 fathoms, at a time when the only cable which could be cited as having worked satisfactorily for any considerable time in deep water, was a short length of the Malta-Alexandria Cable, lying in 420 fathoms of water? To the public, and to many engineers, it did seem hopeless; but the fact that it was precisely those persons who knew most of the subject that risked their reputation and their money, should prepare us to believe, that, although the name of the materials and the form of the insulated conductor remained unchanged, other changes had taken place which fully justified the confidence of the Atlantic projectors. The methods by which the perfection or imperfection of the cables were examined—the methods of testing, as it is called—have in fact made enormous progress, and it is to the discoveries and inventions in this branch of science that we owe both those improvements in the quality of the materials employed, and that certainty of detecting the smallest fault, which led so many practical engineers and electricians to a conviction of the feasibility of the great undertaking now so happily completed. It is on these electrical tests that a reasonable belief may be based of the probable permanence of the two Atlantic Cables, and it is to these improvements that attention will now be directed.

The electrical tests employed for the first cables made were simple enough. It was necessary to ascertain that the copper conductor in the cable was unbroken, and fit to transmit an electric current. This was tested by placing a galvanometer in a simple circuit formed by the battery, the copper conductor of the cable, and the wire of the galvanometer. If the conductor was unbroken, a current passed from one battery pole to the other through the cable, and in its passage through the instrument deflected a needle. The stronger the current, the more the magnetised needle was deflected. If the conductor failed at any point, no current passed. It was also desirable to know that the conductor was insulated, so that no considerable portion of the current entering one end of the cable would be lost before arriving at the other end, where it would be required to produce a signal; to ascertain this the metallic circuit was broken—one pole of the battery remained connected with the conductor of the cable through the galvanometer wire; the other pole was connected with a plate buried in damp earth, the cable was put under water, and its far distant end was insulated. Thus the battery was ready to send a current into the cable, and would do so if the cable were at any point connected with the earth. When the cable was well insulated, no current passed; if there was a fault, that is to say, a connection between the copper inside the cable and the earth or water outside, a current passed and deflected the galvanometer needle. The test consisted simply in trying whether a current would pass through the conductor, and would be stopped by the insulator; the galvanometer being an instrument which showed the presence or absence of a current by its effect on a magnetised needle. Staunch conservatives may still be heard to sigh for the good old times when a cable was good if a needle stood upright, and bad if it leant to one side; when there were neither complications nor calculations to perplex or mislead any one.

These simple tests, when applied to long cables, had serious defects. Sir W. Thomson was the first to insist on the importance of ascertaining not only that some current would pass through the conductor, but that the greatest possible current did pass which could be expected with a conductor of given dimensions and material. The current which a given battery will produce depends not only on the length and size of the conductor, but on the material of which it is composed; roughly speaking, a given battery will produce a six-fold greater current in a long wire of good copper than it will in an equally long wire of iron of the same diameter. The property of the conductor, determining the amount of current which will pass through it under given constant circumstances, is termed its resistance. The greater the resistance the less the current, and *vice versâ*. Each metal and each alloy has its specific resistance, from which the resistance of any given wire may easily be calculated. It further happens that various specimens of commercial copper differ exceedingly in this electrical property, so that one copper wire will transmit double the current transmitted by a second, in similar circumstances, although to the eye the two wires do not differ. To this fact Sir W. Thomson drew attention in 1857. It might seem of little importance what the resistance of a conductor is, since the current can always be increased by increasing the power of the batteries employed; but Sir W. Thomson pointed out that the rapidity with which a succession of distinct currents, such as are required to produce signals, could be made to follow one another through a long submarine cable, was, *cæteris paribus*, inversely proportional to the resistance of its conductor, so that the commercial value of that cable as a speaking instrument depended on this resistance, which could be diminished only by (at increased cost) increasing the dimensions of the conductor and insulator, or, without any sensible increase of cost, by simply selecting that copper which possessed the smallest specific resistance. This point is clearly explained in the following extract from a paper by Sir W.

Thomson, published in the 'Proceedings of the Royal Society,' June 15, 1857:—

It has only to be remarked that a submarine telegraph, constructed with copper wire of the quality of the manufacture A, of only $\frac{1}{21}$ of an inch in diameter, covered with gutta-percha to a diameter of a quarter of an inch, would with the same electrical power, and the same instruments, do more telegraphic work than one constructed with copper wire of the quality D, of $\frac{1}{16}$ of an inch diameter, covered with gutta-percha to a diameter of a third of an inch, to show how important it is to shareholders in Submarine Telegraph Companies that only the best copper wire should be admitted for their use.

As soon as it came to be understood that the value of a cable might be enhanced forty per cent. by a judicious selection of the copper employed, tests were adopted which should not only show that the conductor would transmit a current, but also that it was the best conductor which could be procured of the dimensions and material chosen. In other words, the resistance of the conductor was measured.

Measurement implies comparison with some unit. The resistance of some special piece of wire at a given temperature may be taken as a standard 'one unit,' and the resistance of all other wires or conductors may be referred to this unit. This comparison was rendered possible by the discoveries of Ohm, published in 1827; measurements were made by him and his followers, Lenz and Fechner, in terms of arbitrary units, and Professor Wheatstone in 1843 published an elegant method of making these measurements, and then proposed the adoption of a fixed standard or unit of resistance. When, therefore, it was found desirable to measure the resistance of conductors, the means were not wanting, and were soon very generally adopted. For these measurements 'resistance coils' are required; these consist in a graduated series of fine wires of known resistance, which can be combined at will so as to give any multiple of the standard or unit that may be required; they are arranged in boxes, and fitted with stops, slides, or handles, so that the required additions or subtractions of resistance may be easily made. As early as 1847 or 1848, the Electric and International Telegraph Company in England, and Dr. Siemens in Berlin, used resistance coils for practical experiments connected with telegraphy; but it was not till 1857, during the manufacture of the last seven or eight hundred miles of the Atlantic Cable, that the copper was systematically selected. This example was followed in the Red Sea Cable, when the resistance of the conductor was regularly tested by Mr. Fleeming Jenkin at Birkenhead, and by Messrs. Siemens during the laying. The copper of the first portion of the Atlantic Cable was not selected in this manner, and was of very indifferent quality. Since then the improvement has been continual. Dr. Matthiessen reported to the Joint Committee appointed by the Board of Trade, and the Atlantic Company, in 1858, that chemically pure copper was superior to all alloys, and that the best copper for electrical purposes was to be obtained from Lake Superior and Burra-Burra, the worst from Demidoff and Rio Tinto. The gradual improvement since that date may be gathered from the following table:—

Date	Name of Cable	Specific Resistance at 24° C. in British Association units
1859	Red Sea	0·270
1861	Malta-Alexandria .	0·264
„	Persian Gulf . .	0·247
1865	Atlantic . . .	0·242
1866[1]	Lowestoft-Norderney .	0·240
	Pure Hard Copper .	0·231
	Pure Soft Copper .	0·226

The smaller the figure in the last column the better the material; the last figure represents perfection. The specific resistance is the resistance of a foot of wire weighing one grain. The unit in which it is measured is that selected by a Committee appointed by the British Association in 1861, from whose yearly reports may be learnt the reasons for preferring this to other rival standards—for it is by no means a matter of indifference what unit is employed.

The improvements in the methods and instruments used to measure resistance have far more than kept pace with the prac-

[1] The writer believes that the 1866 Atlantic Cable has better copper than any of the cables in the above table, but he does not know the exact figure of merit.

tical improvement of the material. Resistance coils would now be considered very bad if their normal values were inaccurate to the extent of one part in a thousand; they may be procured ranging from one unit to 100,000. The standards issued by the Committee above named profess to be identical in their resistance, without a greater error than one part in ten thousand. Still greater accuracy could be obtained if required, but the precautions necessary are then very numerous, as may be seen on consulting the various papers by various members of the Committee on Electrical Standards, published in the British Association Reports from 1862 to 1865.

A very wide gulf separates the present practice from the old plan of simply ascertaining the continuity of the conductor. Every hank of copper wire is tested for resistance even before it is spun into a strand. The resistance of the strand is measured by the engineers when covered with gutta-percha, and before being admitted to form part of the cable; for twenty-four hours previous to this test it is kept at a stated temperature. The conductor of the manufactured cable is also daily measured, less for the purpose of ascertaining its electrical properties than to ascertain its temperature from its observed electrical resistance, and also to check the length supposed to be in circuit when other tests are made. These tests are interfered with by variations of temperature, by slightly imperfect connections, by the induction of the wire upon itself, and, after the cable is laid, by earth-currents. But the precautions thus rendered necessary are well understood, and carefully observed in the case of all important lines. The quality of the copper enters into the engineer's specification with precisely the same numerical accuracy as its weight; it is referred to definite units; and no more frequent disputes arise between the contractor and engineer as to these measurements than as to the weights of material supplied.

A further use of these measurements will be spoken of when treating of repairs; but for the present let us leave the tests of the conductor to consider those of the insulator. The conductor may have more or less resistance, and work worse or better in consequence, but if the insulation be defective, the cable may not work at all, and the tests of insulation are therefore the most important of all. The old rough test was defective in many ways. It was found that if large enough batteries were used, and care taken to obtain very sensitive instruments, some current might always be made to pass between the copper and the outside of the insulator; in other words, no insulator offers an infinite resistance to the passage of a current. It was not difficult to judge roughly whether the amount of leakage, as it might be termed, was serious enough to damage a cable; but unfortunately, small faults are apt with time to become large faults, and the rough method was quite useless as a means to detect small faults in long cables. As the cable increased in length, the leakage even through a good insulator became so considerable, that two or three bad places would make no very sensible difference in the deflection observed; and the galvanometers used became less and less sensitive as their deflections increased, so that the addition caused by a moderate fault became imperceptible. Then the galvanometers were not constant in their indications, so that the deflection of to-day was a very imperfect guide as to the deflection to be expected to-morrow. The galvanometers used by different observers were seldom or never compared. Moreover, the batteries used varied, and their properties were not examined; little attention was paid to the temperature of the cable, although this has an immense effect on the leakage to be observed; finally, and worst of all, the cables were not immersed in water, and fifty faults might in that case exist in a cable without producing any sensible effect, either on this old rough test or on any other. Under these circumstances, is it surprising that cables were laid which contained many serious faults, and that, after a short and uncertain period, depending on many circumstances, they ceased to transmit messages? Is it unreasonable to expect that, under a system by which the existence of any sensible inequality in the insulation of a cable is rendered impossible, the cables recently laid may continue in perfect working order for an indefinite period? All experience has shown that sound gutta-percha retains all its valuable properties in deep or shallow water, completely uninjured by use or time. The only decay ever observed has been at bad joints, air-bubbles, or impurities.

It is, again, to Sir W. Thomson that we owe the first suggestion of an accurate method of testing the insulation of a cable.

In 1857, in a lecture delivered to the British Association at Dublin, he pointed out that a so-called insulator was really a conductor of enormous resistance; that this resistance, though large, was measurable in terms of the same units as measured the resistance of conductors, and he then gave an estimate that the gutta-percha of the first Atlantic Cable had a specific resistance twenty million million million times greater than that of copper at about 24° C. At his suggestion Mr. Fleeming Jenkin made systematic measurements of the resistance of the insulating sheath of the Red Sea Cable; and, independently, Dr. Siemens of Berlin had made similar arrangements for those measurements during the submersion of the cable. Unfortunately this cable was not tested under water, and these tests were therefore of little use, except to determine the properties of gutta-percha. Since 1859, every important cable has been tested on a similar system. The methods used have varied, but they have always resulted in determining the resistance per knot of the insulator. Attention has been paid to the temperature, any rise in which rapidly diminishes the resistance of gutta-percha. The necessary allowance for the different dimensions of various cables has also been made, and no test is now counted of any value unless made under water. The result is that definite numerical results are obtained, comparable one with another, whatever be the dimensions, length, or temperature of the cable, and whatever be the variations in the batteries or galvanometers employed. The work of one day is comparable with that of another; the results obtained in various factories, and by various engineers, are all comparable, and no considerable variation in the resistance of the insulator, such as would be caused even by a small fault, can possibly escape detection. The improvements in the tests have here also been followed by a great improvement in the quality of the materials, as well as by increased security against faults. The specific resistance of the gutta-percha of last Atlantic Cable is twelve-fold that of the Red Sea gutta-percha; and at 24° C. may be roughly said to be 200,000,000,000,000,000,000 times that of copper (referred to equal dimensions).

It is difficult to find any comparison which will give a tolerably clear idea of the extraordinary difference between the electrical resistance of these two materials; it is about as great as the difference between the velocity of light and that of a body moving through one foot in six thousand seven hundred years; yet the measurements of the two quantities are daily made with the same apparatus, and the same standards of comparison. This fact is well calculated to give an idea of the range of electrical measurements and the perfection to which the instruments employed have been brought.

Resistance coils and the galvanometer variously combined allow these measurements to be accurately made in many ways. Sir W. Thomson's reflecting galvanometer is now almost exclusively used for this purpose. The simple deflection test is still frequently employed, but it is then reduced by calculation so as to give the results in resistance.

It would be out of place to attempt to explain in detail the modes of testing adopted, but it may be interesting to enumerate the several examinations which each mile of insulated wire undergoes before it is admitted to a cable.

1. The hank of copper wire is tested for resistance.

2. The resistance of the copper conductor of the insulated mile of wire is measured after having been kept for twenty-four hours in water at a constant temperature.

3. The resistance of the insulator is measured under the same conditions, once with a current from the zinc pole, and once with a current from the copper pole of the voltaic battery. The above tests are made by the contractor.

4, 5. The last two tests are repeated by independent observers acting as the engineers of the company.

6. The coil of wire is again tested for insulation immediately before being joined to the manufactured cable.

In addition to these tests, in many cases the insulation is tested in water under a great pressure, to simulate the pressure occurring at the bottom of the sea. This test was patented by Mr. Reid, and is probably of considerable service, although in the vast majority of cases the insulation resistance is increased by pressure. While a cable is being submerged, it is indeed customary to expect an improvement of about 7 per cent. for every 100 fathoms of water, due to this cause only; thus in 2,000 fathoms an improvement of 140 per cent. is expected.

After the cable is sheathed with iron, it lies under water in large tanks; the resistance measurements are repeated daily, and the results compared with those calculated from the length and temperature of the cables. The effects of an increase of temperature in diminishing the resistance of gutta-percha have been separately examined by Messrs. Siemens, Mr. F. Jenkin, and Messrs. Bright and Clark. The results of the various experiments agree very closely. One curious phenomenon deserves mention: the apparent resistance of insulators increases materially while the battery is applied to them, and it is therefore necessary to note the time at which the observation is taken. In the earlier cables even this fact escaped notice. This extra resistance is said to be due to electrification; it ceases gradually after the copper conductor has been discharged by being maintained in electrical connection with the earth, or with the opposite pole of the battery, but in the latter case it reappears as before, increasing as the application of the battery is prolonged. Its cause is not understood. It seems to be a kind of electrical absorption, and is first mentioned by Faraday in experiments on induction.

Enough has been said to explain the care and accuracy with which the insulation of a cable is now measured. The results obtained may be understood from the following facts. Not one-third per cent. of a current entering either the 1865 or 1866 Atlantic Cables is lost by defective insulation before reaching Newfoundland. Such loss as does occur indicates no fault, but is simply due to the uniform but very minute conducting power of the gutta-percha.

Again, if one of the cables be charged with electricity, and its two ends insulated, at the end of an hour more than half the charge will still be found in the cable. The conducting power of the two thousand miles of gutta-percha has been insufficient in one hour to convey half the charge from the copper to the water outside. Those who have tried to insulate the conductor of a common electrical machine well enough to retain a charge for a few minutes will appreciate the degree of insulation implied by the above statement. Contrast these facts with the following extract from the lecture delivered before the British Association by Sir W. Thomson in 1857, at Dublin, and good reason will be seen for believing that the rapid failure of the first cable is not likely to be repeated in the case of those now in use:—

The lecturer proceeded to explain that, when tested by the galvanometer, there was very little difference in the force of a current sent into 2,500 miles of the Atlantic Cable, whether the circuit was or was not completed. This seemed rather hopeless for telegraphing (he continued), where there was so much leakage, that the difference could not be discovered between want of insulation and the remote end. But if there were 49-50ths lost by defective insulation, it would only make the difference between sending a message in nine minutes instead of in eight.[1]

Sir William Thomson did not on this occasion mean to state that there really was no difference when the farther end was insulated or put to earth, but the instruments employed showed very little difference, and on a subsequent occasion only about one-fourth of the current which started was found to have arrived at the remote end. The difference now is not one three-hundredth part, and the current entering the cable when the remote end is insulated is now, under the most unfavourable circumstances, not one-hundredth part of that passing when the remote end is put to earth, or, in other words, when the circuit is completed.[2]

[1] From Professor W. Thomson's lecture before the members of the British Association at Dublin, 1857, as reported in the Glasgow *North British Daily Mail* of September 4, 1857.

[2] The following data, supplied by Mr. Latimer Clark, Engineer to the Anglo-American Company, will be interesting to those who have made this subject their special study. The total insulation resistance of the whole 1866 cable, as it lies at the bottom of the Atlantic, is 1·316 millions of British Association units, or, as Mr. Clark calls them, ohms. This is equal to 2,437 millions of ohms per knot after one minute's electrification. The 1865 cable does not sensibly differ from the 1866 cable. Both lose half their charge in from 60 to 70 minutes The increase of apparent resistance due to electrification is enormous; thus, after thirty minutes' electrification the insulation resistance is more than 7,000 millions of ohms per knot. Mr. Jenkin, in the Red Sea Cable, did not observe a greater increase than 50 or 60 per cent. due to this cause, and a similar amount has been generally observed on other cables. An increase of 200 per cent. for gutta-percha is perhaps unparalleled, although an even greater increase has been observed with india-rubber prepared by Mr. Hooper. While the cable was on board the 'Great Eastern,' it behaved like all other cables as to electrification, rising, for instance, from 681 to 1,051 per knot during thirty minutes, at 18·3° C., so that the increased effect of electrifica-

Probably the imperfection of the old cable was due rather to the joints between the separate miles of wire as manufactured, than to any extreme inferiority in the gutta-percha employed. These joints are even now the weak places in the protection of a cable. When the gutta-percha has been selected and purified with care, and applied by mechanical contrivances of proved excellence, there is little risk of a fault occurring; but this manufacture cannot be so conducted as to produce one unbroken length of wire, and even if it could, convenience in the other processes of manufacture would require the division of this wire into lengths. One-mile lengths are, in practice, usually made without joint, and are joined together by a skilled workman as occasion arises. The copper strands are soldered together with a scarf-joint, two pieces of fine wire are then wrapped over this joint, so that even if it is pulled asunder, electrical continuity will be preserved, and so far the operation is one of no great difficulty. This cannot be said of the next process, the insulation of the wire by hand, and the welding, as it were, of the new sheets of gutta-percha, so applied, with the old sheathing on either side. The gutta-percha is warmed by a spirit-lamp; too much or too little heat is fatal, and the jointer must judge of the temperature by experience; the least moisture will spoil a joint—hence one reason for providing that no moisture can percolate along the metal strand. A very little dirt or impurity will also do much injury—hence the rule that a jointer must do no other work, and that the copper wire must be soldered by one man, the gutta-percha applied by another. A joint may also be spoilt by the presence of air under one of the insulating coats, and as the writer cannot pretend himself to make a joint, other causes of failure probably exist of which he is ignorant, but enough has been said to show the difficulty of the process.

tion must be due to the low temperature and high pressure. Mr. C. W. Siemens, in a paper published in the British Association Report for 1863, arrives at the conclusion that at 24° C. pressure does not affect the change produced by electrification. The resistance of the copper conductor of the 1865 cable is 7,604, that of the 1866 cable 7,209, corresponding to 4·009 and 3·893 per knot respectively. The mean insulation resistance per knot, as measured in the factory at 24° C., was 379 millions, after one minute's electrification. All the resistance measurements are given in British Association units.

Fortunately, joints can now be tested apart from the rest of the cable. In old times when a joint had been made the whole cable was tested; if the leak from the new joint was inconsiderable in comparison with the loss from the whole cable, perhaps some hundred miles long, the joint was supposed to be good, although, perhaps, it may have allowed a greater loss in its few inches of length than occurred from some miles of sound cable. A bad joint seldom does more than this at first, but in time it becomes brittle, cracks, leaves the sound gutta-percha at each side, and, finally, allows the water free access to the strand. Joints of this character have been found in considerable number in old cables, and especially in the old 1857–58 Atlantic Cable. Some of these present an appearance of extraordinary carelessness, even the copper strands being imperfectly joined. It is almost certain that the final failure of the 1858 Atlantic Cable was due to one of these joints in which the copper was imperfectly joined; the wires were pulled asunder when the cable was being laid, they came together again when the strain was removed, but the points of contact soon were oxidised, and all communication ceased. Mere loss of insulation hardly ever entirely stops signals.

The test now employed shows whether a joint is as good as any equal length of the wire, and all joints which do not reach this standard are mercilessly cut out. First the joints to be tested are allowed to soak in water for twenty-four hours, then they are placed in an insulated trough of water connected with a Leyden jar of large surface, the cable is charged with a powerful battery, and a little electricity leaks out through the joints into the insulated trough. If the joint is good, this leakage is so small that the current produced by it could not be shown by the most sensitive galvanometer, but after a minute or two minutes, the insulated trough and Leyden jar will be charged by the gradual accumulation of electricity which has slowly leaked through the joint. If this be now discharged through a galvanometer, it will produce a sensible effect, and can be measured. In fact, the leak which was too small to be directly perceptible is not only perceived, but its amount ascertained by measuring the quantity which accumulates from it in a given time. The test is due to Messrs.

Bright and Clark. Other tests of a similar nature have been proposed, but have been found less convenient. The first test for a joint, distinct from that of the whole cable, was, it is believed, proposed by Mr. Whitehouse. No instance has yet occurred of failure in a joint which has successfully passed the accumulation test above described. There are about two thousand joints in each Atlantic Cable.

Any further description of the various tests would only be wearisome. There are tests of charge, of discharge, of the effects of electrification, of the effects of positive and negative currents, tests with statical electricity as well as voltaic currents; but enough has been said to show that the examination of a submarine cable, as now conducted, is not guess-work, or even a matter of experience and skill; it consists simply of a long and laborious series of exact measurements, so expressed in figures that all electricians can understand the results, and compare them with those obtained from other cables, or by other observers. In this lies our safety.

Granting that the production of a perfectly insulated conductor 2,000 miles long is no longer a matter of chance, can we protect and lay this wire with equal certainty in such depths as the Atlantic presents? or do we here fall back into a region of mere good or bad luck? As to shallow water, the question need not be asked. No serious strains occur, and the submersion of the cable depends on a few simple mechanical arrangements which have long since been perfected. Even in deep water cables have not broken during the laying nearly so often as is supposed. Some very early Mediterranean expeditions, a later attempt to join Candia with Alexandria, and the experimental trip of the first Atlantic expedition, give almost the only instances where a cable parted suddenly during submersion; but it must be allowed that the strains endured in passing over depths of 2,000 fathoms approached far too nearly to the breaking strain of the cables, and it is by no means impossible that some cables may have been injuriously stretched, although they were not broken.

In order to lay a cable of any construction taut along the bottom of the sea, it is necessary to restrain its free exit from the ship by applying a retarding force, nearly equal to the weight of a length of the cable, hanging vertically from the ship to the bottom of the sea. Cables of the old form, in which simple iron wires were laid round its core, would support from 4,000 to 5,000 fathoms of themselves hanging vertically in water. They could, therefore, be laid fairly taut in depths of 2,000 or 2,500 fathoms, such as are met with in the Atlantic, but engineers are in the habit of allowing a very much larger margin than the above. They make all their structures from six to ten times stronger than by exact calculation they need be. This figure 'six' or 'ten' they call the co-efficient of safety. A co-efficient of safety of 'two,' such as was given by these old cables, gave very little safety indeed. When the cables are not laid taut, but with a certain slack, the strain need not be quite so great. The friction of the water tends to relieve the strain, but this relief with the old smooth cables was small.

Sir W. Thomson was again the first to give the true theory of the strains which occur, and the curve assumed by the rope during submersion. The first account of the theory appears in the 'Engineer' newspaper of October 1857.

A much more elaborate investigation was, independently of Sir W. Thomson's theory, made by Messrs. Brook and Longridge, whose able paper was published in the 'Proceedings of the Institution of Civil Engineers' for 1858. Dr. Siemens, of Berlin, independently arrived at similar conclusions; the subject is nevertheless not a very simple one, for the Astronomer Royal was misled more than once in his investigations concerning it.

When the ship and cable are both at rest, the latter hangs in a simple catenary curve, the strains on which are easily computed; but when the cable is being paid out, it lies in an inclined straight line from a point a very little below the surface of the sea to the bottom (provided, however, the cable as it lies at the bottom is not strained); above the water the cable hangs in a short catenary; the angle at which the cable lies in the water depends on the speed of the ship and the specific gravity of the cable; it is independent of the strain on the cable, and is therefore unaltered whether the cable is being paid out slack or taut. As the speed of the ship increases, the angle which the cable makes with the horizon diminishes; the same effect is produced by diminishing the specific gravity of the cable—

that is to say, by increasing its bulk relatively to its weight. The Atlantic Cable, under the water, probably lay at an angle of nearly 7° with the horizon; on leaving the ship the angle was $9\frac{1}{2}$°. In this case, in a depth of two miles, a length of from $16\frac{1}{2}$ miles of cable would lie in the water between the point where it left the ship and that where it touched the bottom. The weight of this cable, weighed in water, would be 231 cwt.; fortunately, as the cable would break with about 154 cwt., only a very small part of this weight is borne by the cable itself as it leaves the ship. Even if the cable were to be laid absolutely taut, a restraining force of 28 cwt. only would be necessary. In practice, 12 cwt. to 14 cwt. was found quite sufficient.

The cable, as it leaves the ship, may almost be said to lie on a long inclined plane of water; if it lay on a solid inclined plane, without friction, it might, by a well-known law of mechanics, be balanced by a length of itself hanging vertically from the apex of the inclined plane to the bottom, and this is almost exactly the strain required to be given by the break on board ship to balance the cable, or, in other words, to prevent it from shooting back along the inclined plane, so as to lie slack in folds at the bottom; but the inclined plane of water is not at rest, it yields under the cable at every instant, at every spot; yet if the cable were pressed through the water, so that the water yielded before it, but did not slip along it at all, the analogy of the inclined plane would be quite perfect. The resistance of the water to displacement would supply the component of the whole force required, perpendicular to the direction of the cable exactly as in the case of a solid plane; but on constructing a diagram, it will at once be seen that the cable, as it descends, slips a little along the plane, and the friction of the water opposing this slip slightly diminishes the strain required to lay the cable taut. If, on board ship, this full strain is not produced by the brakes, the cable slips still faster back along the inclined plane, and with such a velocity that the friction of the water on the cable makes up for the insufficient tension given by the brakes, and equilibrium is again restored, but at the expense of a waste of cable. It will be clear that, with a given depth, the greater the length of cable in the water the less need this waste be, for the friction will be directly proportional to the surface; further, for the same reason, the waste will be less the more bulky the cable and the rougher the surface. With the old iron cables of small diameter and smooth surface, very little advantage was gained by diminishing the strain on the brakes below that due to the full depth of water; a very slight relief of strain was followed by a perfect rush of cable out of the ship, and a loss of twenty or twenty-five per cent. was followed by a comparatively small diminution in the risk of fracture. In the cables of the Atlantic class, the bulk relatively to the weight is very greatly increased by enveloping each iron or steel wire in a separate covering of hemp before laying them round the gutta-percha. These cables lie at a much smaller angle with the horizon, they offer a much larger and rougher surface than the simple iron cable, and consequently the friction, as they run back on the inclined water plane, is very much larger. With cables of that class it becomes practicable and desirable to diminish the strain produced by the brake much below that due to the full depth of water. Slack to the amount of twelve or fifteen per cent. diminishes the necessary strain on the brakes by more than one-half, and the importance of this relief can hardly be over-estimated. It actually becomes practicable to disregard the depth over which the ship is passing. The brakes may be set to give the strain thought desirable, and the cable will then take care of itself. In shallow water less slack will be paid out, in deeper water more, but the amount is never excessive, and can at any time be diminished by increasing the speed of the ship, which, by diminishing the angle at which the cable lies with the horizon, augments the effect of the friction of the inclined water-plane. This effect must not be confounded with the effect that would be produced by a buoyant substance attached to the cable. The hemp is no lighter than water, and does not tend by its buoyancy to carry any part of the weight of the cable, but it increases the bulk, and therefore increases the resistance of the water to displacement, and both directly and indirectly increases the surface friction.

The strain on the new Atlantic Cables during submersion was from 12 to 14 cwt.; their strength is 150 or 160 cwt. Here there is a co-efficient of safety of ten instead of two or four. The first cable out of the water weighed little more than half as much as the new cables; in water, it weighed more than they do. Its strength was 80 cwt., and the maximum strain during its submersion was nearly one ton; the ordinary strains varied from 1,500 to 1,900 lbs.

From these figures we may learn the progress which has been made in the mechanical construction of the cables, and the diminished risk which attends their submersion.

The history of the several attempts to lay the cables helps to show the progress made in the construction, and bears out the conclusions as to the improvements effected. In August 1857 a first attempt was made to lay an Atlantic Cable; 330 knots were laid, starting from Valentia. Then the cable broke, the indicated strain being about 27 cwt. The retarding friction on this occasion was produced by two blocks of wood which were clamped round a small drum. Before the next attempt the Appold brake had been invented, and with the sanction of Mr. Penn, Mr. Field, Messrs. Easton and Amos, Mr. Lloyd, Mr. Everett, and Sir C. Bright, it was applied to the paying-out machinery. This brake is an excellent contrivance, by which the required strain is readily produced and maintained unaltered; the retarding friction being quite independent of the condition of the rubbing surfaces. This brake was successful, and has been used ever since. The 1858 expedition began operations on June 26 by a splice in the middle of the Atlantic, joining the cables contained in the 'Niagara' and 'Agamemnon.' The cable fouled the 'Niagara' and broke. A second splice was at once made, and successfully lowered to the bottom. When the 'Agamemnon' had paid out 37½ miles, and the 'Niagara' 43 miles, the electrical tests showed that the copper conductor of the cable was severed. In technical language, there was a loss of continuity. The 'Niagara' endeavoured to haul in the cable, which shortly broke for the third time. On June 28 another splice was made; but after 111 miles had been paid out the cable broke for the fourth time, with a strain indicated of 2,200 lbs., or nearly one ton. On July 28 another splice was made, and this time the cable did not break, but was laid successfully as a mechanical operation, but unsuccessfully in all other senses. As before stated, a want of continuity did occur.

but it ceased after a few hours, and was passed over as of insufficient consequence to stop the submersion.

Much surprise has been expressed at the rupture of a cable estimated as strong enough to bear four tons, when the indicator showed only about one ton. It has frequently been suggested that the instrument gave false indications; but there is really little reason for supposing this. The cable was covered by 126 small iron wires, spun into eighteen small strands, the whole cable being only 5-8ths of an inch in diameter. The wire was not galvanised, and rusted very readily. It is most probable that in many places its theoretical strength was very much reduced by this cause.

In 1865 and 1866 the same brake and indicator, or dynamometer, as it is sometimes called, were used, but the history of events was widely different. The cable, during submersion, not only escaped fracture, but was not even once strained within a tenth part of its supposed strength. In 1865, the occurrence of a small fault, which would have been far too insignificant to have been detected in 1857 or 1858, caused an attempt to haul back the cable, which was broken by chafing against a projection from the bows of the 'Great Eastern.' The arrangements in 1865 were by no means perfect. The picking-up gear was defective and the system of electrical tests faulty, but the paying-out machinery acted admirably, and the cable hardly admitted of improvement. In 1866 the picking-up gear was good, and the electrical arrangements left nothing to be desired.

The special form of cable adopted, in which each iron wire is enveloped in hemp, presents various interesting peculiarities. It is actually stronger than the sum of the strengths of the hemp and steel employed to make it. This almost incredible paradox was discovered during experiments made by Messrs. Gisborne, Forde, and Siemens for the Government, with reference to a proposed Falmouth and Gibraltar Cable. It seems strange enough that a steel wire can be strengthened by wrapping hemp or manilla round it; but this was soon found to be a fact, and indeed the percentage of elongation undergone by a hempen strand and a steel wire before breaking are by no means so different as most people would imagine. By selecting the best lay of the hemp round the steel, it was repeatedly found that the strength of the two combined exceeded the sum of the strengths of the two separately, and this strange result has been fully confirmed by independent experiments conducted by Mr. Fairbairn and others for the Atlantic and Telegraph Construction Companies. The explanation is simple enough. Neither material is really homogeneous: each has its weak places; it is extremely unlikely that the weak places of both should coincide. When, therefore, the two are combined, we obtain the sum of the average strengths of each material; when they are tested separately, we get the sum of the strengths of the two at their weakest points.

This form of cable was first used in 1860 for a cable between France and Algiers, Messrs. Gisborne and Forde being the engineers, and Messrs. Glass and Elliot the contractors. The cable, after some misadventures, was successfully laid, and behaved well during submersion, but the form fell into some discredit, owing to the discovery that even in 1,500 fathoms the hemp was eaten away by a species of teredo after a few months of submersion. This left a mere cage of loose iron or steel wires, unfit to be lifted, or relaid if lifted. Fortunately it appears that these animals, which in the Mediterranean fasten on every inch of exposed hemp, do not exist in the Atlantic. Where they have eaten the hemp the gutta-percha appears as if marked with the small-pox; but no instance has yet occurred where they have actually penetrated the gutta-percha to any serious depth.

The form has other defects. Many persons think that the two injuries which the 1865 cable received during submersion were not due to malice, but to short pieces of broken wire, which would penetrate the soft sheathing of hemp with much greater ease than the hard mail of the common iron-covered cable. The arguments used in favour of this view are as follows:—The hemp conceals a break in the wire which it encloses; a broken wire may be bent out when being coiled, and penetrate the neighbouring coil; the injury may not occur, or not be fully completed, until the coils are disturbed by the trampling of the large number of men engaged on the coil when it is being paid out. Pieces of broken wire were found actually sticking out in this manner after attention had been drawn to the possibility by the faults which occurred. Probably, however, the great success of the Atlantic Cables will cause their form to be the type for deep-sea lines for some time to come.

Cables on board ship are now almost invariably stowed in water-tight tanks; from these they pass up to a sheave or quadrant over the centre of the coil, and thence to the brake-drum, and over the stern. A turn or twist is put into the rope by every turn which it makes round the tank; that is to say, it is twisted tighter by the mere action of coiling away; but this twist is again taken out when the cable is uncoiled; so that if this operation proceeds with regularity, the cable goes into the sea in the same condition as it left the sheathing machine; but if the cable is stiff and springy, or if it is drawn from the hold by jerks, or if one or two coils stick together and are drawn up at once, the turn in the cable tends to throw it over into a loop, which may easily be squeezed or drawn into an ugly-looking thing called a 'kink.' With circular coils, and experienced men in the hold, this hardly ever occurs, and it is rendered next to impossible if the eye of the coil is filled up by a smooth cone, to which the rope clings in ascending, and which prevents any coil from being drawn into a loop. This cone, together with certain guiding-rings which prevent the cable from flying out under the action of centrifugal force, forms the subject of a patent taken out by Mr. Newall, and first used in 1855 for the Varna-Balaclava Cable. The excellence of the contrivance hardly admits of a doubt; but the action of the Patent Laws receives some curious illustrations from the incidents which this patent has given rise to. The validity of the patent has been greatly contested; substitutes more or less like the thing patented have been devised, but rival manufacturers have seldom consented to use the thing patented and pay the royalty. Although the holds were arranged with contrivances having the same object as Newall's cone and rings, foul flakes, as they are called, twice came up from the hold, once on each expedition. These foul flakes are simply two or more turns of the cable which come up entangled together, and then get jammed into more or less of a tangle on deck, for round the brake-drums they cannot go. The cable has to be stopped at once, the ship's engines reversed, and all hands busied in setting the mischief to rights. The following extract from a speech delivered at Glasgow by Captain Hamilton, who accompanied the expedition as a Director of the Atlantic and Anglo-American Companies, gives a graphic description of the foul flakes which occurred during the laying of the 1866 cable :—

This interruption occurred in consequence of the cable, which was being paid out from the after-tank, bringing up with it a bight from the next lower flake, and also the lead from the inside to the outside of the next layer of the coil, so that five cables were running out from the tank instead of one.

These were carried aft together till they were stopped by the paying-out machinery; when, in a very short time, they appeared like the tangle of a gigantic fishing-line. The ship was immediately stopped, but the night was pitch dark, rain falling heavily, and a fresh breeze blowing, the cable over the ship's stern being only visible by a slight phosphorescent light where it dipped into the water. Sir James Anderson, however, by great skill, contrived so to handle his ship of 23,000 tons, which was riding at single anchor in 2,000 fathoms by a mere thread, that the engineers and sailors had time to reduce this apparent confusion to order, and in about three hours the paying-out was resumed without the perfect testing of the cable having been in the slightest degree interfered with.[1]

160 or 170 miles of cable were paid out daily during the 1865 Atlantic expedition, and from five and a half to six and a half knots per hour may be considered a good speed in cable-laying. In 1866 the speed was rather slower, the distance was generally about 120 miles per diem, and the cable paid out about 135 miles. The 1865 and 1866 cables are 1,896 and 1,858 nautical miles long respectively. The total distance from shore to shore is 1,670 nautical miles. The 1858 cable was 2,022 miles long, and it was paid out as fast as in 1865, but more cable was wasted and the ship went slower. A footnote gives the principal dimensions and weights of these cables.[2]

[1] From the *Glasgow Daily Herald*, November 5, 1866.

[2] *First Atlantic.*—Length as laid, 2,022 knots; copper conductor 7-wire strand, weighing 107 lbs. per knot, diameter 0·083 in.; covered with gutta-percha, weighing 260 lbs. per knot, diameter 0·38 in.; served with tanned hemp, and covered with eighteen strands of seven bright charcoal iron wires

There are some popular fallacies connected with cable-laying which are exceedingly tenacious of life—one is, that inasmuch as the wires are laid round a cable like a corkscrew, they will stretch a great deal before supporting the cable, and so the core will be injured by having to support a considerable part of the strain. In point of fact, nothing of this kind occurs. The iron wires abut one against the other, and form a tube which cannot diminish in diameter as a corkscrew does, or would do, if made of soft wire; and experiment shows that an iron-covered cable stretches very little more than a simple straight iron wire. Cables of the Atlantic class stretch a little more, for the soft strands are compressible; but even in this class of cable the elongation, with half their breaking strain, is quite insignificant, and with the strain actually used it is insensible. Then some people say these cables untwist, and they certainly do a little, but the cables recovered from great depths prove that the number of turns which are thus taken out of a cable is quite insignificant, producing no sensible elongation or change in the lay. Others think the rise and fall of the ship must cause sudden jerks and great changes in the strain on the cable as paid out, and quite a small army of patents stand ready to defend the right of inserting some elastic contrivance by which the cable is to have a certain play. Probably the see-saw which these contrivances might introduce would be far more dangerous than the evil they are designed to remedy, for in truth the strain changes very little even in heavy weather, so long as the ship is going fast enough to let the cable lie at a small angle with the horizon. When the cable hangs vertically the case is different, though even then the change of strain is much less

0·028 in. diameter; total diameter of cable 0·62 in.; weight of cable in air per knot 21·7 cwt.; in water 16·3 cwt.

Second, or 1865 *Atlantic.*—Length when complete in 1866, 1,896 knots; copper conductor 7-wire strand weighing 300 lbs. per knot, diameter 0·114 in.; covered with gutta-percha and Chatterton's Compound, weighing 400 lbs. per knot, diameter 0·464 in.; served with wet tanned hemp covered with ten bright steel wires, each enclosed in five tarred manilla hemp strands, diameter of each wire 0·095 in.; diameter of strand 0·28 in.; diameter of cable 1·125 in.; weight of cable per knot in air, 35¾ cwt.; in water, 14 cwt.

Third, or 1866 *Cable.*—Length as laid, 1,858 knots; similar to 1865 cable, except that the steel wires were galvanised and the manilla strands were not tanned but left white. Weight in air 31 cwt., in water 14¾ cwt.; breaking strain, 8 tons.

ATLANTIC CABLE, 1858.

ATLANTIC CABLE. 1865.

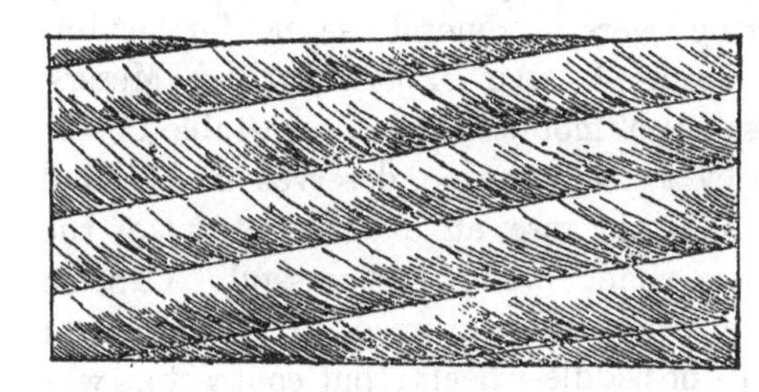

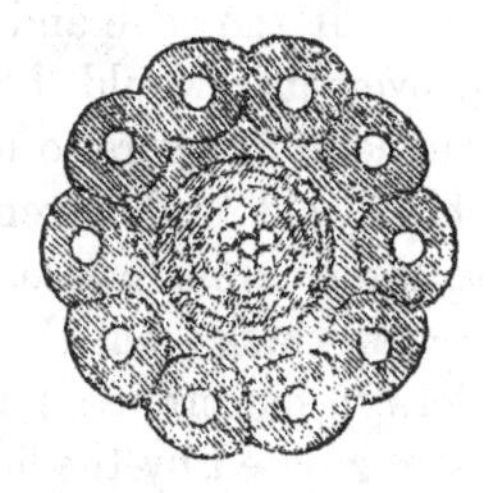

NEW ATLANTIC. CABLE, 1866.

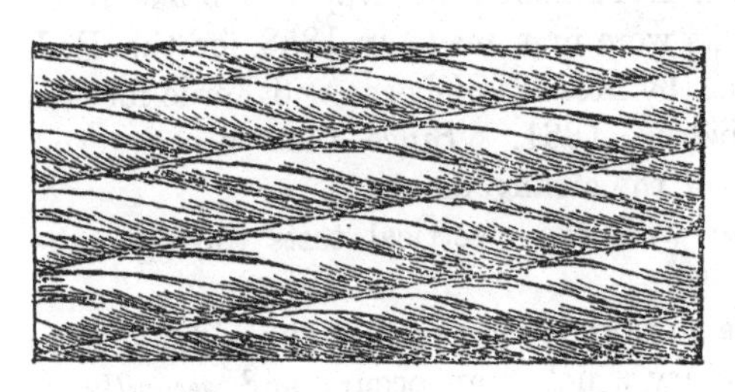

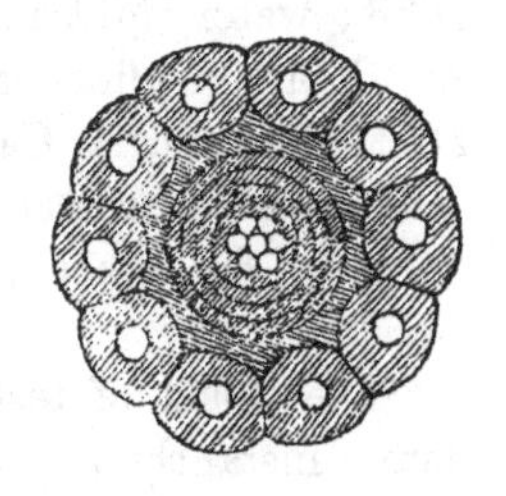

than would be supposed. With the 'Great Eastern' as a *point l'appui* the variation was hardly sensible. Another array of patents defends the privilege of laying a cable through a long auxiliary tube; four patents for this contrivance were taken out in 1857. Other gentlemen wish to tack floats on to the cable; others, parachutes; others, gum and cotton, so as to buoy the cable up for some time; then the gum or glue dissolves and lets the cable down quietly. It is both amusing and sad to read these and many other contrivances. Surely the man who makes a bad invention, and, believing it to be good, spends his life and his fortune in the vain attempt to achieve an impossible success, is almost as fit a subject for commiseration as the real inventor who fails to reap his just reward; and then the former class are much more numerous than the latter.

The machinery now in use for laying cables acts extremely well; if the cone and rings were in general use, no further improvement would be required. An experiment by Messrs. Siemens Brothers to use a reel mounted on a turn-table in the ship's hold and driven by a steam-engine, deserves notice, and to some extent praise, as, at any rate, an experiment out of the beaten track; but the experiment was not successful. Captain Selwyn has proposed a floating reel, the speed of which would be regulated by the floats of paddle-wheels; but contractors who have achieved success by the old plans will be slow to tempt fortune by trying these novel contrivances. It will be seen that very little improvement has been made in the paying-out machinery of late years, simply because it was not wanted. The cone and rings date from 1855; the Appold's brake from 1858; water-tight tanks were first made in 1858 for the Red Sea Cable, but first used by Messrs. Gisborne and Forde for the Malta-Alexandria Cable in 1861. Since then no material change has been made in the arrangements.

It is far otherwise with the electrical tests during submersion.

The object of tests during submersion is twofold: *first*, to detect instantly any injury which may occur; and, *secondly*, to ascertain the position and nature of the injury. Time is of extreme importance in these tests. Faults on board almost always are caused at or near that part of the cable which is in the act of leaving the ship. That is the only portion which is being disturbed, and it is hardly possible that a change can take place elsewhere. If the fault be instantly detected, the ship stopped and the cable arrested as speedily as is consistent with safety, the fault may be retained on shipboard, or if it pass into the sea, only a short length of cable will have to be hauled back before the faulty portion is recovered. As soon as it is quite certain that a fault exists, and the necessary steps have been taken to prevent the cable from running uselessly into the sea, means must be adopted to ascertain where the fault is. One rough method is to cut the cable in the hold near the part being paid out, and then by examining successively the portions in the ship and in the sea, to determine whether the fault is still on board; but electrical methods exist by which, before or after the adoption of this simple examination, the position of the fault can generally be fixed with considerable accuracy. Few statements concerning telegraphy excite more surprise than this does; few people know that accurate measurement of electrical phenomena is possible; some even think that electricity is an agent almost capricious in its action; but those who have learnt that the electrical properties of a conductor or an insulator are susceptible of definite numerical expression should feel no surprise on hearing that, when the electrical properties of a submarine cable of uniform construction are observed to undergo a definite change in virtue of some alteration at some one point, it is quite possible to make such a series of measurements as shall fix the position of that point. There are only two unknown quantities, and whenever by experiment equations can be obtained, including these unknown quantities, they can be determined. Quitting generalities, let us try to show how this is done. We will first suppose that the simple insulation test has shown that the conductor is no longer fully insulated.

A measurement must be made of the resistance of the conductor intervening between the ship and the sea at the fault or earth, as, oddly enough, it is always technically called. If this measurement give 40 units, and the resistance of each knot of the cable is already known to be 4 units, the observer will know that the fault cannot be more than ten miles off. It has already been stated that the electrical resistance of a wire or conductor can be measured with extreme accuracy, and that, as the resistance is proportional to the length, the length in circuit can be calculated from the resistance. Still, from our one measurement, we have not got information enough to know certainly where the fault is—we only know that it cannot be more than ten miles off; it may be less, for the fault itself may have a certain resistance, and about the fault we as yet know nothing. But suppose we can now obtain a similar measurement from the other end of the cable, and this gives 600 units, while the whole length of the cable is 150 miles, we shall then know that the fault is five miles from our end, and has a resistance equal to 20 units; the resistance as measured from our end consists of five miles of conductor and the fault, or 40 units in all, that from the other end consists of 145 miles of cable and the same fault, or 600 units in all, and no other position or resistance of the fault will agree with the two observations made. A comparison with a pipe of water may make this clearer to non-scientific readers. Let us take a pipe 150 yards long, and suppose that we know exactly how much water will run through any given length of a pipe of that diameter from given cisterns at each end. Now, suppose a leak to occur in that pipe: if we stop up the far end and let the water run in from our cistern, we find that as much water runs out as would be allowed to pass by a pipe ten yards long, we then stop up our end of the pipe and let water run in from the far cistern. We find as much water is conveyed away as would be allowed to pass by a pipe 150 yards long; then, as in the electrical case, the leak in the pipe must clearly be five yards from our end, and it must have a resistance equal to that of five yards of pipe. Thus the position of a leak in a water-pipe might be discovered, although the leak itself were buried in the ground. The electrical experiment is quite analogous to this, and is in practice made much more easily than the experiment with water-pipes could be made, for the laws of the flow of water in pipes are much less well understood, and less simple than the laws of the flow of electricity, although we may think we know better what water is than what electricity is.

In cables containing more than one wire, the above test, or something analogous to it, can always be made, for the faulty and good wire being joined together at the distant station, can be treated as one conductor, of which the observer has two ends in his possession. He can then arrange his test so that his observations at both ends are really simultaneous, with the fault in the same condition when added to the two circuits. In this case, a test based on the above principle is quite perfect, and will fix the position of a fault with great nicety. But where the cable has only one conductor, the two tests must be made by different observers at different times. Faults have a disagreeable art of varying very rapidly, so that their resistance is never the same for two minutes or fractions of a minute, and then the test becomes inaccurate, though not actually useless. For instance, the observer in the first case might feel quite sure that the fault was not more than ten miles off, even if he got no information from the other end; if the fault were caused by a nail joining the copper and iron of the cable, it would have no sensible resistance, and the above test would show it was exactly ten miles off. Even if the cable were broken, the observer could guess from the variation of the fault, the current it returned, and other peculiarities, whether it was likely that the fault had much resistance, and thus form by the aid of experience a fair guess at its exact position.

The measurement of resistance is far from being the only test of which the results can be expressed with numerical accuracy; for instance, the statical tension at any point of the wire, its potential, as it is called, can be measured by electrometers, and indirectly by various methods. This statical tension is the quality, in virtue of which one electrified body attracts or repels another more or less strongly. When a current is flowing from a battery through a conductor to earth, the potential gradually decreases from a maximum at the battery to zero at the earth, and decreases according to well-known laws. The observation of this potential at any point gives additional information, therefore, by which the condition of the conductors may be determined. To revert to the analogy of the water-pipe, the potential would be represented by the pressure per square inch, or head, inside the pipe at each point; it would be greatest near the cistern, and gradually decrease to nothing at the mouth of the pipe where the water was discharged.

Another class of fault is more easy to manage. If by accident the pipe got choked up instead of having a hole in it, nothing would be easier than to tell where the obstruction lay, by measuring the quantity of water we could pour into the pipe before filling it. Then knowing the capacity per unit of length, we could calculate the distance by simple division. Exactly so the capacity per unit of length of an electric cable for electricity can be, and is measured, so that if the conductor is broken inside the insulating sheath, without a fault of insulation occurring, the distance of such a fault can be obtained by a simple measurement of the charge which the insulated conductor will take. In short, we can measure current, resistance, potential, and quantity. What is to be measured depends on the nature of the fault observed; but from these measurements or some of them, wherever they can be made simultaneously at each end, the position of the fault can be fixed. Unfortunately, no system of tests on one side of a fault can give its position. A bad fault far off, and a small fault close at hand, cause all the elements which can be observed to vary simultaneously, so as to give no clue as to which has occurred. A bad fault, or one with little resistance, can have its position fixed on the assumption that it has no resistance; but a slight fault absolutely requires the distant test before its position can, even approximately, be determined. Fortunately, signals from the distant end can always be sent past such a fault. We are now in a position to consider the tests hitherto used during laying and the improvements used on the Atlantic expedition.

In very early days people were satisfied if they could speak through a cable whilst it was being laid. Then came the simple insulation test at definite times. Then more complex tests, spaced off into five minutes of this, ten minutes of that, and six minutes of the other, so that each hour was cut up into complex fractions, during which the ship and shore had simultaneously to make more or less complicated changes. If a fault was detected during one arrangement, perhaps half an hour would elapse before the time for speaking and either sending or receiving intelligence would come round. Or, worse still, a fault might occur and not be detected because the connections at the time were arranged for speaking, or for a mere test of continuity, etc. Then blunders would arise from time not being perfectly kept, or from some of the many changes having been incorrectly performed, so that probably this plan was practically inferior to a simple insulation test permanently maintained. It was, moreover, rigid, and could not be readily altered to suit the special tests required when a fault did occur. All these defects were remedied for the first time during the Atlantic expedition of this year. The end of the cable at Valentia was not quite insulated; it was connected with the earth through an enormous resistance, so great that the insulation test of the cable was hardly sensibly affected by the small leakage through it; but this small leakage was easily perceived by an astatic Thomson's reflecting galvanometer. When, therefore, an insulation test was being made on board the 'Great Eastern,' the current used was perceived at Valentia, where the observer could further judge of the tension or potential produced by the 'Great Eastern's' battery by observing the current it would produce through his enormous but known resistance. Any fault would lower that potential, and reduce this current at Valentia. More than this, the 'Great Eastern,' by slightly decreasing or increasing their battery, could cause such small changes in the current observed at Valentia as should serve as signals, and this without intermitting their insulation test. Conversely, Valentia, by drawing off little charges, or adding them, could produce effects similar to slight changes in the insulation of the cable, and those effects could be used as signals from the shore to the 'Great Eastern;' being of short duration, and definitely arranged, they could not be mistaken for faults. Thus simultaneous and continuous tests could be made on ship and on shore. Nevertheless, conversation could be carried on in either direction at any time. No fault of insulation would escape detection, even during conversation, and as soon as it did occur the instruments were ready arranged to make those simultaneous tests by which alone its position could be determined, and then to transmit that intelligence from one end to the other. The merit of this admirable invention is due to Mr. Willoughby Smith. The details of the arrangement actually adopted were worked out by him, in concert with Sir W. Thomson and Mr. Cromwell F. Varley, whose valuable assistance had been given to the Atlantic Company from the time of the failure of the 1858 cable.

The above description of Mr. Smith's invention is not strictly accurate as applied to the arrangements used during the expedition, but the leading idea remained unaltered. Thus the Wheatstone balance was used to measure the insulation resistance in definite units, instead of the simple deflection insulation test. The bridge was arranged with what Sir William Thomson calls a potential divider, a set of resistance coils giving 10,000 equal subdivisions by the mere sliding of two contact pieces. Continuity is never lost, nor the resistance irregularly changed, in these slides—a considerable practical advantage. A special galvanometer was introduced to test continually the constancy of the ship's battery, without which constancy the potential tests would have been much diminished in value. On shore the potential produced by the ship's battery was measured by two methods perhaps more accurate than the deflection through Mr. Smith's galvanometer and large resistance. One method, also suggested by Mr. Smith, was by discharges taken from a condenser charged by the conductor of the cable; the second by an electrometer reading, which could compare the potential of the cable with that of each of the 10,000 subdivisions of a slide similar to that used on shipboard. The battery producing current through the coils of the slide was on shore also maintained constant, or corrected by observations on a special galvanometer. By these arrangements the observer could obtain, in a simple form, the various elements required for the immediate calculation of the distance of a fault had one occurred.

The speaking arrangements were also modified. Charges were not actually withdrawn from the cable or put in at the shore. The withdrawal of a succession of charges would have produced an appearance alarmingly like a fault. Mr. Varley suggested the use of a condenser attached to the cable on shore, by which he induced slight positive or negative charges, which transmitted the signals to the 'Great Eastern.' He, as it were, instead of at each signal withdrawing a few drops of fluid from our typical pipe, pushed the water a little way back in it, or pulled it a little way on, and signalled by these impulses without withdrawing one drop of the fluid. When messages were being received on board the 'Great Eastern,' they simply caused the slight necessary oscillations in the marine galvanometer (an invention of Sir William Thomson's, dating June 1857), which were insufficient to disturb the insulation test. When the signals were being sent from the 'Great Eastern,' the rush of current in and out of the cable would have disturbed the galvanometer unduly, so it was shunted; that is to say, a part of the current was derived by a little sliding arrangement —at the end of each word the slide was moved and a perfect insulation test made. These various practical improvements can only be understood by professional men, but the leading idea of Mr. Willoughby Smith's plan may be grasped by all. The arrangements worked as well in practice as they were admirable in theory. Fortunately no fault occurred.

When a fault does occur, stopping the cable is a very trying and hazardous proceeding. It can only be done gradually. The ship is perhaps running at six miles per hour, or a mile in ten minutes. She will not lose her impetus for a considerable time, even if the engines are reversed; and when the ship is stopped, the cable cannot be instantly checked—if it were, the strain would rapidly become far too great for it to bear. The twelve or fifteen miles which lay straight on the inclined water-plane, as before described, would quickly fall into the common catenary curve, of which the whole weight would have to be borne ultimately by the cable at the ship's stern; for when the cable ceases to sink the resistance of the water ceases to buoy it up. The strain caused by a flat catenary of this length is enormous; thus, if in a depth of two miles only ten horizontal miles intervened between the ship's stern and the point where the cable lay on the ground, the strain due to the catenary would, with the Atlantic Cable, be fourteen tons. In practice, therefore, the cable is generally restrained by such a force as is thought safe, and then allowed to run out until it lies in a catenary short enough to produce only this small strain, or if the cable must be held, the ship must go astern over the cable. When the foul flakes occurred during the 1866 Atlantic expedition, the 'Great Eastern' was stopped in two minutes after the signal was given, and only 130 fathoms of cable paid out during that time. The only time which can be safely saved is the time between the occurrence of the fault and the alarm; and, secondly, between the detection of the fault and the decision of the electrician as to its probable nature and position. The arrangements of 1866, in both these respects, were greatly in advance of all that had been previously attempted.

By far the most remarkable recent achievement in submarine engineering was the recovery of the 1865 Atlantic Cable from a depth of two miles. Cables obviously could be laid in deep as in shallow water—this was a mere question of mechanical arrangement—but very few persons possessed an imagination sufficiently hardy to allow them even to conceive the possibility of recovering a rope which had sunk to the bottom of the Atlantic Ocean. It is not true, as is now frequently asserted, that no one but those engaged in the expedition had any hope of success; for so soon as it appeared from the attempts made in 1865 that the cable could be hooked, Mr. Henley, Mr. Fleeming Jenkin, and a little later Mr. Latimer Clark, publicly expressed their conviction of the probable success of the undertaking; but it is certain that the public, and some even of the directors of the companies concerned, entirely disbelieved in the possibility of success, and put no faith in the assurances given, that the cable really had been found in 1865. Success had attended similar attempts in considerable depths—this was known to engineers—and calculation showed that what had been done in 600 fathoms by Mr. Henley was possible in 2,000 fathoms. Still the greatest credit must be given to Mr. Canning, now Sir Samuel Canning, for his courage in making the attempt in 1865. Few men would have had the nerve to begin an apparently hopeless search at the very moment of failure in a great but comparatively simple undertaking. The admiration is due not so much to the means adopted either then or in 1866—they were simple enough—but to the resolution which prompted the attempt at a moment of great depression. The result will have, or ought to have, a greater effect in promoting the establishment of deep-sea cables than the successful submersion of a dozen cables across the Atlantic. It had been thought and said that men sharing the risk of a deep-sea cable were embarked in a desperate or gambling venture;

one accident, and their money was irretrievably lost. This view had been especially advocated by Mr. Francis Gisborne with great show of truth. He contended, and many approved of his opinions, that it was madness to venture across a deep sea, when a cable could be laid in shallow water, simply because in shallow water the cable could always be repaired, whereas in deep water it could not, and one fault involved the loss of the whole capital embarked. This argument, if not entirely swept away, is very much weakened. Deep-sea cables are no longer gambling ventures, but legitimate speculations.

Nothing can be simpler than the means by which the result was attained. A grapnel or small anchor with five prongs, hung to the end of a hemp and steel rope two and a half miles long, was slowly dragged along the bottom of the sea across the line where the cable was supposed to be. The strain on the steel rope was watched; sometimes it rose and sometimes it fell, as the ship went a little quicker or slower through the water, or as the prongs bit more or less deeply into the sand.

Presently the strain rose from 42 cwt. to 80 cwt., and this strain did not again decrease; but, had the ship been allowed to drift further, would have continued to increase. Surely this increase of strain was due to the cable as it lay on the bottom. The ship's head was allowed to come round so as to face the supposed cable; the steel rope was hauled in; the ship brought vertically over this rope. Still the strain increased, instead of decreasing, even when the length of rope still out of the ship could not reach to the bottom, and then those on board knew that the cable hung on the grapnel. If the cable were not there, the strain would decrease as the weight of steel rope hanging to the bow decreased, but an increase of strain surely proved that more and more weight was being lifted as the grapnel approached the ship, and what conceivable object could produce this effect except the cable, of which a greater and greater length was every minute being lifted from the bottom? This was the reasoning which, in 1865, proved to all on board that really on more than one occasion the cable had hung upon the grapnel. It is needless to say much of the failure to bring the cable to the surface—a failure caused by weak shackles and insufficient machinery—but it is quite worth while to attend to the reasoning of many persons who, in 1865, wrote to prove that, even if the cable were found again, it could not possibly be brought to the surface by mere hauling. The argument used was, that such an enormous length of cable must be lifted, stretching east and west on either side of the grapnel, that it would break under its own weight long before coming to the surface; as one gentleman put it, there was not a sufficient length of cable to reach to the surface. This argument had a certain amount of truth in it, but those who urged it did not generally take the trouble to make accurate calculations, and some made erroneous calculations. Stretch a piece of fine chain, 100 inches long, across a floor, lay it straight, and fasten down the ends; try to raise it in the middle, and you will find that, unless it has been pulled very taut indeed, it will rise an inch or two without difficulty. Even when a cable is supposed to be laid taut, it can be raised to a surprising distance; but the 1865 Atlantic Cable was not so laid; it contained 12 per cent. of slack cable; that is to say, 112 miles of cable lay on about 100 miles of ground. Now, lay the 112 inches of chain on 100 inches of floor, and fasten the ends as before. The middle of the chain can now easily be raised 21½ inches from the ground. The chain will then hang on each side of the point of suspension in catenary curves; the weight supported by the string used to lift the chain will simply be the weight of the chain that is off the ground; the strain on the cable at the point of suspension will be equal to 83½ inches of chain. This strain is less than the whole weight of the cable lifted, so long as the angle made by the chain at the grapnel is less than 120 degrees, as will always be the case when more than 6 per cent. of slack exists; it is a minimum, and equal to half the weight lifted when the cable hangs vertically down on each side of the grapnel, or when the slack is infinite. The more the slack the less the strain, for less cable will be lifted before the grapnel reaches a given height, and the angle at the grapnel will also be more favourable. With 12 per cent. slack, nearly 9¾ miles must be lifted from the ground to reach to a height of two miles. The weight of Atlantic Cable so lifted would be about 6¾ tons; the strain on the cable near the grapnel less than 5½ tons. As the cable would bear 7¾ tons, it was clearly possible to lift it by mere hauling. Moreover, the strain at the bottom would in this case be four tons, tending to pull in more slack from either side, and thus diminish the whole length lifted and consequent strain. By increasing the length of our experimental chain on the floor, and omitting the fastenings at the end, which in the actual cable only exist as friction in the sand, this effect may be clearly seen, and if, instead of tying the chain to the string used to lift it, the experimenter will fish for the chain from a table with four bits of wire, bent into a fish-hook shape, with their shanks bound together, making a mimic grapnel, the illusion will be complete, and the dredger will be surprised to find with what certainty he can hook the cable on a moderately smooth carpet. In 1866 the cable was once fairly hauled to the surface by mere brute force. Calculated from the weight ($6\frac{1}{2}$ tons) then on the grapnel, $9\frac{1}{4}$ miles of cable must have been hanging by the rope. The catenary formed was such as to require 15 per cent. of slack. The strain at the bottom, of about $3\frac{1}{2}$ tons, had therefore pulled in an extra 3 per cent. along the sand on either side. The strain on the cable was about $4\frac{1}{2}$ tons, but with this strain, or little more, it parted shortly after being brought to the surface, and before, owing to rough water, the necessary stoppers could be fixed to it from the ship. This simply affords another instance of the very well-known fact that no engineer should ever depend on obtaining, at all points, the full theoretical or experimental strength of a cable or other structure. There must be a margin. The cable was finally secured by the plan recommended almost unanimously by all who had had experience in similar undertakings. Suppose the chain on the floor laid too taut to come up to the level of the table without breaking; if we with a pair of nippers cut the chain a few feet to our right, a little will slip over the grapnel, the two ends will hang down vertically, and we shall easily land our prize. Just so when the 'Great Eastern' had hold of the cable. She directed the 'Medway' to find it, and lift it three miles nearer to America, she then told the 'Medway' to haul away as fast as she could to break the cable, and the 'Great Eastern' hauled in more slowly, but fast enough to keep hold of the cable with her own grapnel. When the 'Medway' got the cable within 400 fathoms of the surface, it broke at her grapnel. The end fell down, and the loose bight was easily hauled on board by the 'Great Eastern.' The strain on this occasion was six tons. Owing probably to the hemp covering, the cable did not slip along the grapnel after being cut by the 'Medway.' The only chance of failure was that the cable might have rusted so much that, even when hanging vertically, it could not bear its own weight in two miles of water. On the contrary, little or no signs of rust were observed, and there is no reason to suppose that the 1865 cable had materially lost strength during its year of submersion. Considering the perfect success of this simple method of recovering the cable, it really is unnecessary to discuss the many ingenious plans suggested; probably the use of a holdfast grapnel in one ship, and a cutting grapnel in the other, would avoid a few mischances; but it is clear that even these appliances are unnecessary. A few accidents from broken chains, weak swivels, stoppers that slipped, bent grapnel-prongs, etc., did occur and always do occur, even when cables are repaired in shallow water; and, indeed, the repairs of cables in the English or Irish Channel often last longer than the three weeks occupied in recovering the Atlantic Cable. A rocky bottom causes more difficulty and delay than 2,000 fathoms of water.

The cable, when recovered, proved perfect; and on September 2 Sir Samuel Canning telegraphed to Sir Richard Glass, the able manager of the contracting company, that he had much pleasure in speaking to him through the 1865 cable. It was a noble triumph, well earned. The 1865 cable was completed on September 8, 1866. It lies about thirty miles north of the 1866 cable. Those who wish to learn more of the history of this enterprise will find accurate and clear information in 'The Atlantic Telegraph,' by W. H. Russell, LL.D., illustrated by R. Dudley; and in the diary of Mr. Deane, the secretary of the Anglo-American Telegraph Company, published in the 'Times.'

The success of the Atlantic Cable was not gained by the effort of a single genius, but resulted from the co-operation of many minds and divers kinds of men. Some have followed the undertaking from first to last; for instance, Mr. Cyrus Field's unflinching faith has carried him on from first to last as an ad vocate whose zeal never flagged. Sir Richard Glass was a member of the firm of Glass and Elliot, which made half of the first cable, and he is the manager of the company which has successfully completed the task. His work is known to all who practically were connected with the undertaking. He is the recognised chief of all, and willingly recognised. Sir Samuel Canning and Mr. Clifford accompanied the first expedition. Sir William Thomson was on board the 'Agamemnon' in 1858, and has freely spent time and money in forwarding a work in which he saw a means of worthily employing the powers of a mathematician, the experimental skill of a naturalist, and the inventive faculties of a man of genius. His name has already been frequently mentioned as first to make this and that invention or improvement; and not only has he reaped with his own hand a meet harvest of scientific discovery, but he has the satisfaction of having prompted others whose work has been a supplement to his own; and indeed he may be said to have founded a new school of practical electricians in England. Mr. Varley came later into the field, but he too worked hard, and his assistance during the long period of depression from 1858 to 1865, and at Valentia during the last expedition, together with his additions to the testing and speaking instruments, give him strong claims. Mr. Willoughby Smith, of whose beautiful system of testing it is difficult to speak too highly, has only lately been placed in high command, but indirectly, as electrician to the Gutta-Percha Works, from first to last he has helped, and helped effectually, in improving the materials employed. Mr. Chatterton, the manager of those works, should not be passed by in silence. To these must be added the well-known names of Captain Sir James Anderson and Commander Moriarty, C.B., as well as those of the early pioneers, Sir Charles Bright, Mr. Whitehouse, and Mr. de Sauty, and last, not least, the Directors and Officers of the Atlantic Company, the Anglo-American Company, and the Telegraph Construction and Maintenance Company. The difficulties these gentlemen have had to contend with do not admit of being scientifically stated. They can indeed only be known within a very narrow circle, but those who have been similarly placed, and have had to administer the affairs of a company heavily involved in a dangerous undertaking, requiring continually large supplies of fresh capital, will be able to guess at the work and anxiety they must have undergone. A pair of baronetcies among them all is really a moderate reward.

But now that the rewards are distributed, and the cables are laid, how are they being used? Alas! very little as yet. Perhaps no one even believes that a long submarine cable ever will be laid, and so preparations are never made to meet it. The Persian Gulf Cable was laid nine months before the land lines were completed which allowed it to transmit messages. Until very lately they were so wretchedly bad, or badly managed, that messages often spent a week in the overland journey, and arrived so much out of condition as to be unrecognisable by their friends.

The Atlantic Cables end at Heart's Content, Newfoundland, and the journey of the messages often ends there too. It is said that up to the beginning of November, since the line was opened, the lines from Newfoundland to America have been interrupted for thirty days, or nearly one-third of the whole time. Through large tracts of desolate country a single wire was expected to do all the work, and no due arrangement for its management seems to have been made. The short submarine line from the island to the mainland, laid in 1856, and eighty-five miles long, was also out of order when the 1866 cables were completed, and so we have hitherto reaped comparative little benefit from those cables, looked on as a commercial speculation. The high price of 20*l*. for a twenty-word message, only recently lowered to 10*l*., has been justified by the fact that, if many messages had been obtained from the public, they really could not have been sent. So in practice one or two hours' work per diem has been sufficient to send on one cable all the American and European continents had to say in a hurry. This cannot last, but it is almost amusing as a commentary on the lively disputes which occurred on the power of the cables to transmit a large amount of work. In time the outlet from Newfoundland will be completed, and the cables will then surely be flooded to such an extent as to test their utmost capabilities.

Engineers and electricians, half alarmed at their own audacity, gave certificates that seven or eight words per minute might be sent along the new cables. New and complex instruments were devised to insure even this result, and now eighteen or twenty words per minute have been obtained; with the omission of many of these inventions, twelve words per minute is the fair average speed. Nevertheless, the engineers and electricians were not to blame. On the contrary, they deserve praise for their moderation. To explain how their estimate was formed, a sketch of the theory of the transmission of submarine signals is required; and here again Sir William Thomson must be named as the first to state that theory, and draw the main conclusions from it. In a letter to the 'Athenæum,' dated November 1, 1857, Sir William Thomson pointed out, in opposition to Mr. Whitehouse, then electrician to the Atlantic Telegraph Company, that the number of words which in a given time could be sent through a long submarine cable varied inversely as the square of the length of that cable; that when the length of a cable was doubled, only one quarter the number of messages per diem could be sent through it. In a paper published in the 'Proceedings of the Royal Society,' Sir William Thomson gave the complete theory, showing that on all lines a limit existed to the speed of transmission, and giving an estimate of the probable speed through the 1858 Atlantic Cable as three words per minute.

The speed of electricity used to be given as 288,000 miles per second, but in reality Professor Wheatstone's beautiful experiments only proved that this speed might in given circumstances be attained. Electricity seems to have no proper speed, in the usual sense of the word. The speed depends in each case on the condition of the conductor,[1] and may on certain conceiv-

[1] This fact, and the increased retardation observed in underground wires, and therefore in submarine cables, is guessed, or rather foreseen, in a very curious proposal for an electric telegraph, by Francis Ronalds, published in 1823, containing an account of experiments made in 1816, long before the days of Gauss or Cooke and Wheatstone, before even the discoveries of Oersted and Ampère, which have rendered our present system of telegraphy possible. The writer is indebted to Mr. Latimer Clark for the knowledge of this fact. Mr. Ronalds' proposal, based on actual experiment through eight miles of wire, deserves to be better known than it is. His book is called *Descriptions of an Electrical Telegraph*, and was published for R. Hunter, 72 St. Paul s Churchyard, in 1823.

able conditions be treated as infinite, though we have no proof that the laws now known hold good up to or nearly up to this limit. When the Valentia end of the Atlantic Cable is joined to the signalling battery, a current rushes into the cable without any perceptible loss of time, but no effect whatever can be perceived in America for at least one-tenth of a second; after say fifteen-hundredths of a second, the received current begins rapidly to increase, according to a definite law, and if the battery contact at Valentia is continued, the current entering the cable there, and the current flowing out of the cable at Valentia, will be sensibly equal after say two and a quarter seconds. After this the currents would remain equal so long as the battery remained in action. When the battery contact is broken at Valentia, and the cable put to earth, the current flows on at Newfoundland for say one-tenth of a second, as if nothing had happened; it then begins rapidly to decrease, and sensibly ends say two and a quarter seconds after the contact was broken.

Thus the current arrives in gradually increasing waves, and dies out in a precisely similar manner. (The numbers given are not the result of direct experiment, but are probably not far from the truth.) On an average three waves, or arrivals of waves, are required to indicate a letter of the alphabet, and five letters are required for each word, so that if on each occasion the wave had to rise to its maximum and fall to its minimum, each letter would require twelve seconds for its completion, and one word per minute only could be sent. With the ordinary Morse instruments used on land and short submarine lines, probably this result would be nearly the limit of the working speed. On the Malta-Alexandria Cable, which has a larger core, and one, therefore, better adapted for speed than the Atlantic Cable, only 3·18 words per minute were obtained through 1,330 knots. Calculated from this, and allowing for the difference of the cores, the speed on the Atlantic Cables would be little more than 1½ word per minute; and be it remarked that until the present year no other instruments than these ordinary Morse instruments were in practical use on submarine lines. Our engineers were therefore bold when they promised seven or eight times this speed by means of new instruments. Moreover, the New Atlantic, even allowing for the difference in length and in copper, can only be about two and a half times as good a speaking instrument as the 1858 Atlantic, on which only two and a half words per minute had been obtained.

In order to produce a succession of distinct and legible signals, it is not necessary that the wave should reach its maximum and fall to its minimum at each signal. If the sending battery contacts are changed or reversed before the full height of the wave is reached the wave is not obliterated, it is simply diminished; if battery contacts, alternately with one and the other pole of the battery, succeed one another with considerable rapidity, say three reversals every second, or ninety dot-signals per minute, the waves will be reduced to say 10 per cent. of their maximum; but if we can render these little waves visible they may be interpreted as legible signals. The old Morse system, which simply indicated a blunt Yes or No, could not show these little waves or follow them in any way. Sir William Thomson's reflecting galvanometer does render these little waves legible, even when they are no larger than 1 per cent. of the maximum current. The received current deflects a tiny magnet to and fro. A little mirror swinging with the magnet reflects a spot of light on to a distant scale, where by its oscillations the spot of light indicates movements of the magnet too small to be directly seen. The little swinging magnet follows every change in the received current, and every wave, great or small, produces a corresponding oscillation of the spot of light on the scale. These little oscillations, produced in due order, are easily read by a practised clerk—no one knew how easily. The sending arrangements were designed to produce perfect regularity in these little waves, making them as sharply defined as possible; but just as by a practised eye handwriting can be read in which there is no single clearly formed letter, so it has been found that a clever clerk will instinctively, as it were, disentangle most irregular oscillations of this little spot of light into the letters and words they should represent, and from this cause the greater part of the sending gear has been found unnecessary, and yet the bold estimate of eight words per minute has been exceeded. The addition of a single simple little instrument allows four times the number of words to be sent through the Atlantic that could be sent through the Malta-

Alexandria Cable, of little more than two-thirds the length. Here, then, is at least as great an advance as in the other branches of submarine telegraphy. In future long cables the speed may be calculated on this new basis, which has long been advocated by a few, but which had not received practical confirmation till now. The speed at which messages can be sent through a given length of cable is simply proportional to the quantity of copper and gutta-percha used, provided the relative proportions of these materials remain unchanged, as is now practically true in most cases. The new experiment would therefore allow the engineer to adopt a core of one-eighth the weight which he could have adopted upon the old system of telegraphy to obtain the same speed; or if he be not bold enough to adopt so small a core as this would sometimes lead to, he may at least choose the smallest core which on mechanical grounds he thinks safe to adopt. Here we may catch a glimpse of possible cheap cables hereafter.

So long as the cable was coiled on board the 'Great Eastern' it was not possible to transmit more than five or six words per minute through it, even with the best appliances. The difference between a coiled cable and a straight one, as a means of signalling, has long been known, and that difference appeared from Mr. Jenkin's experiment on the Red Sea Cable ('Transactions of the Royal Society, 1862') to be possibly great enough to halve the speed, or even reduce it to a still smaller fraction of that obtained on the straight cable; but crucial experiments on this point were wanting. The extra retardation is produced partly by the induction of the current on various parts of itself in neighbouring coils, and partly by the magnetisation and demagnetisation of the iron sheathing, which forms a sort of huge electro-magnet. The effect produced by the coiling is analogous to giving the electric fluid an inertia, and consequent momentum, an analogy long since pointed out by Sir William Thomson, in a paper by him and Mr. Jenkin on the discharge from a coiled cable, published in 'Phil. Mag.,' 1861. A new illustration of this analogy was discovered on board the 'Great Eastern' by Sir William Thomson, in the fact of an oscillating current flowing in and out of the insulated cable when first charged. This phenomenon was described by Mr. Varley in a paper read at the meeting of the British Association this year at Nottingham, and published in the 'Athenæum' of September 15.

The signalling arrangements on board ship and on shore present this peculiarity, that there is no voltaic circuit. The current is received at Valentia in a Leyden jar or condenser, which acts as a sort of elastic reservoir. When receiving, it is alternately charged by the cable, and discharged back into the cable; while the galvanometer placed between the condenser and the cable indicates the alternate forward and backward tide in the current. Similarly, when Valentia wishes to send signals, it charges the condenser positively or negatively by induction from the battery, and thus causes corresponding movements in the charge of the cable. This arrangement, already alluded to in the testing arrangements, is due, we believe, to Mr. Varley, and prevents to a great extent the action of earth-currents, which would otherwise be found troublesome with so sensitive a receiving instrument as the mirror galvanometer. Much has been said concerning these earth-currents, and some people thought they would render signalling across the Atlantic impossible.

Different parts of the earth and sea are found to be at different electric potentials. One part is electro-positive or electro-negative to another. There is, that is to say, the same difference between two parts of the earth that exists between the two poles of a battery. If, then, these two points are joined by a wire, a current will flow through that wire as if from a battery, and this current is termed an earth-current, to distinguish it from a current produced by an ordinary voltaic battery. This difference of potential between two given spots, such as Newfoundland and Valentia, is not constant, but continually varies. So does the current it produces. The current, and the variations in the current, interfere with the signalling current, disturbing the distinctness of the signals. When no voltaic circuit exists, no direct current will flow from one end of the cable to the other, except that caused by the discharges into and out of the condenser; but a change in the potential of either station will still have some disturbing effect, by changing the charge in the condenser. When very rapid changes take place in the electric condition of the centre, a magnetic storm is said to be taking place, and this, on all lines, will occasionally put a stop to signalling.

Very little is yet known as to the cause of earth-currents or their laws. The electro-motive force producing them does not seem to increase with the distance between the two ends of the line after the first few miles. No greater force than that due to ten ordinary Daniell's cells is reported by Sir W. Thomson and Mr. Varley to have been observed at any time between the two ends of the cable. Much the same may be said of the Malta-Alexandria Cable; but Mr. Varley has spoken of a force equal to 400 cells on short land lines in England. Thus submarine lines appear to have in this respect an advantage over air lines. The latter are further subject to induction from changes in the atmosphere, producing effects similar to earth-currents. The conducting mass of the sea should screen submarine circuits from all these effects; but Mr. Varley has informed the writer that currents were observed from the Atlantic which seemed to be of this nature. One most singular phenomenon was also communicated to the writer by Sir W. Thomson and Mr. Varley. Owing to changes in the potential of the sea, the capacity of the cable for a statical charge varied. The immense Leyden jar formed by the cable at times therefore poured out a current at one end, while the other was insulated, giving apparently more than infinitely good insulation. Not only did the battery used to test insulation then fail to force any current through the gutta-percha of the insulated cable, but a current was actually forced back on the battery, as if coming through the gutta-percha into the cable. From this cause, even if the Atlantic Cables were joined in a metallic circuit, continual currents would fluctuate to and fro between them, owing to changes in the difference of potential of the sea in their respective tracks, thirty miles apart. The two cables afford an unrivalled opportunity of studying earth-currents, about which really little is known; and it is to be hoped the opportunity will not be neglected. There is not the slightest reason to fear that they will prove any obstacle to the transmission of messages through any submarine cable, of whatever length it may be. One method of avoiding all disturbance from earth-currents is to use so powerful a battery as to overpower their effects; but this plan is not to be recommended, since the action of a powerful battery has been known to change small faults into great ones, and though not even a small fault is believed to exist in either

Atlantic Cable, it is well to avoid so powerful a decomposing agent as is furnished by a large voltaic battery. 400 cells were used in 1858. For the signals sent in 1866, 12 cells are sufficient, but 20 or 30 is the number in daily use. Mr. Latimer Clark sent signals through the two cables joined in one, being a circuit of 3,754 geographical miles, with one small cell formed in a thimble. In connection with the subject of signalling, it is interesting to remark that perhaps Sir William Thomson's connection with the Atlantic Telegraph Cables was due to a controversy between him and Mr. Whitehouse on the subject of signalling; the letters are published in the 'Athenæum' from August to November 1856, and are extremely curious. Mr. Whitehouse misinterpreted some careful experiments, and remarks in one place that to lay such a cable as he thought Sir William Thomson's theory demanded would require Mr. Scott Russell's 'Leviathan.' It is needless to add that subsequent experiment has confirmed every part of Sir W. Thomson's theory, although the constants he used have been somewhat modified by experience.

Attention has so far been chiefly directed to the Atlantic Cables, because in connection with these almost every late improvement has been adopted or invented. Lines in shallow water remain much what they were in general construction ten years since. Those of later design are heavier on the average than the earlier cables, for experience has shown that a saving of weight and strength results in great ultimate loss. The average life of a shallow-water cable, weighing less than two tons per mile, is about five years, whereas no limit can as yet be assigned to the life of cables weighing eight or ten tons to the mile. The iron wires are now almost always galvanised, and frequently covered with hemp, and Bright and Clark's silicated bituminous compound, which seems very efficiently to protect the cables from rust, and to prevent broken wires, during submersion, from fouling any part of the machinery, a frequent occurrence some years since, producing what was called a brush, formed by the one broken wire remaining on board in a constantly increasing coil or tangle round the axis of the rope, while the rest of the cable went overboard. The Persian Gulf Cable made by Mr. Henley, and tested and laid by Messrs.

Bright and Clark for the Indian Government, under the superintendence of Colonel Stewart. was thus covered. The excellence of this cable, 1,176 miles long, laid near Kurrachee, in a sea with the temperature at the bottom of 24·2° C., and at the top of 26° C., is a proof that gutta-percha may, with due precaution, be used in tropical climates.

The gutta-percha resistance per mile of this cable varies from 575 to 268 millions of British Association units per mile, according to the temperature at the bottom in the various sections. It was laid from sailing vessels towed by a steamer. The diameter of the main cable, covered with compound, is $1\frac{1}{8}$ inch, and its weight is about 3·7 tons per mile. The Lowestoft-Norderney Cable, 240 miles long, laid in September last, from England to what was Hanover, is the heaviest, on the whole, yet laid. It weighs $10\frac{1}{2}$ tons per mile, is 2 inches in diameter, contains four insulated conductors, is covered with Bright and Clark's Compound, and would bear a strain of twenty tons. It has twenty miles of shore end, each mile of which weighs twenty tons, and would bear a strain of forty tons. The insulation resistance of each mile, as it lies in the North Sea, is 1,100 millions of British Association units, and the four wires are remarkably uniform. This cable was laid for Messrs. Reuter's Telegram Company, under the superintendence of Messrs. Forde and Fleeming Jenkin. The contract was let to the Telegraph Construction Company, and the cable made and laid by Mr. Henley. The England-Holland Cables are shorter examples of equally colossal proportions. There are now seven cables at work between England and the Continent, and three between England and Ireland. The Malta-Alexandria Cable, 1,330 miles long, laid for Government in 1861, under Mr. Forde's superintendence, by Messrs. Glass, Elliot and Co., also deserves mention. Although not designed for shallow water it has done good service; but the frequent interruptions which occur will serve as a warning not again to use a cable weighing less than two tons per mile in shallow seas. Those who wish for fuller information concerning the less important lines may consult the references given in the course of this article. The most important fact to be stated about shallow sea lines is that the Dover-Calais Cable, laid sixteen years ago, is still working, and likely to continue to work for many years to come.

It is extremely difficult to obtain accurate statistics as to the number of miles of cables laid, lost, and now at work. Many are in the hands of distant Governments, who give no information; and some are so frequently under repair that it is difficult to know in what category to place them. In 1862, in Mr. Jenkin's report to the Jurors of the Great Exhibition, 5,345 statute miles of cable, and 9,456 miles of gutta-percha wire, were said to be in working order; 9,406 miles of cable were classed as having been successful for some time, but not then working; 557 miles were classed as total failures. These numbers were avowedly mere approximations.

Mr. Francis Gisborne published statistics in 1865, in which he put the numbers for working and non-working cables at 5,066 and 11,261 statute miles respectively. Dr. Russell gave 6,842, and 9,407, as the length of cables at work and abandoned. Since then 3,754 miles have been added to the successful list by the Atlantic expedition. The Gutta-Percha Company claim now to have supplied insulated core for 12,100 miles of cable, which are still at work. Whatever the actual numbers are, it is incontestable that a large proportion of lines laid have failed from time to time. The Red Sea, and Batavia-Singapore Cables, upwards of 4,000 statute miles long, failed from their want of weight and strength; they rusted rapidly, and could not be repaired when faults appeared, or when they chafed through; or rather the expense of the continual repairs was such that they were abandoned, perhaps somewhat prematurely. The failures of some deep-sea cables have in all probability been due to lightning. An unaccountable apathy has in many cases led to the neglect or actual removal of the lightning dischargers attached to submarine lines, and the writer has seen neglected dischargers with points fused and burnt away, proving that the line had been struck repeatedly. Other failures have been attributed to the tautness of the cable when laid, to friction at the bottom, to volcanic action, etc.; but not much is known about these causes, which are rather hypothetical than proved. Neglected faults have certainly, in some cases, been much enlarged, by the use of more and more powerful batteries, added by ignorant clerks.

These failures need alarm no one; they simply prove what

should be known without proof, that there is a real difference between ignorance and knowledge, between care and neglect; and that supervision after submersion is not less necessary than during manufacture. The main object of this article has been to show that the success obtained in late years, as compared with early failures, is due to no chance, but to a real advance in every branch of Submarine Telegraphy. If the reader does not understand or believe in this advance, the writer has failed in his object. He prefers to think that he has shown good reason for believing that the success is likely to be permanent. Much might be written on the proposals for still further improvements, real and imaginary. Mr. Hooper has succeeded in preparing india-rubber, so as to be apparently permanent, while it certainly surpasses gutta-percha in all the electrical properties which are required for the insulation of cables. The Indian Government has taken the initiative in employing this material, which is eminently suited for tropical climates. The other preparations of india-rubber have very generally been found subject to decay; and various newly proposed materials, such as parkesine and balata, can hardly be said to have been fairly brought before the public.

Messrs. Siemens have employed for some portion of their lines lately laid with success in the Mediterranean a very novel form of cable, formed of hemp bound with strips of copper, which they believe will be much more permanent than the old-fashioned cables. The forms proposed, but not yet tested, are very numerous, and little is known of their merits. The cost of an experiment is so great that engineers hesitate to recommend even what they believe to be well worth a trial. It is certain that the old forms answer well, but it is equally certain that their expense will preclude their adoption, except on the main lines. A cheap light cable, durable in deep water, would lead to an immense extension of the telegraphic system. It is by no means certain that a simple gutta-percha-covered wire would not answer as well as the most elaborate cable. However this may be, the next advance must be towards cheapness; efficiency is attained.

The importance of telegraphic communication is often claimed on very narrow grounds. The advantages in war and diplomacy are to some extent counterbalanced by very obvious disadvantages. Even the gain to individual merchants admits of doubt. By diminishing risks, telegraphy is sometimes thought to diminish profits. The mere convenience of sending a message quickly is outweighed in many minds by the annoyance of receiving, at all odd hours, scraps of news, often unintelligible from their conciseness. But on the broad ground that with the assistance of the telegraph the wants of one country can be supplied from the excess of another, in little more than half the time required for the purpose without the telegraph, we may claim for that invention a recognition that it is useful, in the sense that free trade, good roads, or fleet ships, are useful. The measure of that good is a problem in political economy which it is not now our business to solve ; it is certainly out of all proportion with the price paid for the information sent. Up to the present time full advantage has not been taken of the power we possess. From a want of organisation and some political difficulties we cannot at this moment send a message to any distant part of the world with a certainty that it will be delivered without considerable delay and probable mutilation. Mr. Reuter, to whom we owe the organisation of the despatch of public news, has begun to organise a system by which, in time, the great capitals of Europe may really be placed in instantaneous communication. A Parliamentary Committee has been considering proposals of a similar kind extended to the East; but, meanwhile, a message sent through the Atlantic Cable may be delayed five or six days in the wilds of Nova Scotia, and mutilated messages continue to arrive after a fortnight's journey from India. So long as this is the case, no calculation can be made of the employment which would be found for our great submarine cables if worthily worked. But the fact that much remains to do, even after the Atlantic Cable has been laid, need not prevent a just pride in a really great victory, achieved, not by chance, but by a knowledge resulting from the patient efforts of many minds for many years.

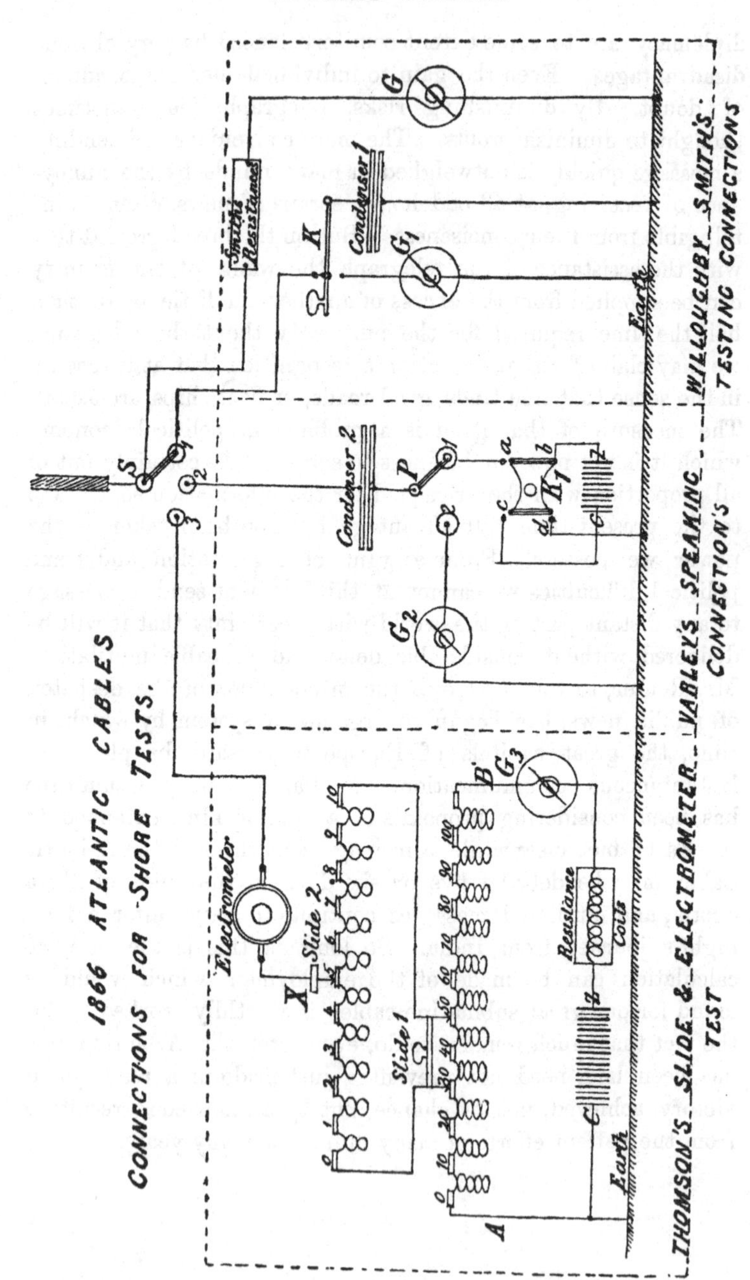

1866 ATLANTIC CABLES
CONNECTIONS FOR SHORE TESTS.
Electrometer
Slide 2
Slide 1
X
A
B
Resistance
Coils
Earth
Condenser 2
Condenser 1
Smith's Resistance
S
P
K
K1
G
G1
G2
G3
THOMSON'S SLIDE & ELECTROMETER TEST
VARLEY'S SPEAKING CONNECTION'S
WILLOUGHBY SMITH'S TESTING CONNECTIONS

ADDENDA.

Explanation of Diagram of Shore Tests—Atlantic Expedition.

S is a switch by which the cable can be connected at pleasure with the studs 1, 2, and 3.

The cable is always connected with Smith's resistance, through which a feeble current passes, deflecting galvanometer G, and showing by its magnitude the state of insulation of cable, or, more strictly speaking, the potential at S.

When S is connected with stud 1, as drawn, condenser 1 is charged by the cable. It can be discharged through galvanometer G_1. This discharge is a second test of the potential at S. It will cause an onward movement in the current applied by the ship's battery, and a consequent momentary change of deflection in the ship's galvanometer. This will be a continuity test for the ship, proving the copper unbroken. This test might have been used as an arrangement for speaking from shore to ship. Two condensers of different sizes would have given two signals, from which an alphabet could be formed.

When S is connected to 2 and the second switch P to 1, as drawn, the cable will charge one plate of condenser 2; and conversely, if the potential of the other plate be changed by being connected through the reversing key, K_1, alternately with each pole of the battery, C Z, this will induce alternate positive and negative charges in the cable, which, being superadded to or withdrawn from the original charge produced by the ship's battery, will cause alternately backward and forward currents which will show as signals on the ship's galvanometer. Similarly, if switch P connect the second plate of condenser 2 to the galvanometer G_2, any changes of potential in the first plate of condenser 2 caused by the ship's battery will induce corresponding changes in the charge of the second plate, and therefore currents to and fro through galvanometer G_2, which currents may be used as signals.

The cable at S may be connected through stud 3 with the electrometer, and its potential directly measured by comparison with the potential of some point of a set of slide resistance coils, through which a current is flowing from a local battery. This current is maintained constant by reducing the resistance in the resistance coils, so as to maintain a constant deflection in the galvanometer G_3. The potential at the successive studs of the slide 1 may be called in an arbitrary unit, 10, 20, 30, ... 100; and if the electrometer were directly connected with these studs, it would show with an accuracy of ten parts in a hundred what was the potential of the cable in the

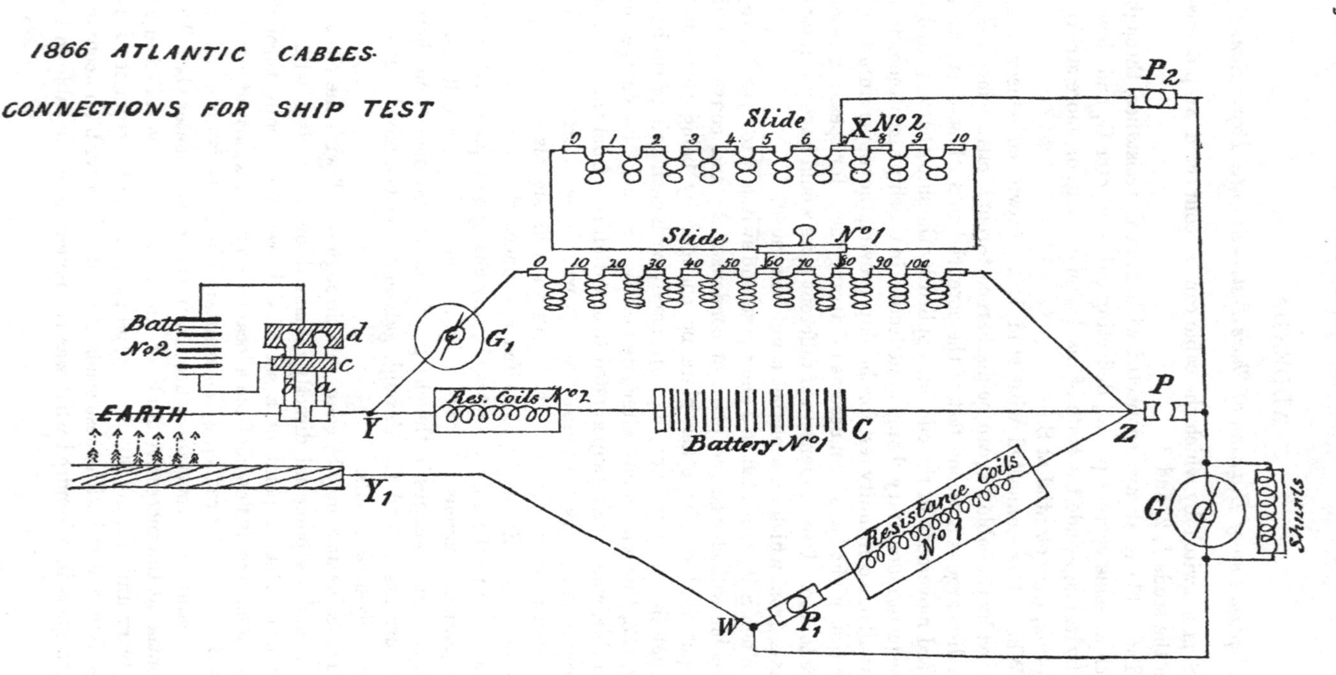

1866 ATLANTIC CABLES.
CONNECTIONS FOR SHIP TEST

same arbitrary unit. To obtain further subdivision, a second set of resistance coils, each equal to one-fifth of each of those of the first slide, are joined so as to bridge over any two of the first set. The first set consisting of eleven equal coils, the total resistance between A and B will always be ten units, and the ten studs of slide 2 will subdivide into ten equal parts the potential between the two studs of slide 1 which they bridge. Thus, as drawn, calling the total potential of A zero, and of B 100, the potential of X will be 35. Slides 1 and 2 during the test are moved till the potential is found, which is exactly equal to that of the cable. In reality, there were 101 equal coils in slide 1, and 100 in the second slide, each equal to $\frac{1}{50}$th part of those in the first. Thus the potential between A and B could be subdivided into 10,000 equal parts.

Explanation of Diagram of Tests on board Ship.

The diagram, when connection is made by plugs at P_2 and P_1, but broken at P, represents a Wheatstone's balance, the four arms of which are, 1st, a set of resistance coils No. 1 between Z and W; 2nd, part of the slide resistance coils between X and Z; 3rd, the remainder of the slide resistance coils between X and Y; and 4th, the gutta-percha sheath of the cable between the conductor, Y_1, and the earth, Y. G is the galvanometer of the balance, arranged with shunts to diminish its sensitiveness when required. The resistance of the gutta-percha = resistance coils No. 1 multiplied by $\frac{XY}{XZ}$. Thus if the slide reading had the value 67, as shown in the diagram, we should have gutta-percha resistance = resistance coils No. $1 \div (\frac{100}{67} - 1)$.

In the actual slides with 100 coils each, the resistance in coils No. 1 was divided by 10,000, divided by the slide reading, and a unit subtracted. The same formula gives the copper resistance if the remote end of the copper conductor is put to earth. Resistance coils No. 2, and galvanometer G_1, were used to maintain a constant current through the slides as described for the shore tests. Y was usually connected to earth by the bar C which joins *a* and *b*, but when the keys *a* or *b* were depressed, they sent signals to the shore from battery No. 2, but these signals only slightly raised or lowered the potential of the cable, and did not much disturb the electrification. When signals were being sent, the sensibility of the galvanometer G was diminished by shunts, which were removed during the pause after each word to take a perfect insulation test. If anything had gone wrong with the Wheatstone balance arrangement, the plugs P_2 and P_1 would have been withdrawn, and connection made at P, which would have established the old direct simple insulation test from C to Z, P, G, W, and Y_1.

TELPHERAGE.[1]

In the first place, it is necessary that I should define what is meant by this word 'telpherage,' and, perhaps, that I should defend its formation. The word is intended to designate all modes of transport effected automatically with the aid of electricity. According to strict rules of derivation, the word would be 'telephorage;'[2] but in order to avoid confusion with 'telephone,' and to get rid of the double accent in one word, which is disagreeable to my ear, I have ventured to give the new word such a form as it might have received after a few centuries of usage by English tongues; and to substitute the English sounding 'telpher' for 'telephore.' In the most general sense telpher lines include such electric railway lines as were first proposed by my colleagues, Messrs. Ayrton and Perry. The word would also describe lines such as I have seen proposed in the newspapers for the conveyance of small parcels at extremely rapid rates. But to-night I shall confine myself entirely to the one specific form in which the telpher line first presented itself to my mind, and which it has fallen to my lot to develop. In this form telpher lines are adapted for the conveyance of minerals and other goods at a slow pace, and at a cheap rate. The problem which occurred to me was this: Was it not really possible to send vehicles, by means of electricity, along a single suspended wire or rod—in fact, to telegraph goods and passengers instead of messages? The idea is familiar as a joke, but, on consideration, it appeared that there might be good grounds for supposing that the idea was both practicable and useful. I am now able to show you the realisation of that idea, and the result of experiments on a large and practical scale has, I think, justified the arguments

[1] A paper read before the Society of Arts, May 14, 1884.

[2] The 'o' is found in all compounds of this root, as in *λαμπαδηφόρος*, *κριοφόρος*, or, to take a more familiar example, phosphorus.

which have induced me to devote much time and labour to telpherage.

[Here the model was shown in action. This model consisted of two concentric octagons of wire, the length of each outer span being 5 feet. On each octagon there was a single locomotive and train, equal in length to that of the span. These trains ran well and steadily in opposite directions round the lines.]

These arguments may be stated as follows. We could not with steam employ a vast number of little one-horse engines to pull along a number of small trains or single waggons. There would be waste in the production of power and great cost in the wages of the men employed at each engine. But an electric current of, let us say, 50 horse-power, will, as it circulates through a conductor of moderate size, drive thirty small engines each of one horse-power, which require practically no supervision, and can be made nearly as economical in their action as a single electro-motor of 30 horse-power could be. But if the power can be distributed economically along a line, say, ten miles in length, this allows us to employ thirty small trains, corresponding each to a waggon pulled by one horse, instead of a single train such as might require 30 horse-power. If we further distribute the weight by making each train of considerable length, we are able to employ an extremely light form of road, such as a suspended rope or rod of, say, $\frac{3}{4}$-inch diameter. Later on in the paper I will show the amount of traffic which such a rod can practically convey. Meanwhile, I simply draw your attention to the general principles of the subdivision of power and the subdivision of weights. In distributing the power by means of electricity it was clear that considerable waste must be incurred, but the amount of that waste is easily calculated, and is by no means prohibitory. Moreover, the power, being obtained from stationary engines, or in certain cases from falls of water, could be produced at a cheaper rate in comparison with that obtained from locomotives or traction engines. When I examined the various forms of possible road by which the distributed power and distributed load could be conveyed, it seemed to me that the single suspended rope or rod offered great advantages. The smallest railway involved embankments, cuttings, and bridges, fencing, and the purchase of land. A single stiff rail, with numerous supports, from which the train might hang, seemed better, and may, in some cases, be employed, but the supports would require to be numerous—say, one post every ten or fifteen feet—and even with these spans the girder required to carry vehicles weighing 2 cwt. each would be costly. With a single suspended rod or rope we may have supports 60 or 70 feet apart. A $\frac{3}{4}$-inch rod, thus supported, will carry five vehicles, each bearing 2 cwt., without excessive strain. No purchase of land is necessary, no bridges, earthworks, or fencing. The line can be so far removed from the ground that it will not be meddled with, either by men or animals. A single wheel-path gives the minimum of friction, and the rolling-stock can be much more easily managed than if we attempted to let vehicles run on double swinging ropes. On all those grounds it seemed well worth while to devise means by which trains could be electrically and automatically driven along the single suspended rod.

Before proceeding further, I had better state how far this idea has been realised. The Telpherage Company, Limited, was formed last year to test and carry out my patented inventions and those of Professors Ayrton and Perry for electric locomotion. On the estate of Mr. M. R. Pryor, of Weston, two telpher lines, on my plan, have been erected. One of these is a mere straight road, with spans of 60 feet, and various form of rod and rope. The first full-sized train was run on this line with a locomotive which we call the 'bicycle-wheel loco' (Figs. 8 and 9, p. 260). The line was found inconveniently large and high, and the experiments were continued on a line $\frac{5}{8}$-inch diameter, of round steel rods, with 50 feet span. This line is continuous, that is to say, it re-enters on itself. It is 700 feet long, and we have run a train of more than one ton at a speed of five miles per hour on this line with complete success. The insulation has given no trouble. It need hardly be said that we see our way to great improvements in details. Thus, we can make the road more uniform, and stronger for its weight; we can lessen the quantity of material used, and greatly diminish the amount of skilled labour required in erection. We can improve the design of the posts. We can improve the trucks and locomotives, so that they will go round sharper angles, and so forth; but the main object has been practically carried out. We have had trains on a scale as large as I am prepared to recommend, running at the highest speed I have contemplated.[1] I trust it will be clear to you, from this description, that what I have contemplated and realised is not an electric railway destined to compete with steam railways in conveying goods and passengers at high speeds, neither is it a new form of communication destined for small parcels and high speeds; it is simply a cheap means of conveying heavy goods which, like coal or grain, can be carried in buckets or sacks, each containing two or three hundredweight. The speed on a telpher line will be that of a cart, and the object we aim at is to cart goods at a cheaper rate and more conveniently than with horses. I assume that you all know that an electric motor is a machine which will run so as to exert power whenever an electric current is passed through it. You also know that a machine called a 'dynamo,' driven by a steam-engine or other source of power, will produce an electric current which may be conveyed along a suspended and insulated rod, and used to drive an electric motor. In describing the details of my system, the first point to be explained is how the current produced by the dynamo, and conveyed along a single line, is taken from that line and directed round the motor.

In endeavouring to realise this idea, the first thought which occurred to me was that of dividing the line into lengths, equal to the length of the train, so that, using the train to bridge over a gap between two sections at different potentials, the current could be conveyed from the leading to the trailing wheels of the train round the motor. This idea is employed in the model now shown; but, in the first form which suggested itself, the gaps between the sections were opened by a switch worked by the front of the train and closed by a switch worked by the end of the train. The first model, which may have been seen by some present working in Fitzroy Street, was made on this plan. Trains driven in that way would all be coupled in series. The present model is differently arranged; there are no working

[1] A form of points for sidings and branch lines has been constructed, and acts satisfactorily. The trains in the sidings are, at Weston, driven electrically, but when a dangerous electromotive force is used this would be worked by hand.

parts or switches. Let the successive sides of the polygon be called the odd and even sides; the odd outer sides are connected with the even inner sides; and the even outer sides with the odd inner sides. We thus have two continuous conductors each going right round the model, but not joined to each other; these are connected to the two poles of a battery. So long as no train bridges a gap no current flows, but whenever the train bridges the gap a current flows from the positive to the negative pole round the motor. This plan is called the 'cross-over system;' all the trains are joined by it in parallel arc, and the current is reversed each time a train passes a gap. This reversal does not affect the working of the motor. This is the plan which has been carried out on a large scale at Weston. Its simplicity leads me to believe that it will be the plan most usually adopted, but several other methods of driving have been devised. A spark passes between the wheels and the line each time the current is stopped, but this spark occurs between large masses of metal, where it appears to be harmless; it has given no trouble whatever at Weston. Moreover, it has been found very easy to make connection between the line and the train. The ordinary truck wheels answer admirably, so that no complicated brushes are required. There are some absolute advantages in having interruptions at regular intervals, but the discussion of these would lead me too far for my present purpose. Only one of the two continuous conductors requires to be insulated; this results in alternate insulated and uninsulated sections all along each line. Fig. 1 shows a saddle, as we call it, with an insulated attachment, B, at the one end, and an uninsulated attachment, A, at the other, as used for a short sample line which has just been sent to Peru for the Nitrates Railway Company. The line itself is a $\frac{3}{4}$-inch steel rod with forged ends; and Fig. 2 sufficiently shows the mode of attachment. The insulation is given by a vulcanite bell insulator, D, carrying a cast-iron cap, C. All the parts are designed to stand 2·2 tons strain; the vulcanite is secured between two layers of Siemens' cement. The experiments at Weston have shown that vulcanite answers perfectly, but the material is rather expensive I have here a small porcelain insulator, which has been subjected to 2·2 tons strain. I believe porcelain will answer well in all respects, but it has not yet been subjected to the test of actual traffic day by day. At Weston, the vulcanite was used between layers of Portland cement, the only objection to which is, that it takes some time to set. The simple steel rod has been found preferable in all ways to rope. We find that there is less friction, and less jar with the rod, and ample flexibility; it is also much easier to secure. Moreover, a solid rod, with welded ends, can be made

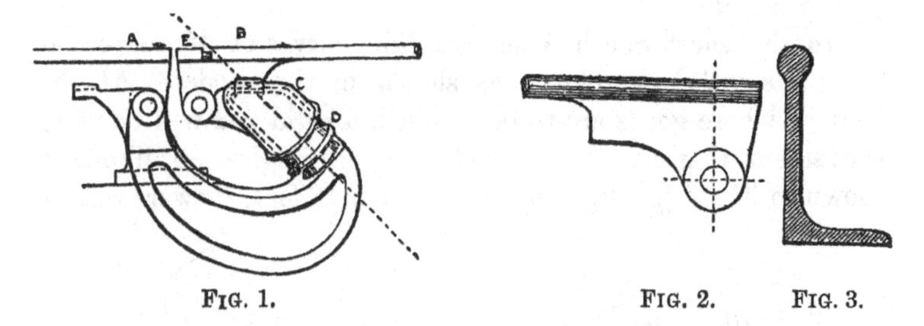

Fig. 1. Fig. 2. Fig. 3.

so that the ends, where supported, are, to some extent undercut, as is shown in the corresponding bulb angle-iron (Fig. 3) used for rigid parts of the road; this undercutting allows much greater freedom of rolling than would be compatible with the horizontal gripping wheels, especially when gripping wheels are used which, like those in the model, actually hold on to the line,

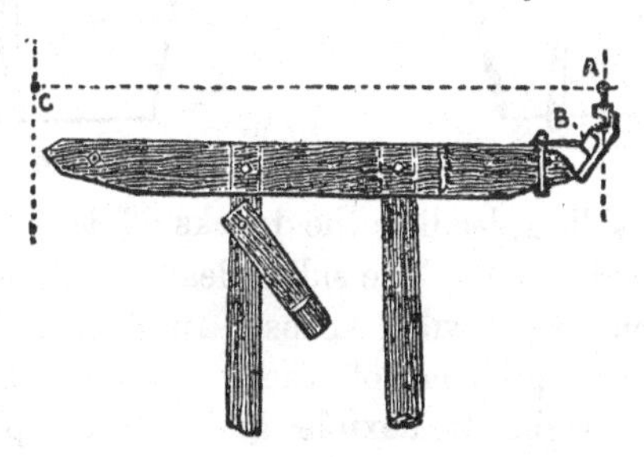

Fig. 4.

so as to resist being lifted. A short piece, E, slightly insulated, prevents the sections from being short-circuited by the wheels. Fig. 4 shows the post and crosshead supporting the line. In the 1-inch example this design was fully carried out, and the posts stood the cross strain due to the overhanging load perfectly. In the five-eighth-inch line an attempt was made to cheapen the construction, but the posts, in wet weather, work at the foundations; it is well that we are put on our guard against this danger. In the first design a sort of rocking saddle was employed, to allow the strain to be transmitted from one span to the next, but the flexibility of the posts provides amply for this object. Abutment posts are required at intervals, and these can be made use of to provide compensation for changes of temperature, and to limit the stress on the rods. In straight lines I reckon four abutment posts per mile.

In the short South American line, curves of 45 degrees at the posts will be employed, as shown in the model. At the stations where goods are to be handled, a rigid rod will be more convenient than the flexible rod. A bulb angle-iron, like that shown in Fig. 3 (p. 257), supported every 10 feet, answers well at

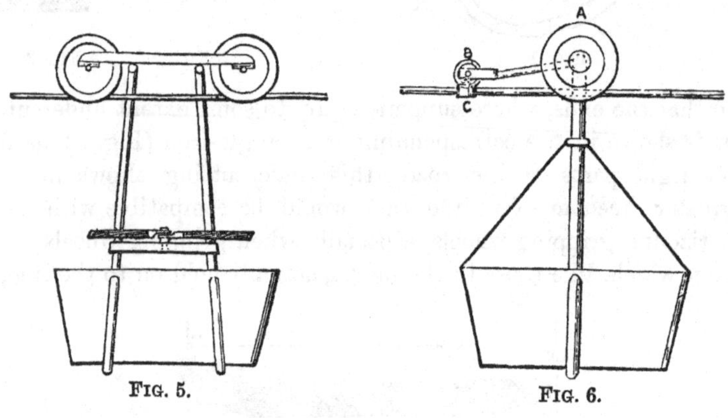

FIG. 5. FIG. 6.

Weston, and a siding, leading the trucks off this line, has been satisfactorily carried out. The siding leads back to the line at a point between two flexible spans. In fine, it may be said to-night that the problem of the continuous line, whether straight, curved, rigid, or flexible, has been completely solved. Drawings and specifications can be put, without further delay or experiment, into the hands of contractors. Trucks used on ordinary rope lines are designed to be pulled by ropes on a road which is necessarily straight. When trucks of this description, with wheels 8 inches diameter and 22 inches wheel-base (Fig. 5), were tried at Weston, arranged in trains, some new difficulties presented themselves. Any sudden check to the motion was followed by a rearing action, throwing the truck off the line; similar results followed the application of any sudden pull. Moreover, trucks, with two rollers, on a rigid frame, even with so great a wheel-base as 22 inches, require curves of considerable radius if we are to avoid serious binding at the flanges. Notwithstanding these difficulties, the trains at Weston, with a little care, run well and lightly, but the trucks which have gone to South America are on the plan adopted in the model; they run much more safely, and turn much sharper curves. They have two peculiarities—first, each wheel, 7 inches diameter (Fig. 7), is pivoted on an axis, B, vertically over the centre of the wheels, A; this allows the truck to run with the

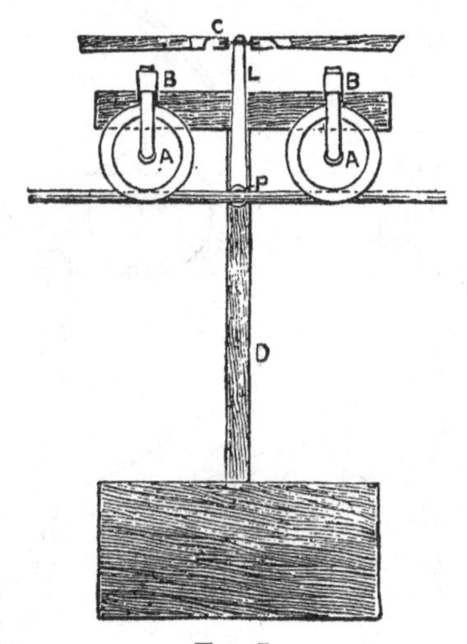

Fig. 7.

freedom of a bicycle round curves; secondly, the weight carried is hung on a swinging arm, D, pivoted to the frame at a point, P, on a level with the line. The result is that any force applied in a plane containing the line acts as if applied at the line itself, and will neither lift the wheels in front nor behind. In the model, the coupling, as you see, is on one line attached to the top of the swinging arm (L, Fig. 7), where the coupling rods are well out of the way. In the other line, the coupling is below the road. The swinging arm relieves the locomotive from all jerk at stopping or starting. The truck is completed by a small hook or catch embracing the rod. In case of any accident causing the wheels to leave the line, this hook will prevent the truck from falling. The weight of the two-wheel stiff truck, shown in Fig. 5 (p. 258), with wrought-iron buckets, is 75 lbs. The weight of the two-wheel pivoted trucks, with wooden bucket, is 63 lbs. They are both adapted to carry 2 cwt. Fig. 6 (p. 258) shows a one-wheeled truck tested; the results were not favourable. A special form of bucket must be designed to suit each kind of traffic. Simple iron hooks for sacks will, in many cases, be available, and these hooks can be so contrived that on being struck they will drop the sack.

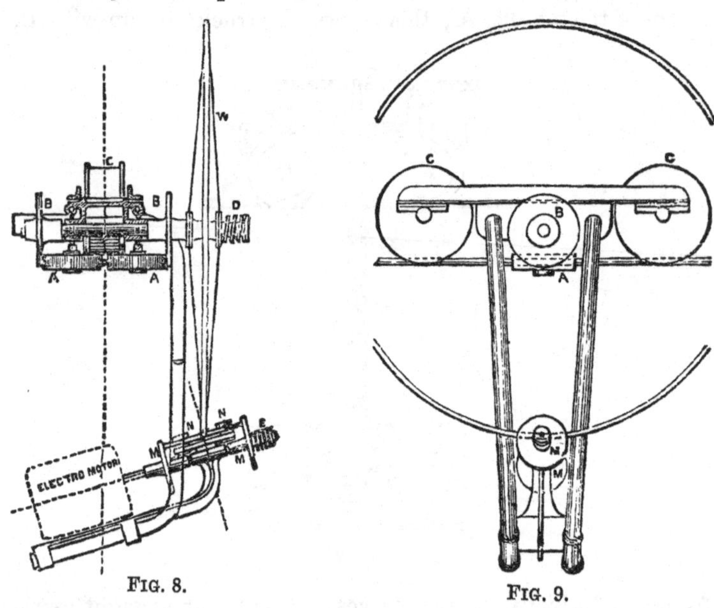

FIG. 8. FIG. 9.

The first type of locomotive which was tried on a large scale is shown in Figs. 8 and 9. The motor lies nearly horizontally across the line, and is connected by a form of frictional gearing, which I term 'right angle nest gearing,' with the edge of a bicycle wheel, W. The shaft of the bicycle has on it two discs, B B, one of which is fixed on the shaft, while the other can slide longitudinally on the shaft. These two discs are pressed together by a spring, D. Their edges bear on the horizontal gripping rollers, A and A, which seize the line. These rollers are supported in such a way as to be free to come together under the pressure of the spring transmitted by the discs A and B. By tightening the spring, any required grip can be obtained with no injurious friction, either on the cross shaft or on the spindles of the rollers. This grip is a form of right angle nest gearing. The weight of the locomotive was taken by wheels, C C, fore and aft. The following defects were observed :—The frictional surfaces, both in the upper and lower nests, were too small, and the materials too soft, so that rapid wearing resulted with a consequent increase of friction. Moreover, the grip was so powerful, that the rollers, A A, were capable of supporting the weight, and thus a small inclination of their vertical axis was enough to cause the

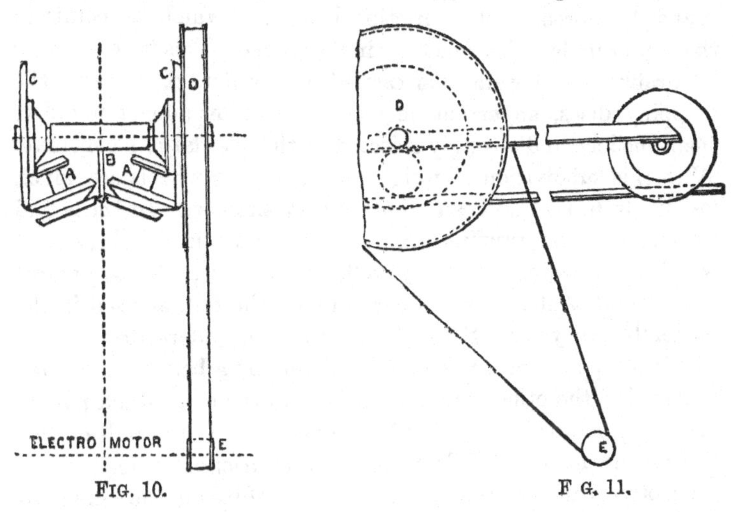

FIG. 10. F G. 11.

locomotive to rise and even run off the line; moreover, the vertical curvature in the rope, or at the posts, required the rollers, A A, to be deep, thus limiting the extent to which rocking was admissible; moreover, very broad pulleys, fore and aft, would be required even for moderate horizontal curves. Nevertheless this locomotive ran sufficiently well on the 1-inch line during an exhibition to the shareholders last autumn. The weight for a five-eighth-inch line of a somewhat improved form of this type, to exert one-half horse-power, on the average, is 200 lbs.; an extra half-hundredweight would give one horse-power. The driving wheels, A A, of this example are 6½ inches diameter. The motor makes 9·23 revolutions for one of the driving wheels.

One mile per hour corresponds to 473 revolutions per minute of the motor. 35·21 inch pounds at the motor spindle are required for a pull of 100 lbs. at the rail. Figs. 10 and 11 show a locomotive designed by Mr. A. C. Jameson, when I was personally unable to attend to work. This locomotive, which is called the 'belt locomotive,' shows a great advance on its predecessor. The general arrangements of the upper nest grip is retained, but a most ingenious modification has been introduced, by which the discs, C C, run on one path on the rollers, A A, while the rod runs on another. In this way the dirt from the line is never conveyed to the driving disc surface between A and C. Moreover, these frictional surfaces, which are points in the first form, have become lines in the second. This type answers admirably. The weight is carried by a roller, B, between the gripping discs, an arrangement contained in one of my first small models, and wrongly rejected in the first large locomotive. With this subdivision of weight, the gripping wheels are much less likely to rise, and can be made very shallow. In the actual locomotive, these gripping wheels are of an open inverted Λ shape, which has certainly run very well, although I prefer at present the upright V shape, which closes under the rail, as used in the model before you.[1] Both of the gripping rollers drive, as in the first type. The cross shaft is driven by a belt on a 20-inch pulley, D; the other end of the belt runs on a 2-inch pulley, E, on the motor spindle. The friction due to the pull of this belt on the motor spindle is relieved by friction rollers. This locomotive runs extremely safely and steadily on the line; indeed I am not aware that it has ever been thrown off. The following are particulars of its construction:—Weight, with a 96 lb. 1 horse-power motor, 269 lbs.; wheel-base, 2 ft. 6 in.; diameters of driving rollers, 6 in.; 4·94 revolutions of motor per one revolution of driving wheel. A couple of 60·6 inch lb. on the motor is required for 100 lbs. pull at the rail; 276 revolutions of motor correspond to one mile per hour on the rail. The only improvements I have to suggest in this design are—first, the addition of gear which will give a higher speed of motor for the normal

[1] It has been suggested in an article in the *Engineer* that this grip is less powerful than the form first described. The suggestion can only be due to some misunderstanding; the gripping action is identical in the two cases.

speed of four miles per hour, which we contemplate; secondly, the addition of a swivel or bogie arm, such as is used in the model before you; thirdly, improvements in the belt connection. Moreover, the machine requires strengthening in some places. It will, however, be seen that none of these points touch the essential features of the design, which might at once be adopted in practice. Worked with motors of the Gramme type, the additional gear would not be required.

Before the belt locomotive had been completed it was necessary to design a locomotive for the South American line, which I have several times mentioned. I had, meanwhile, constructed the model which is now before you; and this little locomotive

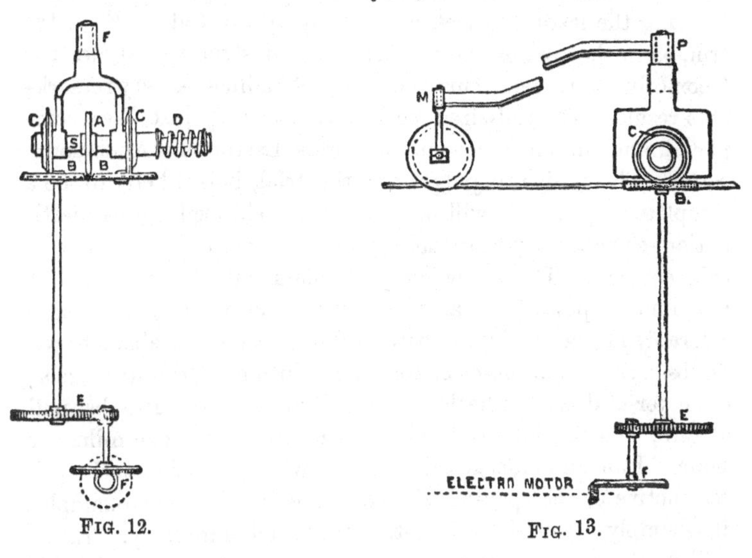

FIG. 12. FIG. 13.

in which the power is transmitted by ordinary spur-wheels, ran so extremely well that I adopted the general arrangement for the next example on a large scale. This arrangement is shown in Figs. 12 and 13. The grip (C C and B B) is a third variety of the right angle nest, simpler than that in the belt locomotive. In this form also we have line contacts, and two paths for the discs and rod. Where it is desired to drive from both sides, this arrangement is less powerful than that in the belt locomotive. In the South American locomotive I drive from one side only, leaving the off-side roller free to revolve as it pleases. This avoids grinding at rapid curves, and the adhesion given by one wheel will be ample in a dry country, such as that where this locomotive is to work. The arrangement of the gearing, E and F, is obvious. It allows the locomotive to lie fore and aft instead of across the line, and this design has some advantages in the adjustment of the weights. The surface of the gripping wheels are arranged like an upright V, so as to hold on under the line. This makes it very difficult for the wheels to leave the line, both because of their absolute hold, and because the inclination of the V is such as to favour the action of gravitation in overcoming the friction of the grip, instead of opposing it, as in the inverted V. Another feature of this machine is the arm pivoted at P, and carrying the leading wheel, which is again pivoted at M in the arm, as in the case of the trucks. The construction allows the locomotive to traverse curves of six-feet radius—a very remarkable result. The full-sized locomotive has only just been completed, and run on three spans at Messrs. Easton and Anderson's. So far as I am able to judge from the trial, it is likely to be a complete success. It will be immediately shipped for its destination, so that its performance cannot be more fully tested in this country. The following particulars will show that it is much more powerful than the belt locomotive, but it is considerably heavier :—Wheel base, 2 ft. 6 in.; weight, about 3 cwt. 14 lbs.; 15 revolutions of motor per revolution of driving-wheels; diameter of driving-wheels, 10 in.; 33·3 in. lb. per 100 lbs. pull at rail; 504 revolutions of motor per minute for one mile per hour. I am in doubt at this moment whether to adopt the belt locomotive or the spur-wheel locomotive for the next example; it is simply a question of cost, weight, and durability. Either will do the work. In all the arrangements it is essential that the second bearing-wheel at M (Fig. 13, p. 263) should lead, not follow, the drivers in regular work. The reverse arrangement lets the rope lead on at an angle with the plane of the roller, causing an injurious grinding action. Details of couplings have been well worked out, but space fails for their description.

As general features of the train running on the line, I may mention that the deflection of the rod within reasonable limits has very small influence on the resistance. When the deflection on a 50-foot span was about 2·4 feet, the resistance for a train of trucks, weighing in all 1,260 lbs., was 22 lbs.; and no sensible difference could be detected when the deflection was materially reduced. This resistance was measured by pulling a train along, span after span, by one end of a rope passing over a pulley on the leading truck, and having a weight hanging vertically from the other end of the rope. The weight thus limited the pull. This pull differs extremely little as the train moves along, for when one part of the train is descending the curve the other part is ascending. It should be noted that during this experiment no special care had been taken to oil the bearings, and I have no doubt this pull can be materially reduced. I have ventured to dwell at some length on the mechanical problems involved in this form of telpherage, because the experiments made so far have chiefly borne on questions of mechanics. The makers of dynamos can put at our disposal apparatus which will generate day after day, with perfect certainty and regularity, currents of electricity such as will transmit the horse-power generated by powerful steam-engines. These makers have already solved the chief electrical problems which present themselves in connection with telpher lines. They can give us at will constant current or constant electromotive force, high or low, as we may choose. They are now able to arrange their apparatus so that any number of incandescent lamps may be turned off or on, without disturbing the regularity with which other lamps are supplied; and by the same arrangement we are enabled to start or stop any number of telpher trains without disturbing the running of others. The electrical problems of the telpher line and those of electric lighting run in absolutely parallel lines. The electric motor, although it may be termed a mere inversion of the dynamo, has not as yet been brought to equal perfection; but month by month improved designs, proportions, and materials are being introduced, and the result already attained is sufficient for our purpose. It is all the more encouraging to feel that these results will certainly be surpassed—and far surpassed—in the immediate future.

The following short summary of the problem of the transmission of power by means of electricity may interest those who have not studied the subject. There are three steps in this transmission—1st, we convert mechanical power into electricity by means of a dynamo; in doing so we incur a loss of from 10 to 20 per cent.; 2nd, this electricity, in flowing along a conductor, generates heat, representing a further loss, analogous to that resulting from friction in mechanical gearing. This loss, depending on the distance of transmission, the size of the conductor, and the electromotive force employed, is easily computed. 3rd, we re-convert the electricity into mechanical power by means of an inverted dynamo, which we term an 'electric motor.' With motors in which large weights of iron and copper are employed, the loss in re-conversion need not exceed 20 per cent., but with light motors, weighing from 70 lbs. to 100 lbs. per horse-power, such as we must employ in the locomotives, I could not undertake with certainty at this moment to effect the re-conversion without a waste of one-half. The effect of all these sources of loss is, that at the stationary engine I must exert about 3 horse-power for every single horse-power which is employed usefully on the line. I look forward confidently to the time when 2 horse-power at the engine will be sufficient to give 1 horse-power to the motor. To put these conclusions in a more scientific form, I may assume the efficiency of my dynamo as 80 per cent., that of my small motor as 50 per cent.[1] The waste by heat expressed as horse-power is equal to $\frac{C^2R}{746}$ where C is the current in amperes, and R the resistance in ohms. The horse-power represented by the current is equal to $\frac{E\,C}{746}$ where E is the electromotive force in volts and C the current in amperes. It follows from the last expression that I may increase the horse-power in three ways, by increasing either E or C, or both. If I increase E, leaving C the same, I do not increase the loss during transmission along the line (except by leakage), no matter what horse-power the given line may transmit. A practical limit is set to the application of this law by the difficulty met with in dealing with electromotive forces above 2,000 volts. Marcel Deprez, taking advantage of this law—first pointed out by Sir William Thomson—has transmitted seven or eight horse-power over seven or eight miles, through an ordinary telegraph wire, and he obtained a useful duty of 63 per cent., taking into account all three sources of loss which I have enumerated. With small motors I cannot yet promise a result so good as this, and I

[1] The last motor tested, weighing 117 lbs., made by Mr Reckenzaun, gave 1·72 horse-power with 54 per cent. efficiency.

merely mention it to let you understand that in speaking of three horse-power at the generator for one at the locomotive, I am leaving a very ample margin. Quitting generalities, I will give some details as to the electrical and other conditions necessary, in two examples, for what may be considered as typical telpher lines.

First Line.—Length, five miles. Length of circuit, out and in, ten miles. Twenty-five trains running at once, spaced one-fifth of a mile apart; speed four miles per hour. Let each require 1 horse-power on the average; let the motor take on the average two amperes of electric current; let the electromotive force near the stationary engine be 840 volts; the electromotive force at the end of five miles will be about 746 volts.[1] The total current entering the line will be 50 amperes at the near end of the line. Fifty amperes and 840 volts represent 56·5 horse-power; of this, 6·5 horse-power will be wasted in heating the line; the remaining 50 horse-power will do work in the motors equivalent to 25 horse-power. In order to give this current of 50 amperes with 840 volts, the stationary engine will require to exert $\frac{10}{8} \times 56{\cdot}5$ horse-power, or roughly, 70 indicated horse-power, or somewhat less than three times the useful horse-power. Let us now examine the economical results to be obtained from such a line as this. Mr. Dowson, in an interesting comparison between the cost of horse-power obtained from coal and gas, reckoned the cost per horse-power for a 100 horse-power engine, at the rate of 3*l*. 6*s*. 9*d*. per annum, to include wages, coal, oil, and depreciation. Mr. Dowson would naturally be led to put the cost of steam power obtained from coal rather high than low. I will, however, adopt a much higher figure, and assume that the power may cost as much as 6*l*. 10*s*. per horse per annum; this gives 455*l*. as the cost of the 70 horse-power required for my telpher line. Let the twenty-five trains each convey a useful load of 15 cwt. In a day of eight hours the line will have conveyed a traffic which we may express as 600 ton-miles—i.e. it will be equivalent to 600 tons conveyed one mile, or 60 tons on each line conveyed from end to end daily. If we count 300

[1] It has been suggested that this fall of potential would be inconvenient. For electric lighting, it would be too great; but in telpherage it would simply cause a decrease in the speed of the train at the far end, and as all trains would be equally affected, none would overtake the other, and no inconvenience whatever would arise.

working days in the year, the sum of 455*l.* gives 1*l.* 10*s.* 4*d.* per diem, and the 600th part of this is about 0·61 of a penny as the cost of the power required to carry a ton one mile. In Great Britain we ought easily to be able to reduce this below a half-penny per ton per mile, which proves that the apparent great waste, even of two-thirds of the power in transmission, does not involve prohibitory expense. In calculating the whole cost of transport, we must further take into consideration the cost of installation. Taking the spans at 70 feet, I estimate this cost as follows:—

Line 500*l.* per mile	£2,500
Engine, boiler, and shed, at 20*l.* per indicated horse-power	1,400
Dynamo and fittings	1,000
Twenty-five trains	2,500
Contingencies	600
Total cost	£8,000

Allowing 12½ per cent. for interest and depreciation, this represents an annual cost of 1,000*l.* Allowing 100*l.* as the salary of an electrician or young engineer, and adding 455*l.* as the cost of the power, this gives a total annual expenditure of 1,555*l.* for a daily duty of 600 ton-miles. If we continue to assume the year as containing 300 working days, the total cost of conveying one ton one mile will be found equal to 2·07*d.* If goods are to be transmitted for long distances, the same calculation applies. We should simply have stations ten miles apart, working lines five miles long on each side of them. This, then, is the practical outcome of the general principles stated at the beginning of this paper. We may expect with great confidence that it will pay investors to convey goods for any distance at the rate of 2*d.* per ton per mile, by the agency of the suspended telpher line.

Second Line.—Matters are somewhat modified when the traffic is smaller. Making similar calculations for a line one mile long instead of five, with only four trains running at once, we might employ an electromotive force as low as 100 volts; the loss by heating would be insignificant; we should require about 12 horse-power; the work done in eight hours would be 96 ton-miles. I estimate the cost of installation at 1,600*l.*, and the annual cost of working 344*l.* without the annual salary of an electrician.

This corresponds to 2·875*d.*, or less than 3*d.* per ton per mile. One very important feature in respect to the cost of telpher lines is the fact that the larger part of that cost is due to plant, such as locomotives, trains, and dynamos. This plant can be increased in proportion to the work required; thus there is a very moderate increase of cost in the rate per mile for a small traffic as compared with a large one, and, on the other hand, a line laid down for a small traffic will accommodate a much larger traffic with no fresh outlay on the line itself.

There are numerous minor electrical problems involved, but time does not permit me to enter into the consideration of these to-night. It will be sufficient for electricians when I say that I see my way to governing, blocking, and breaking the trains, without ever interrupting the current used to work the motor, except between the line and rolling wheels. We already know that the interruption at this point, although accompanied by a spark, does no injury whatever. I have often been asked whether the frequent reversals involved in the cross-over system do not tend either to injure the dynamo or the motor. I made special experiments on this very point lately with a compound wound Crompton dynamo and Mr. Reckenzaun's motor with thirty-six coils. I was unable at the commutator of the motor to detect the smallest change in the motion due to the most rapid reversal. At the dynamo commutator I could just see when the reversal occurred, but there was no change of a character to cause the smallest alarm. At the same time I may state that, when from any cause reversals may be thought undesirable, we are in possession of apparatus which we call 'step-overs,' which, without diminishing the simplicity of the permanent way, enable us to send a continuous and unreversed current. These and similar electrical questions, such as the performance of Messrs. Ayrton and Perry's excellent motors, might possibly have had greater interest for electricians than some of the mechanical details discussed to-night; but I have felt that the main point to establish, in bringing this invention before the public, is that we have in telpher lines a means of conveying goods in an economical manner, by lines, locomotives, trucks, dynamos, and motors, which have undergone their preliminary trials with success, and can be at once applied to the more searching test of performing work for the public. If I have established this fact, I think you will have no difficulty in believing that the subsidiary electrical problems have been, or will be brought before you in detail on many occasions by many men. In conclusion, I will enumerate some of the uses to which telpher lines may be put. They will convey goods, such as grain, coals, and all kinds of minerals, gravel, sand, meat, fish, salt, manure, fruit, vegetables; in fact all goods which can be divided conveniently into parcels of two or three hundredweight. If it were necessary, I should feel no hesitation in designing lines to carry weights of 5 or 6 cwt. in each truck. The lines will carry even larger weights, when these, like planks or poles, can be carried by suspension from several coupled trucks. The lines admit of steep inclines; they also admit of very sharp curves. Mere way leaves are required for their establishment, since they do not interfere with the agricultural use of the ground. They could be established instead of piers, leading out to sea, where they would load and unload ships. With special designs, they could even take goods from the hold of a ship and deliver them into any floor of a warehouse miles away. When established in countries where no road exists, the line could bring up its own material as a railway does. Moreover, wherever these lines are established, they will be so many sources of power, which can be tapped at any point, for the execution of work by the wayside. Circular saws, or agricultural implements, could be driven by wires connected with the line, and this without stopping the traffic on the line itself. In fine, while I do not believe that the suspended telpher lines will ever compete successfully with railways, where the traffic is sufficient to pay a dividend on a large capital, I do believe that telpher lines will find a very extended use as feeders to railways in old countries, and as the cheapest mode of transport in new countries. In presenting this view to you, I rest my argument mainly on the cost of different modes of transport, which may, I believe, be stated approximately as follows:—Railway, 1*d.* per ton per mile; cartage 1*s.* per ton per mile; telpher lines, 2*d.* per ton per mile; and let it be remembered that, in taking the cost of cartage at 1*s.* per mile, the first cost and maintenance of the road is left wholly out of account; whereas, in my calculations for the telpher line, allowance has been made both for establishment and maintenance.

ON THE APPLICATION OF GRAPHIC METHODS TO THE DETERMINATION OF THE EFFICIENCY OF MACHINERY.[1]

PART I.

1. THE object of the present paper is to show how, by graphic methods, we may find the relation between the effort exerted at one part of a machine in motion, and the resistance overcome at another part; the solution found is rigorous for all motions in one plane, and takes count of the friction, weight, and inertia of the parts. It also takes into account the stiffness of ropes or belts.

The paper shows that we may represent any machine, at any given instant, by a frame of links, the stresses in which are identical with the pressures at the joints of the machine. This self-strained frame is called the *dynamic frame* of the machine, and may be so drawn as to represent the machine either rigorously, taking into account friction, weight, inertia, and rigidity, or approximately, omitting some of the conditions under which the machine works.

Moreover, it is shown that for all machines (in which the motions can be represented as in one plane) the dynamic frame is of one type, either simple or compounded. The dynamic analysis of machinery into parts represented by this simple frame is believed by the author to be novel. It is consistent with the kinematic analysis of REULEAUX.

The driving effort and the resistance are in the frame represented by stresses in links, and reciprocal figures afford an easy method of determining the relation between those stresses so soon as the frame has been drawn. Incidentally, the method also gives the resultant pressure at every joint in the machine, a

[1] From *Transactions of the Royal Society of Edinburgh*, vol. xxviii., 1877

result of considerable practical value. When the relation between the stresses due to effort and resistance has been found, it is easy to calculate the *efficiency* of a machine by taking into account the spaces through which these stresses act in equal times, so as to show any loss of energy which may arise from deformation, such as may be due to the stretching of ropes.

W. J. MACQUORN RANKINE introduced the word 'efficiency' to denote the ratio of the *useful* work done by a machine to the *whole* work done. He showed that the efficiency of a train of mechanism is measured by the continued product of the efficiencies of all the successive pieces or combinations, and he gave methods by which the efficiency of certain elementary pieces and modes of connection could be ascertained. These methods assume that in each case the effort is known in magnitude, position, and direction; also that the position and direction of the resistance are known. The weight of the piece is taken into account, but no mention is made of the resistance due to inertia, which could, however, be treated in the same manner as the forces due to weight.

RANKINE did not carry his explanation of the subject so far as to show how to find for any actual complete machine the direction of the successive efforts and resistances; nor does he draw attention to the fact that they are interdependent. We cannot determine the efficiency of a whole machine by calculating the efficiency of each part separately without regard to its position in the machine, for it is this position which determines the directions of the driving and resisting efforts. These directions cannot be found by mere kinematic analysis, but depend not only on the form of the neighbouring parts, but also on their friction, weight, mass, and rigidity. In many cases the chief difficulty in determining the efficiency of a machine consists in determining the direction and point of application of the effort and resistance to the motion of each part. These directions are found by the *dynamic frame*. The writer has endeavoured to take up this subject where RANKINE left it, and to give a general method by which the efficiency of the great majority of actual machines can be practically calculated. In doing this, he has found the graphic method convenient.

Since the method adopted is based on a novel analysis of

machines, it has been found necessary to begin by some elementary definitions.

2. *Elements of Machines.*—All machines consist of parts so joined that any change in the force with which one part presses against another will produce some change in the force with which the other parts are pressed or held together. This condition obviously holds good when the parts are rigid, and it constitutes the test whether flexible ties or elastic fluids form part of a given machine.

A loose coil of rope cannot by this definition form part of a machine, but a rope is part of a machine, whenever a change in the tension of the rope changes the pressures or tensions between other parts of that machine. Similarly, steam or water forms part of a given machine whenever a change in the pressure of that steam or water changes the force exerted between other parts of that machine. This relation between the parts of a machine is dynamic, and is more general than the kinematic relation which exists between the rigid or inextensible parts of a machine. A definite position or motion of one part of a machine is not in every case accompanied by a definite motion or position of another part, irrespective of the forces in action, neither does the same relative position of all the parts of a machine necessarily imply the same forces between the parts, but a change in the force exerted between any two parts always implies a change in the forces between the other parts.

The word *element* will in this paper be used to designate the continuous parts of machines. Each *solid element* is a part which is continuous in respect that no portion can slide upon or break away from another portion in immediate contact with it. Any rigid bar or structure in a machine is an element, as, for instance, the connecting rod of an engine, the cylinder and bed plate, the crank and fly-wheel. A belt between two pulleys, or a rope on an axle, when these form parts of machines, are separate elements. A slack belt or a slack rope forms no part of a machine, since it does not fulfil the first condition of altering the force at one end when the tension is altered at the other. A continuous fluid forming part of a machine will be called a *fluid element*. A portion of fluid is continuous when a change of pressure at one place is transmitted by the fluid so as to change the pressures throughout the element. Thus the steam in the cylinder of a steam-engine is a fluid element, and, strictly speaking, this element comprises the steam in the pipes and in the boiler, as well as the water in the boiler. Generally, we need only consider the steam enclosed in the cylinder cut off and bounded by the side valve. Wherever discontinuity of motion occurs, a solid element will be considered as terminated or bounded. Thus two rigid bars joined by a flexible tie will be treated as three distinct elements. The flexibility of the tie replaces the sliding motion at a joint which would otherwise be required. The discontinuity of motion here insisted upon as indicating the surface of separation between elements is the property by means of which we shall be enabled to determine the relative forces with which separate elements press on one another. Two elements are treated as separate when at any part they are discontinuous, and also in certain limiting cases when the relative motion at the surfaces of contact is infinitely small.

The elements of a machine as now defined correspond closely with Professor REULEAUX' kinematic links. All Professor REULEAUX' links are elements, but the definition now given of a dynamic element would embrace certain parts of machines which can hardly be called kinematic links, as, for instance, the steam and water mentioned above. Each element of a machine will in what follows be designated by a single small italic letter.

3. *Joints.*—The name of joints will be given to the surfaces of separation between two elements. When the elements are rigid, joints can occur only where there is sliding or rolling contact between the elements. A surface of separation between a flexible tie and a rigid bar will also be treated as a joint, inasmuch as discontinuity of motion may occur at this surface similar to that which occurs at a service of separation between rigid elements. A joint in the sense in which the word is here used is essentially a joint between *two* elements. There can be no common joint between three or more elements. In ordinary phraseology, we speak of three or more members of a frame as jointed at one place, and speak of the joint there as a single joint, meaning that they all abut against a single pin; but, in the present paper, each portion of the surface of such a pin as bears against a single element will be spoken of as a single joint.

We shall always assume that a pin of this kind is fixed relatively to one of the elements, so as to reduce the elements of the machine to the smallest number. A joint will be designated by two italic letters, being the names of the elements between which it occurs. Thus in Fig. 1 (p. 279) there are two joints, *ab* and *ac*, the pin being fixed on *a*. All actual joints are surfaces of greater or less extent. They may be divided into three classes:—1st. Joints between a solid and hollow sphere; this case includes that of a plane resting on a plane. 2nd. A solid prismatic cylinder inside a hollow cylinder. When the cylinder has a circular cross section, relative turning and sliding are both possible. 3rd. A solid screw inside a hollow screw. When the pitch is infinite we have one form of case second in which only translation is possible. When the pitch is zero we have a joint which only allows of rotation and is the joint or bearing most commonly met with in machinery. The third case is the most general, and may be considered as including the two others as special forms. The solid and hollow cylinder, which allow simple rotation of one element relatively to another, will be called the pin and eye of a joint.

A dynamic joint occurs wherever pairing occurs, the word 'pairing' being used in the sense given it by Professor REULEAUX. The kinematic pair as defined by him differs from the dynamic joint in respect that complete pairing implies closure or complete restraint except in one direction. Moreover, Professor REULEAUX is able to classify pairs with reference to the mode of closure. A dynamic joint implies no restraint other than that given by the force exerted in the line of bearing pressure. All portions of the pair might be cut away, except a surface of sufficient strength round the bearing point; distinctions between force closure, pair closure, link closure, have no dynamic signification; and restraint, such as that afforded by collars on bearings, is of no importance dynamically when the machine is properly designed. A short distinction may be made between a dynamic joint and a kinematic pair, by saying that the latter might be self-strained by imperfect fitting, while the former cannot be self-strained.

In certain cases the surface in contact may approximately be represented by lines, as in what Professor REULEAUX calls the higher pairs, and we may even conceive a class of joint which requires only points of the elements to be in contact, as where a sphere is pressed against a plane. In connection with geometrical diagrams, the word 'joint' will be applied to *points* round which intersecting lines may be said to turn relatively to one another. Geometrical joints of this kind will generally be designated by a single capital letter.

4. *Definition of a Complete Machine.*—The object of machinery, considered dynamically, is the application of energy, or, in more popular language, power, to the performance of useful work, and the name *complete machine* may be given to any combination of elements so joined that the energy developed in one element, or between two elements, is, by the relative motion of the elements, enabled to do useful work in overcoming a resistance exerted either by one element or between two elements. A complete machine is self-contained, and the internal action between its parts can change neither its momentum nor its angular momentum. Most actual machines have one portion fixed relatively to the earth, which then becomes part of one element of the machine.

5. *Lines of Bearing Pressure.*—The elements of a complete machine are so held together at the joints by the forces which are in play when the machine is in action, that each element of the machine occupies a determinate position relatively to all the others, and presses against its neighbours at each joint with a force determinate in magnitude, direction, and position. A given position of the parts does not necessarily imply constant forces at the joints, but it does imply a determinate relation between the forces at the joints. A line indicating the direction and position of the equal and opposite resultant forces at a joint will be called a *line of bearing pressure.* In consequence of friction, a line of bearing pressure where it cuts the joint, must, when the machine is in motion, make an angle with the common normal to the surfaces equal to the angle of repose, or angle of which the tangent is equal to the coefficient of friction for the surfaces at the joint, and must be so inclined that the force exerted by the one element on the other has a tangential component directly opposed to the sliding of the second element relatively to the first. This condition will hereafter be

frequently referred to, and will, for brevity, be spoken of as the condition that the bearing pressure shall make the *stated angle* with the joint.

6. *Driving Element and Resisting Element—Driving Link and Resisting Link.*—The source of power in a machine is often contained in a single element, which is usually so combined with the others as to exert equal and opposite forces in two directions, and only in two directions. The steam in a cylinder of a steam-engine affords one example of this kind of element, which will be called a *driving element.* If the directions of the equal and opposite forces exerted by the driving element lie in the same straight line, the power tends to lengthen or shorten the element in a definite direction. The two forces in this case tend to separate or draw together the other elements with which the first is jointed, as a straight elastic link would do if extended or compressed so that its line of action coincided with the direction of the forces exerted by the driving element. We may therefore conceive an actual driving element as replaced by an ideal elastic link producing the same force at the above-named joints. Some machines have no material driving element, but are actuated by an attraction or repulsion between two of their elements, as, for example, when a machine is driven by a weight. In this case, as in the former case, we may conceive the machine as driven by an ideal elastic link between two elements. This link is as well suited to replace the equal and opposite forces due to gravitation as the equal and opposite forces due to steam. If the two opposite and equal forces exerted by a driving element do not lie in one straight line, these forces exert a couple between the elements with which the driving element is jointed; a similar couple may be exerted by the forces of attraction or repulsion between two elements. In either case, the machine may be considered as actuated by an ideal couple due to the elastic reactions of two links. What has been said of the driving power applies *mutatis mutandis* to the useful work done by a machine. In some cases the work is actually done in lengthening or shortening a separate resisting element, as when a bale of cotton is compressed. In other cases an attraction or repulsion is overcome between two elements, as when a weight is lifted. In both these cases we may conceive the resistance

as represented by an ideal elastic link exerting definite, equal, and opposite forces in a definite position and direction. This ideal resisting link is equally suited to express the resistance by which cohesion, friction, or inertia opposes the relative motion of two elements. When the resisting element, or the resistance due to the united action of two elements, gives rise to a resisting couple, this resistance may be represented by two ideal links between two elements. Thus, we see that in every machine where there is only one source of power, we may represent that source by one ideal link (or two ideal links) connecting two elements, and acting on these elements as one member (or two members) of a strained frame acts on the rest of the structure. The name of *driving link* will be given to ideal links of this kind. The name of *resisting link* will similarly be given to ideal links used to represent the useful resistance in a machine. In all cases the cause of motion and the useful resistance must be considered as an action between two elements. We cannot properly speak of a driving point or resisting point, but only of driving or resisting links.

7. *Links.—Dynamic Frame.*—Any actual material element might at any instant be removed from a machine without altering the stresses on the other elements, if at each joint thus laid bare, force could be applied corresponding in position, magnitude, and direction with the pressure supplied at that instant, and at that joint by the element removed. The series of forces supplied by any one element are necessarily such as would, when balanced by equal and opposite forces, leave the element in equilibrium. If, therefore, an element has only two bearing joints and is without weight or inertia, the forces it supplies must lie in one straight line, and either push or pull, as the member of an actual frame does, when under simple tension or compression; the element, in fact, acts like one link of a frame, as shown in Fig. 2, where the element a might be replaced[1] by the link 1, shown by the line on which arrow-heads are placed. The ideal link replacing the actual element does not necessarily or generally lie in the geometrical axis of the element. When

[1] The writer has in this paper ventured to use the verb 'to replace' as it is usually employed by writers on chemistry, namely, as the translation of the French word *remplacer*.

there are three bearing joints in an element having no weight or mass, the three lines of pressure must lie in one plane, and either be parallel or intersect in one point. Any one of the

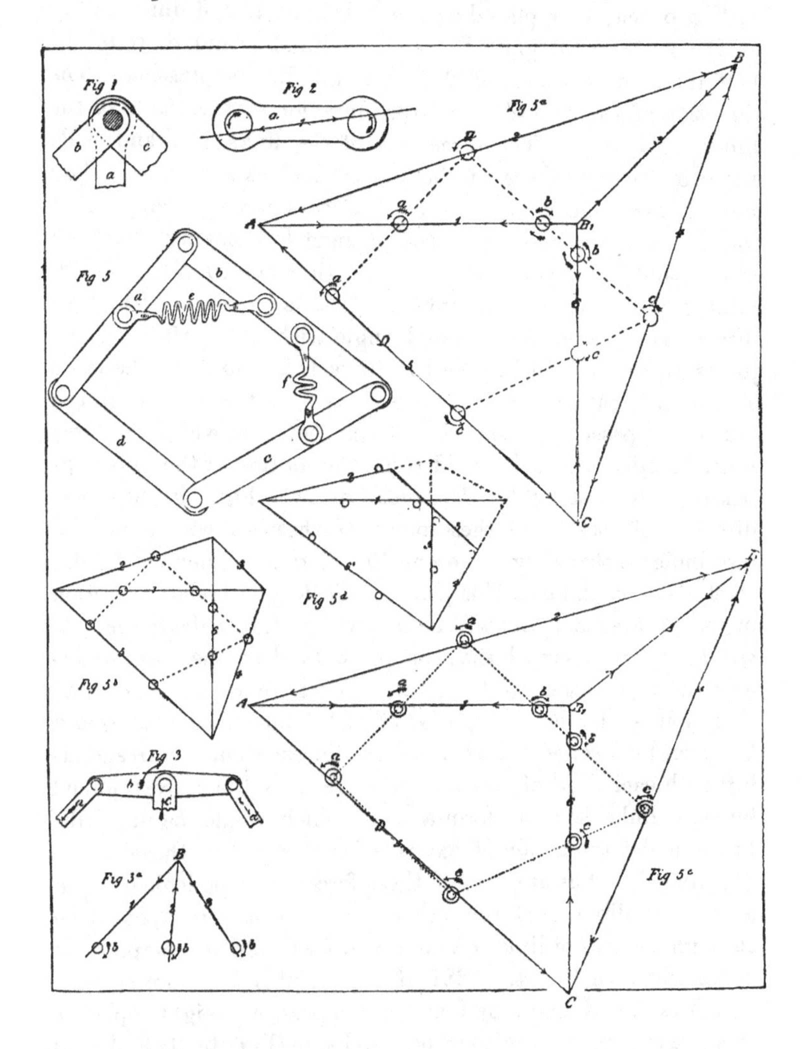

forces supplied may be looked upon as the equilibrant of the two others. The forces which this element supplies might therefore be replaced by three ideal half links, each coinciding in position and direction with the line of bearing pressure at the three joints, and all connected by an ideal geometrical joint without friction at the point of intersection (which, in the case of parallel forces, will be at an infinite distance). Thus element *b*, Fig. 3, may be replaced by the half links 1, 2, 3, intersecting at the geometrical joint B, Fig. 3*a*; links 1 and 3 would be half links in tension, link 2 a half link in compression. The direction of the arrows shows the direction of the stress in the links replacing *b*. The directions of the links 1, 2, and 3 do not coincide with those of the geometrical axes of *a*, *c*, and *d*; indeed, these elements may be stiff bars having many other joints; but if we know that the element *b* is moving relatively to *a*, *c*, and *d*, as shown by the arrow, then one condition determining the directions of links 1, 2, 3 is given us, for these directions must make the stated angle with the surfaces of the joints *ba*, *bc*, *bd*. In Fig. 3*a* the links 1, 2, and 3 are therefore shown, not passing through the centres of the circles at the joints, but passing on that side of the centre on which the force represented in the link would resist the motion of the pins supposed to be fast on *b*. The small arrows, Fig. 3*a*, show the direction of rotation of these pins, and these arrows are lettered *b* to indicate that they represent the motion of element *b* relatively to *a*, *c*, and *d*. This plan of indicating the relative motion of the surfaces at joints will be followed in future diagrams. It will always be assumed that the pin is *fixed* in the element indicated by the letter at the arrow. When there are more than three joints, the forces supplied at each joint are such as would be given by a series of half links, one for each joint, corresponding with each line of bearing pressure, and themselves joined by other links, so as to form a frame which would be in equilibrium under the action of external forces equal to those in the half links.[1] Thus any one of these forces may be looked upon as the equilibrant of the others, and as acting upon them through a series of links which are subject only to compression or tension. In Fig. 4, p. 281, if the member *b* is jointed with members *a*, *c*, *d*, and *e* by four parallel pins, we might replace *b* in any machine by the four half links, in Fig. 4*a*, 1, 3, 4, and 5 and a complete link 2, joining the intersection of 1 and 5 with

[1] In this frame it might be necessary to include at least one stiff bar or frame to meet opposite and equal couples.

that of 3 and 4; this last link will be in equilibrium under the action of the forces acting on b, as shown in the half links. In a more complicated example the link 2 might be replaced by a

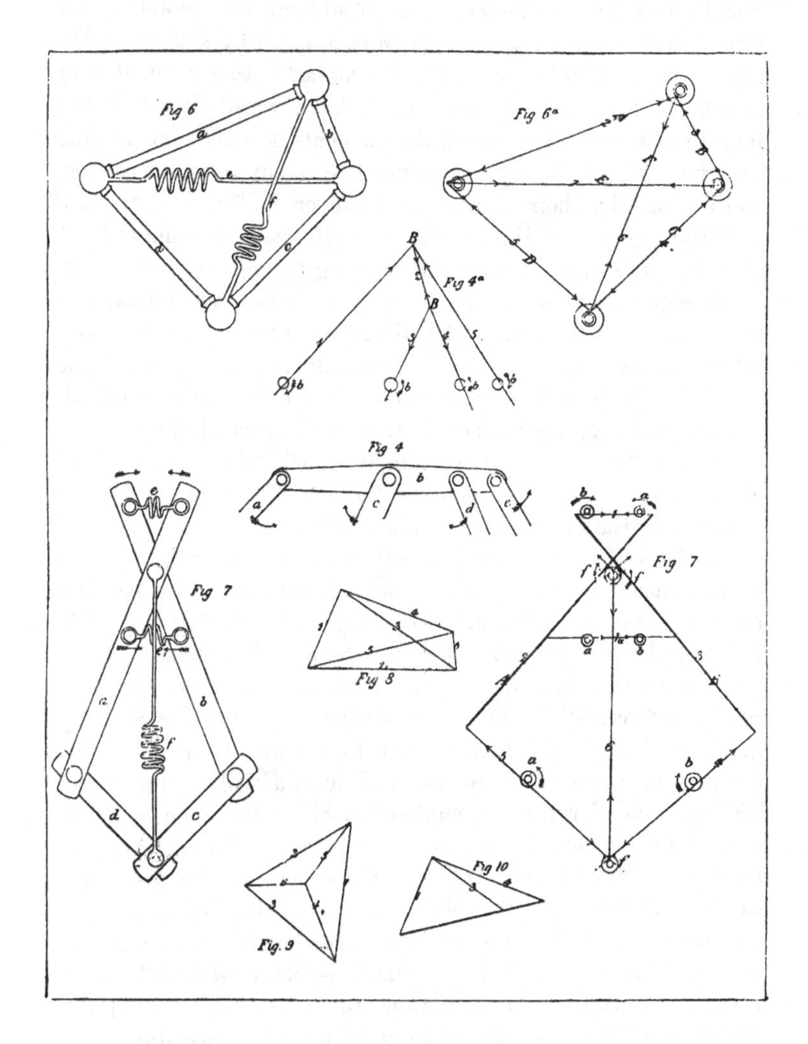

complete frame, or by a rigid plate. Since the forces acting on one element either intersect at one point or in a series of points which may be joined by other links, we may always designate the point or points in question by the same letter of the alphabet as is used to denote the element. The geometrical intersections will be marked by capital letters, whereas the elements are marked by italics. When there are several geometrical joints for one element, these joints will be denoted by the same letter, but distinguished from each other by suffixes. The substitution of links or half links for an actual element may be effected even when forces are parallel, if we admit joints at an infinite distance. If, now, all the elements of a machine, in their relative positions at any one instant, be removed in succession, and replaced by their equivalent links or half links, we shall substitute for the original machine a self-strained frame of links such that the stress in each link passing through a joint will be in all respects equal to that on the joint, while the stresses in the driving and resisting links will represent the effort and useful resistance in the machine. Each half link at a joint of one element is necessarily met and completed by the other half link in the same line, due to the reaction of the second element.

The self-strained frame, composed of links, as described above, will be called the *dynamic frame* of the machine with its elements in the given relative position.

8. *Example.*—An example will probably assist in showing what is meant by the dynamic frame. Let a machine be composed as the six elements, *a*, *b*, *c*, *d*, *e*, *f*, joined as in Fig. 5, p. 279. Let *e* be the driving element, and *f* the resisting element. We will suppose in this and in the following examples that the resulting forces all lie in one plane, although the figures may not show the split joints necessary to ensure this result. For the present, the effect of weight and inertia will be neglected. The machine shown is a complete machine; the element *d* has two joints, the elements *a* and *c* have three joints, and the element *b* has four joints. The dynamic frame may be drawn assuming the friction at the joints to be insensible, or it may be drawn taking friction into account. In the former case it will be represented as in Fig. 5*a*, which is obtained as follows:—Link 1 may first be drawn through the axis of *e*, for we know that the half links at the joints *ae*, *eb* must lie normally to the surface of these joints (being frictionless), and must therefore lie in one straight line, passing through the centre of the pins at *ae* and *eb*. For similar reasons we draw 5 and 6 through the

axes of elements *d* and *f*. The forces exerted by the elements *d* and *e* on the element *a* must be balanced by the force due to the third joint *ab*, therefore the direction of this balancing force must pass through the intersection of 1 and 5, and must be normal to the surface of the pin at *ab*. We therefore are now able to draw the link 2. For similar reasons we must draw the link 4 through the centre of the pin at *cb*, and through the intersection of the lines 5 and 6. The element *b* is in equilibrium under the action of the four forces acting at the four joints; in other words, the resultant of 2 and 4 must be equal and opposite to the resultant of 1 and 6, so that we may complete the dynamic frame by drawing the link 3 as shown; this last link, however, would be equally well placed as shown in Fig. 5*b*, where it joins the intersection of 2 and 6 with that of 1 and 4. Both the frames shown in Fig. 5*a* or Fig. 5*b* are kinematically equivalent to the actual machine shown in Fig. 5 in the following sense: A given small contraction of the element *e* would, supposing all the other elements to be rigid, produce a definite extension of the element *f* in the actual machine. A small contraction in link 1 equal to that in element *e* would, supposing all the other links of the frame inextensible, produce an extension in link 6 equal to that produced in *f* in the machine. We may calculate the relation between the stresses in *e* and *f* by the relative rates of their contraction and extension, that is to say, by the principle of virtual velocities, or we may calculate the relative stresses between links 1 and 6 of the frame by the ordinary principles of statics, for instance, by a 'reciprocal figure.'[1] The ratio between the stresses in *e* and *f* and that between the stresses in 1 and 6 would be identical, whichever method were adopted. It need hardly be said that the method by virtual velocities would be much the simpler. It is not until we wish to take friction into account that the utility of the dynamic frame becomes apparent. In order to take friction into account we have to change the form of the frame only in this respect, that the links, instead of being normal to the surface of each joint, must be inclined so as to make the angle of repose with the normal to that joint, and must be so placed that the reaction due

[1] *Vide* J. Clerk Maxwell on Reciprocal Figures, *Phil. Mag.*, April 1864; nd Fleeming Jenkin, *Trans. Roy. Soc. Ed.*, vol. xxv. 1869

to the elasticity of the link—or, in other words, the stress in the link—may oppose the motion of rotation of the pin in the eye of the link; in brief, the link must make the *stated angle* with the surface of the joint. In the present example, where all the joints are made by circular pins and eyes, this is readily done by making the directions of the links tangent to, and on the proper side of, circles drawn with their centres at the centres of the pins, and having each a radius equal to $r \sin \phi$, where r is the radius of the pin in question, and ϕ is the angle of which the tangent is equal to the coefficient of friction for the surfaces in question.

These circles will, in what follows, be called *friction circles.* The following mnemonic rules will be found useful in selecting the side of the circle to which any given link must be tangent. Case 1. When the link represents an element with only two joints, and therefore does not end in any geometrical joint named by the same letter as an element. Consider the link as terminating in an eye which rotates on a pin fastened to the other element. Mark by an arrow the direction of rotation of the pin inside the eye. Mark also by an arrow the direction of the force exerted by the link at the joint, then place the tangent so that the force indicated by the arrow in the link appears to oppose the motion of the pin. It must be remembered that the arrow indicating the direction of the force exerted by the link may, in the diagrams, frequently be found at the end of the link furthest from the joint. Case 2. When the link ends in a geometrical joint marked with the letter denoting the element in which any pin is fast, then the arrow on the half link next that geometrical joint must point as if opposing the motion of the pin relatively to the other element.

Fig. 5*c*, p. 279, shows the dynamic frame when the friction at the joints has been taken into account, or, as it may be called, the *dynamic frame with friction.* Eight friction circles are first drawn, arrows are placed on the links, as in Fig. 5*a*, to indicate whether these are in tension or compression; the directions of stress are by hypothesis known in links 1 and 6, and we easily see what must be the direction of stress in the other links to keep links 1 and 6 in equilibrium. We next put arrows at each friction circle in Fig. 5*c*, showing the motion of each pin relatively to its eye when *e* contracts, and *f* is lengthened; before doing this, we must choose in which element the pin is to be fixed. The links of the dynamic frame are then drawn tangent to the friction circles, so as to make the stated angle with the surface of each joint. The point where the pin bears against the eye, marked by a dot on the surface of each pin, will aid in showing the meaning of the diagram, as explained in § 7. The italic letters near the arrows denote the element in which the pin is fast, and the arrows show the direction of rotation of that element relatively to the other element of the joint. If *f* were made the driver the direction of these rotations would all be reversed, and consequently the links would all have to cross over to the other sides of the friction circles. This would materially alter the form of the diagram, and in many machines would result in an impracticable diagram, showing that the machine cannot be driven backwards.

9. *Modification of the Dynamic Frame—Couples.*—When any element of a machine is in equilibrium under the forces applied at two joints the dynamic link will in direction and position correspond more or less closely with the direction in which the element of the machine lies between the joints. Thus, when the machine constitutes a material frame in one plane, as in Fig. 6, p. 281 the dynamic frame, Fig. 6*a*, will have a general resemblance to the machine. The similarity between the actual frame composed of material links and the ideal dynamic frame in this case must not lead the reader to expect that he will always be able to identify a link in the dynamic frame as corresponding to an element in the actual machine. Half a dynamic link corresponds to each *joint* of the machine, and there may be dynamic links which, like link 3 of Fig. 5, have no corresponding joint in the machine, but the dynamic link will only correspond with an actual element when the number of joints in the latter is limited to two. Since a two-jointed element has a link resembling it, we may sometimes name this link by the letter of the element; thus, all the links of 6*a* might be called indifferently by the letter of the element or the number assigned to the link; but in Fig. 5*b* or 5*c*, we can name only the link 5 by the letter of an element. In fact, the capital letter always signifies

an ideal joint, but in cases like that of link 5 this ideal joint is any point in a given straight line. When the dynamic joints are at an infinite distance owing to the parallelism of certain links it is convenient to substitute ideal stiff frames, bars, or plates joining the parallel links and acting as the actual rigid elements or stiff frames would do, with the exception that the joints between the ideal bars and links are frictionless; this substitution may also be made when the links are nearly parallel. It is clear that, by the ordinary method of statics, we might calculate the relative stresses in all the links of the frame in Fig. 5, if we were to imagine links 1, 2, 5, and 4 joined by a stiff bar or triangular frame to which they were jointed without friction; this gives a modified dynamic frame, as shown in Fig. 5*d*. The position of this stiff bar is unimportant, but when used to connect parallel links, it is conveniently drawn perpendicular to these. When the links are not all in one plane, these bars become imaginary rigid plates or stiff frames. The rigid bars will be shown in the diagram by thicker lines than the links.

A couple in an actual machine can only be exerted between *two* elements which are acted upon in opposite directions. There is no such thing as 'a solitary couple' in nature; we always find a pair of equal and opposite couples, as we find a pair of equal and opposite forces. Two equal and opposite couples require two rigid elements between which they are exerted, and these elements appear in the modified dynamic frame as two rigid bars, perpendicular to the forces producing the couples. Fig. 7 shows a simple machine in which the driving link of Fig. 6 is replaced by a driving couple between the elements a and b. The driving couple is indicated by the two springs e and e_1, which it is assumed are producing exactly equal and opposite stresses between a and b in two parallel directions. Fig. 7*a* shows the dynamic frame for this machine with friction. First we draw the links 1 and 1*a* tangent to the friction circles for the joints ea, eb, e_1a, e_1b. The distance between the links 1 and 1*a* shows the arm of the driving couple as diminished by friction; next we draw links 4 and 5 by the rules already given, then, remembering that the force in link 5 at joint ad must produce an equal and parallel bearing pressure on the pin at the joint af, we draw this bearing pressure tangent to the friction circle, so as to make the stated angle with the surface of the joint; we also draw the bearing pressure due to element *b*, on another part of the same pin. The intersection of these lines of pressure gives the dynamic joint through which the pull of link 6 must be exerted; this link is now drawn, and the diagram completed by the two bars drawn perpendicular to links 4 and 5 respectively. Let P_1 be the force of the original couple, and D the distance between links 1 and 1*a*. Let A be the distance between the lines of bearing pressure on element *a*; then P_5, the stress in link 5, is given by the expression $P_5 = \frac{P_1 D}{A}$. Similarly $P_4 = \frac{P_1 D}{B}$. The stresses in links 4 and 5 being known give the stress in link 6 by the simple composition of forces. The portion of Fig. 7 referring to the material means of producing the two couples does not necessarily belong to the diagram; the couple between *a* and *b* may in certain cases be produced by some other machine, being, in fact, the resisting couple of that other machine, and in that case the efficiency of the means of producing the couple must be determined by an examination of the first or driving machine. This case is, in fact, one case of compound machinery, and will be treated hereafter. In what follows, a driving couple may be occasionally described as existing between two elements, without reference to the mode in which it is applied; a resisting couple may be spoken of in the same manner.

10. *Assumption that the Links of Frame lie in one Plane.*—One object of our investigation is to find a means of ascertaining the efficiency of any mechanical arrangement—the word 'efficiency' being understood in the sense given it by RANKINE—as the ratio of the useful work done in a machine to the whole work or energy expended. Now RANKINE ('Millwork,' § 371 A) has pointed out certain conditions which must be fulfilled to give the highest efficiency in any design, viz.: First, that the useful resistance to the motion of any element, the effort to move it and the force due to the weight of the part must lie nearly in one plane, or else act in directions parallel to one another; and secondly, that the acting parts must not overhang the bearings. Injurious couples are introduced if these conditions are not

fulfilled. RANKINE'S conditions are, however, generally fulfilled in all important designs of machinery, and when this is the case we may, without serious error, assume all the actions to take place in one plane parallel to the plane of action in the machine, and this hypothesis will be adopted in all that follows when the contrary is not stated.

11. *Simple and Compound Dynamic Frames.*—The dynamic frame of a complete machine must contain a driving link or couple and a resisting link or couple, together with the links necessary to connect the driving and resisting links or couples in such a way as to form a self-strained frame. A complete machine may be either simple or compound. *Simple machines* are those having dynamic frames which cannot be decomposed into two or more self-strained frames, such that the resisting link or couple of the one becomes the driving link or couple of the other. *Compound machines* have dynamic frames so formed that they can be decomposed into the frames of simple machines so connected that the resisting link of one becomes the driving link of the next. If we exclude rigid bars as members, there is only one frame which can be self-strained, and which is yet incapable of analysis into two distinct self-strained frames. This frame has been already described, and consists of a quadrilateral, with two diagonals, as shown in Figs. 8 and 9, p. 281. There is no essential difference between those two figures, which each consist of a quadrilateral figure 2, 3, 4, 5, having opposed angles joined by the two links 1 and 6. This simplest self-strained frame will be shown to be the dynamic frame of many elementary combinations in machinery. The driving and resisting links may in this frame be arranged in two ways. 1. They may, as in the examples hitherto given, be represented by two links, such as 1 and 6, 2 and 4, or 3 and 5, which are not joined together; any one of these pairs may be considered as diagonals of a quadrilateral joined by the four remaining links, and each pair may be called conjugate links. For the convenience of description, when these link are placed as 2 and 4, or 1 and 6, Fig. 8, they may be called opposite links. 2. The driving and resisting links may be adjacent; that is to say, they may, as in the case of 1 and 4, or 2 and 6, have a common intersection or joint. When the driving and resisting links are adjacent, as 1 and 4, those links which

do not abut at the intersection of the two adjacent links[1] need not represent bearing pressures at working joints, as will be seen by considering an actual machine corresponding to the dynamic frame, as, for instance, that of Fig. 11, p. 292; the three elements corresponding in this case to links 2, 5, and 6 will then simply constitute a stiff system or frame, which might be replaced by a single rigid element. When this is the case the machine will belong to class 2, the dynamic frame of which is that of Fig. 10, p. 281, in which a bar is substituted for links 2, 5, and 6. When treating of compound machines, we shall find reason to consider machines of class 2 rather as half machines than complete simple machines. When couples are admitted in place of driving and resisting links, the dynamic frame necessarily includes two stiff bars, between which the couple or couples act. We then have three cases:—1. The resisting and driving couples may act between the same pair of stiff bars. This gives a dynamic frame of merely two bars, with the two pairs of links by which the couples are exerted. 2. A driving or resisting couple between two bars may be combined with a resisting or driving link between the same bars. 3. A resisting couple or a driving couple may, in the quadrilateral of Figs. 8 and 9, be substituted for any link, the couple being exerted between two bars replacing two links, which, together with the link removed, form a triangle. When we examine the usual combinations of elementary parts forming actual machines, we shall in all cases find that these combinations may be represented by a dynamic frame of one of the classes described.

12. *Efficiency of Elements.*—The relation between the energy exerted and the useful work done in a machine is affected by a loss of energy in transmission through the elements, as well as by a loss in transmission past the joints. At each joint we may say that only a certain fraction of the energy received is transmitted, the remainder being wasted in overcoming useless

[1] In the frame of the machine shown in Fig. 45, p. 323, links 1 and 6 might be drawn so as to appear adjacent, by placing link 4 so as to join the intersection of 1 and 6 with that of 2 and 5. The links 2, 3, 5, do not, however, in this example, form a stiff frame, and the machine belongs to class 1. This is obvious when link 4 is placed so as to join the intersection of 1 and 5 with that of 2 and 6. Machines of class 2 have only 5 working joints.

friction; the fraction transmitted is the measure of the efficiency of the joint (RANKINE). Let this fraction be called J. Similarly, let the ratio between the energy received and that transmitted by each element be called e, then the efficiency of the whole machine consisting of a linear train of joints and elements will be the product $J_1 J_2 J_3 \ldots . \times e_1 e_2 e_3 \ldots .$ This formula is, however, of little practical use, because the values of $J_1 J_2$, etc., are materially influenced by the directions of the forces at each joint, and these cannot be assumed for one joint independently of the others. In other words, the values of J are not independent of one another. The value of the product $J_1 J_2 J_3$, ... etc., can only be found by solving a large number of troublesome simultaneous equations, or by means of the dynamic frame. With respect to $e_1 e_2 e_3 \ldots .$ we must distinguish between two cases. 1. Those in which the element is in equilibrium under the external forces independently of any progressive change in its own form. 2. Those in which the element is not in equilibrium under the forces applied at the joints. As an example of the first class, I may take a straight rope used to transmit power. Although the rope stretches, yet the whole pull at one end is transmitted to the other end, but there is a loss of work, because the distance traversed by the driving end is greater than that traversed by the following end. In all cases of this kind the values of e are not only independent of one another, but do not affect the values of J. They do not alter the relation between effort and resistance, and their aggregate effect is easily taken into account; e for each element is a constant fraction, which can be independently determined, and the fraction expressing the efficiency of the machine will be the product of two factors—first, the efficiency as found by the dynamic frame, and secondly, a coefficient obtained by multiplying together all the values of $e_1 e_2 e_3$, etc., for the elements concerned. When the energy thus employed acts against a reciprocating resistance in the elements, as where it bends the beam of an engine, it is without influence on the whole efficiency for a complete cycle of operations, such as a whole revolution of the crank. It simply alters the relative efficiency at different periods of the stroke, and may, therefore, generally be neglected. Where, however, the lost work is done against a non-reciprocating force, as in stretching a rope, it cannot be safely neglected, and leads to sensible diminution of efficiency, as where power is transmitted by belts and pulleys. Coming to the second class of cases, it is clear that when a heavy element is being lifted, lowered, accelerated, or retarded it is not in equilibrium under the action of the external forces at the joints calculated in the manner hitherto described; we shall, however, hereafter include the forces due to these causes in calculating the forces at the joints, and there remains only one mode in which a loss of energy occurs in the course of its transmission by an element, namely, its dissipation in overcoming an internal couple. This case finds an illustration in the case of a rope wound on to a pulley, or unwound from one; the pull on the rope is not transmitted in a direct line, as we have hitherto supposed, but in consequence of the couple required to bend or unbend the rope, the line of action is shifted sideways through a length $\frac{m}{F}$, where m is the moment of the couple, and F the force transmitted. This translation of the force affects the values of J for all the subsequent joints. It can be represented in the dynamic frame by showing the line of action of the force shifted parallel to itself in a disadvantageous direction. If in the given problem we know the useful resistance, we must, in constructing the diagram, shift the force at the driving end; *vice versâ*, if the driving effort is known, we must shift the force at the resisting end. The fraction expressing the loss of efficiency due to this cause is not, like that due to friction or stretching, independent of the magnitude of the forces involved, but will, on the contrary, always involve complete inefficiency when the force is very small, and implies a gradually increasing efficiency as the force transmitted increases. Thus, a small force exerted on a stiff rope passing over a pulley produces no effect on the further side, because it is insufficient to bend the rope. The loss by an internal couple always diminishes the resistance which a given driving effort can overcome, whereas the loss of internal work done in overcoming a single force has not this effect. The case of the transmission of power by fluids in pipes will be examined in a subsequent paper, after machines composed of solid parts have been analysed.

13. *Simple Machines.—Lever.*—Let a, Fig. 11, be a lever to

which a driving effort is applied by the spring *e*, and a resistance by the spring *f*. Let the lever have a fulcrum or bearing in the element *b* to which the elements *e* and *f* are jointed,

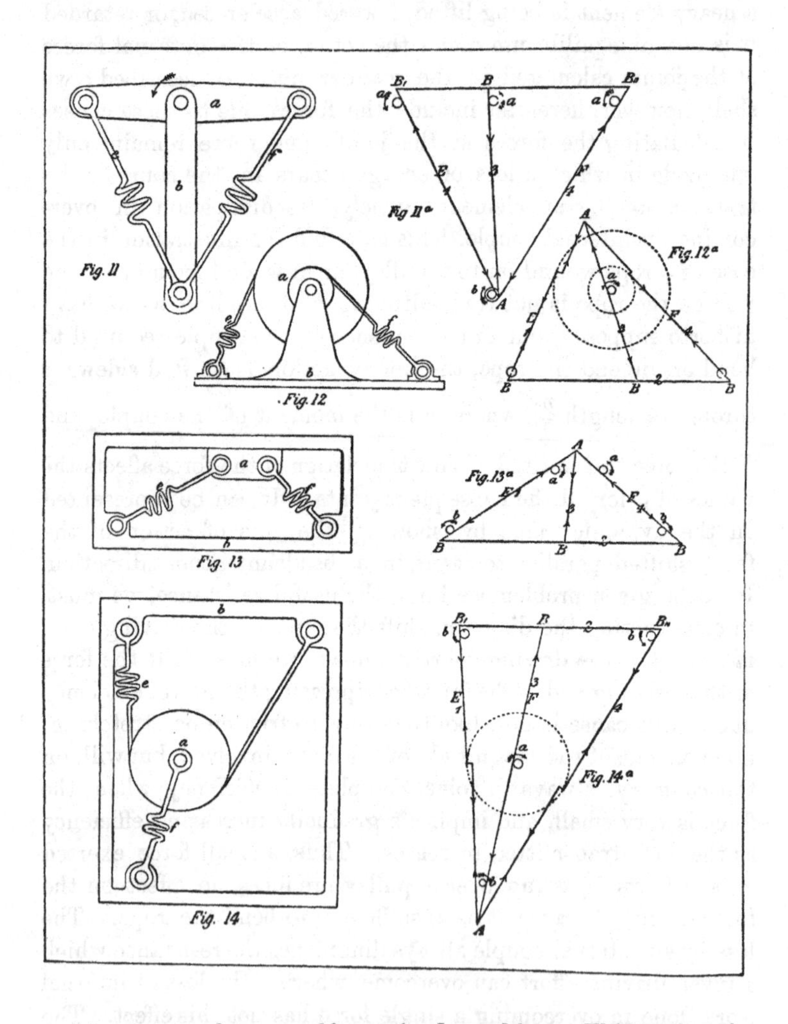

making a complete or self-contained machine. This system is a self-strained frame, with one stiff bar, namely, the lever. The bar in this, as in the other drawings, may be regarded as the symbol of a stiff frame, the form or design of which is unimportant in the given question.[1] The relation between the longitudinal stresses in the elements *e*, *f*, *b*, is given by the dynamic frame, Fig. 11*a*, which takes the friction into account at all the joints. The friction circles are drawn for joints *ae*, *ab*, *af*, *be*, and *bf*; the circle for the joint *bf* is, for clearness in the diagram, supposed to be a little larger than that for *be*. Arrows are placed at each friction circle to denote the motion of one part relatively to the other at the joints; each arrow is marked with the letter of the element, the motion of which it denotes; thus, at the joint *ae* the arrow marked *a* shows that, when the driving element *e* moves the lever, the rotation of *a* is left-handed relatively to *e*. Similarly, the arrow marked *b* at the joints *be* and *bf* denotes that, relatively to *e* and *f*, the rotation of *b* is right-handed. The letter *b* also denotes that the pin is fixed in *b*; (it is not a matter of indifference in which element this pin is fixed). Links 1, 3, and 4 can now be drawn, each tangent to their two friction circles. We choose the side on which to draw them as follows:—The forces acting on A balance one another, and therefore meet in one point marked A (Fig. 11*a*); the directions of the forces acting on *a* are marked by three arrow-heads near A, and the equal opposite forces by opposite arrow-heads near B. The links 1 and 4 appear as compression links in the dynamic frame, whereas they are tension links in the machine. The direction of the stress is also reversed in link 3. In explanation it must be remembered that the point A represents the lever, while the bar B_1BB_{11}, which may be drawn anywhere between the links, represents the element *b*. The links must be placed on that side of the circles where the arrow-heads of the forces acting on *a* (shown near A) oppose the motion of the arrows *a*, while the arrow-heads of the forces acting on *b* (shown near B) oppose the motion of the arrow *b*. The manner of drawing the figure for this example has been described in fuller detail than will in future be thought necessary. The relation between the driving effort in link 1 and the resistance in link 4 can be found from Fig. 11*a* by the ordinary graphical or trigonometrical methods.

[1] When the method of reciprocal figures is used to find the stresses in the links, it will be necessary in all cases to substitute a stiff frame of 3 links for the bars shown in the diagrams.

The conception of a complete machine has not been recognised by any writer on mechanics as necessary for the statement of problems connected with the lever. If, however, these problems are to be practical, and not confined to abstractions, such as 'forces applied to points,' they do require the consideration of a complete machine, as here drawn. The friction at *ab* is usually taken into account; that at *ae* and *af* is more generally neglected, and the friction at *be* and *bf* has perhaps never been thought of as an essential part of the problem. The reason of the neglect is clear. The forces represented by links 1 and 4 are in many problems due to attraction between some parts of elements *a* and *b*, as, for instance, when these forces are due to weights actually forming part of the element *a*, and attracted by the earth which supports, and is indeed part of, the element *b*. In this case the joints *ae*, *af*, *be*, and *bf* are frictionless, or may be said to disappear as joints. When the weights are hung by pins at *ae* and *af*, the friction at those pins must be taken into account, and whenever the forces represented by links 1 and 4 are due to another machine, to springs, or any other material element, the problem requires all the circumstances to be taken into account which are indicated in the dynamic frame as shown.

14. *Wheel and Axle.*—The wheel and axle, with its driving element, resisting element, and bearing, forms a complete machine when the parts are connected, as shown in Fig. 12. The wheel and axle constitute the element *a*, and the other elements have names given to them corresponding to those for the lever. The dynamic frame is shown in Fig. 12*a*. When the wheel and axle are circular, there is no friction at joints *eb* and *fb*; moreover, the pins and eyes which form the joints at *ea* and *fa* are replaced by the flexible rope. There is no friction at the joint *eb*, since *e* does not rotate relatively to *b*, and we may therefore assume that the force in the tie *e* is uniformly distributed relatively to its cross section: the resultant force, therefore, will pass through the centre of the pin at *eb*, and similarly the resultant of the resistance will pass through the centre of the pin at *fb*. If the ropes were perfectly flexible we might, in Fig. 12*a*, draw links 1 and 4 from the centres of the pins at *eb* and *jb*, tangent to the dotted circles drawn with the effective

radii of the wheel and of the axle; from their intersection link 3 would be drawn tangent to the friction circle for the joint *ab*. The stiffness of the rope must, however, be taken into account, and this can be done by drawing links 1 and 4 as broken links, of which the lower halves are drawn as above described, while the upper halves represent the lines of action of the forces shifted sideways. The driving link is brought nearer the centre of *a*, and the resisting link removed further from this centre. The distances d and d_1, by which each half link is shifted, are given by the expressions $d=\frac{m}{P}$, $d_1=\frac{m_1}{P_1}$, where m or m_1 is the moment of the couple required to bend the given rope to the given radius, and P or P_1 is the stress in the link. It must be remembered that this stress is not the same in the driving and resisting links, and that if we are given the stress in *e* we must proceed, by trial and error or simultaneous equations, to find the stress in *f* before we can determine exactly the distance by which the link 4 is shifted. When the ropes are long, their efficiency must also be taken into account, when our object is to compare energy exerted with work done. When we simply wish to compare effort and resistance, the loss of energy due to the stretching of the rope may be neglected. Inasmuch as the axes of elements *e* and *f* are assumed to lie in parallel planes, perpendicular to the axis of *a*, the forces in the elements *e* and *f* (unless parallel) give rise to an injurious couple on the bearings, which, except when these are very far apart relatively to the distance between the planes of *e* and *f*, sensibly diminishes the efficiency of the machine.

15. *Inclined Plane.*—The idea involved in problems on the 'inclined plane' is that one element, sliding on another with a plane joint between them, shall be employed to maintain equilibrium between forces applied to the sliding element in a plane perpendicular to the joint. We may embody this idea in a simple complete machine, as shown in Fig. 13, where *b* is a fixed element, *e* a driving element jointed to *b* and *a*, the sliding piece having a plane joint with *b*; *f* the resisting element jointed with *b* and *a*; the axes of *e* and *f* are in a plane perpendicular to the joint *ab*. We have here a self-strained combination fulfilling all the required conditions. The dynamic frame with

friction is shown in Fig. 13*a*. Links 1 and 4 are drawn tangent to the friction circles, and link 3 is drawn from their intersection A, making the stated angle with the plane joint *ab*. The bar 2 may be drawn anywhere, but is conveniently shown parallel to the joint *ab*. When link 1 coincides with link 3 or makes a greater angle with the joint *ab* than link 3 does, the mechanism will not work.

16. *The Hanging Pulley.*—The hanging pulley becomes a complete simple machine, when the driving element and resisting element are attached to a common element, as shown in Fig. 14; *b* is the fixed element, *e* the rope by which the effort is exerted, *a* the pulley, *f* the element on which useful work is done. Links 1 and 4 are drawn for the dynamic frame in two parts as shown; the moment of the couple dividing the parts being that required to bend and unbend the rope, and its force the force exerted on the rope. The pulley will take up a position in which the link 3 drawn tangent to its two friction circles cuts the intersection of 1 and 4 at A.

It is curious to observe that while Professor Reuleaux has very properly rejected the lever, inclined plane, hanging pulley, and wheel and axle, as elements of kinematic analysis, nevertheless these so-called mechanical powers do furnish the characteristic features of four simple machines of class 2 in which the number of elements is restricted to four. We shall find that the wedge is a characteristic feature in a simple machine with six dynamic links of class 1. These considerations show that dynamical and not kinematical reasoning guided the mechanicians who selected the so-called 'powers.'

17. *Example of a complete Machine having a Dynamic Frame of Six Links.*—A steam-engine with an oscillating cylinder, steam piston and piston rod to represent element *e*, and a pump to represent element *f*, as sketched in Fig. 15, affords an example closely approximating to a simple machine of class 1; the typical dynamic frame of this engine has already been shown in Fig. 6, and is here repeated with the slight variation that the pins are supposed to be fastened to *b* and *d* instead of *e* and *f*. It must be observed that the stress cannot be axial either in the driving or resisting elements; indeed, these parts are not true elements, for there is a joint between two elements in each of

them. They, in fact, with the links of the quadrilateral, constitute a machine working a machine such as will be discussed when treating of compound machines; similarly, when we re-

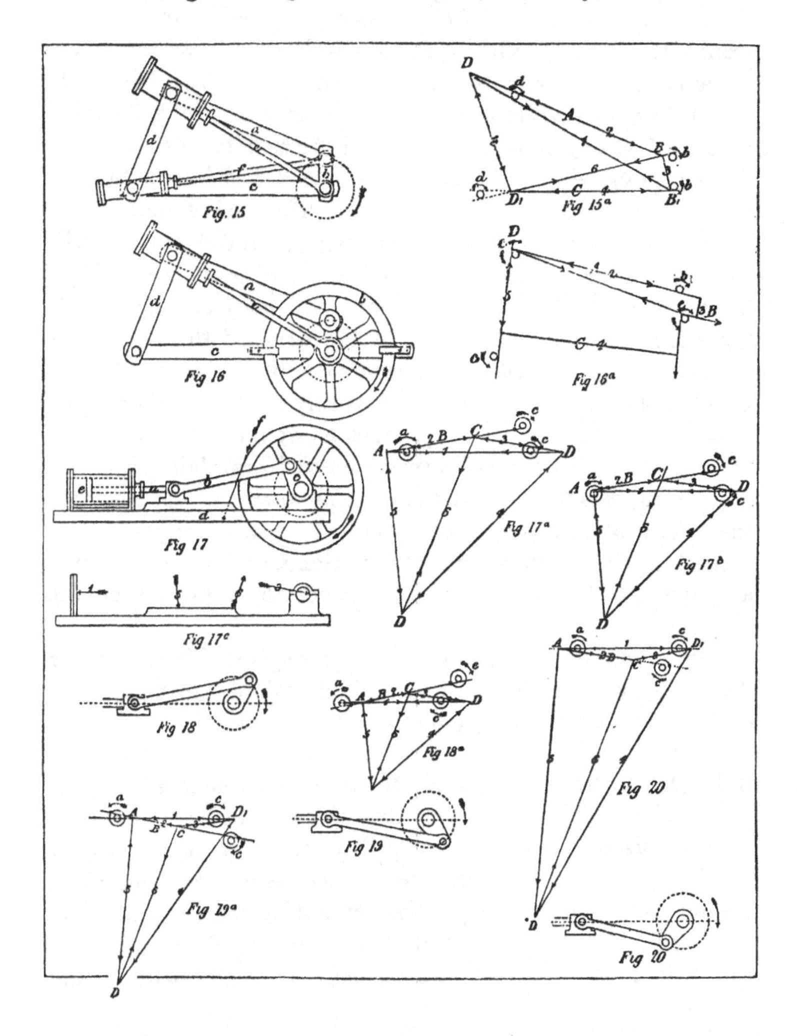

present the driving and resisting element by springs, as in the foregoing cases, we might have observed that in an actual spring the stress would not be strictly axial. In most practical cases, however, the stress will be so nearly axial that for the present we may neglect these considerations, and assume that we know the stress in the driving or resisting link.[1] We then have the frame shown in Fig. 15*a*.

If the same engine, Fig. 16, were employed to overcome or transmit a couple substituted for link 6 or element *f*, the dynamic frame would become that shown in Fig. 16*a*. The engine is shown with the piston rod in tension; links 2 and 5 must be first drawn tangent to their friction circles, then the bearing pressures parallel respectively to 2 and 5, and tangent to the friction circles for *eb* and *ec*. The line of pull in the element *e* is given by the line joining the intersection of the bearing pressures with that of 2 and 5; the diagram is completed by the bar perpendicular to 5 and that perpendicular to 2; these bars are mere indications of the arms of the equal and opposite couples exerted on elements *c* and *b*. These elements must be stiff, and their rotation relatively to one another is, by hypothesis, resisted by a couple such as would be produced by friction exerted on a wheel forming part of *b* revolving between clips or rubbers forming part of *c*. The diagram supposes that the frictional resistance thus obtained is so exactly equal at opposite ends of a diameter of the friction wheel as to constitute a couple which does not directly affect the pressures on the joints *eb*, *ec*.

18. *Ordinary Direct-acting Steam-Engine.*—The ordinary direct-acting steam-engine with a single cylinder gives another example of a simple complete machine. Fig. 17 shows a sketch of an engine of this type with the resistance exerted as if by a link between the periphery of a fly-wheel and the fixed element or bed-plate. We will assume that the position and direction of this resistance is known, being that shown by the arrow on *f* in Fig. 17. The elements are: *a*, the piston rod and block sliding on the guide bars; *b*, the connecting rod; *c*, the crank, axle, and fly-wheel; *d*, the bed-plate, including the cylinder; *e*, the steam in the cylinder. This element is jointed with *a* and

[1] The steam, piston, and cylinder constitute, with the resisting link, a simple inclined plane machine, *vide* § 15, and this machine drives the second machine, constituted by the piston rod, connecting rod, crank, bed plate, and resisting link; the piston rod and bed plate are common to the two machines, *vide* §§ 24, 25.

d; the position and direction of the force exerted are in this case determinate; for the cylinder does not oscillate and the piston with its rod are subject to no stress that is not axial; the element *a* is in equilibrium under the force due to *e*, and those due to the joints *ab* and *ad*. In Fig. 17*a* link 1 is drawn coinciding with the axis of the cylinder, and represents the bearing pressure produced by *e* on its joints; the link 2 may next be drawn tangent to the friction circles for *ab* and *bc*. The third force under which *a* is in equilibrium is that due to the joint *ad*; the link 5 is drawn through the dynamic joint A, making the stated angle with the guide bars. This diagram corresponds to an engine in which the slide block is as usual fast on the piston rod.[1]

We next observe that element *c* has three joints—first at *cf*, secondly at *bc*, and lastly at *cd*. The resisting link must, in order that the machine may be complete, abut at its other end against the bed-plate *d*; it may be due to actual friction, as when the fly-wheel is, for experimental purposes, fastened between two friction blocks secured to *d*. We know, by hypothesis, the place and direction of its application, and therefore the position and direction of link 6. The intersection of 6 and 2 gives the dynamic joint C; the third force acting on *c* must pass through this joint, and make the stated angle at the joint *cd*; we therefore draw link 3 from C tangent to the friction circle for *cd*. Lastly, we observe that the element *d* is in equilibrium under the following forces:—1st, that due to the joint *ed*, equal and opposite to that on *ae*; 2nd, that due to the joint *ad*; 3rd, that due to the resisting link *f*; 4th, that due to the joint, *cd*. We have already, on the Fig. 17*a*, the position and direction of all these forces. They do not, however, all meet in one joint, and we must, therefore, to complete the dynamic frame, add a link which shall receive the equal and opposite resultants of the forces compounded in two pairs; this we may do by joining the intersection between links 5 and 6 with that between 1 and 3,

[1] If, however, there were a joint between these parts, such that the pressure from the guide bars must pass very near the centre of the pin at that joint, then links 5 and 2 would be first drawn, and link 1 drawn cutting their dynamic joint; this arrangement would cause the effort exerted by the piston to pass a little way from the axis of the cylinder, as shown in Fig. 17*b*.

giving the complete diagram of Fig 17*a*. The diagram Fig. 17*c* will help to explain the significance of the several links. The element *d*, which is the bed-plate, is here shown by itself; it is in equilibrium under four external forces, indicated by arrows numbered as the links in the frame are numbered; and as this plate is itself in equilibrium, the resultant of 1 and 3 will be opposite and equal to the resultant of 5 and 6. The forces at the joints may therefore be represented by the stresses in the four half links, 1, 5, 6, and 3, joined as in the diagram by a link 4, in which the stress will be that due to the resultant of 1 and 3, or 5 and 6. The frame, if drawn on the hypothesis of no friction, will be kinematically equivalent to the actual engine; that is to say, the infinitesimal compression of link 6, resulting from a given infinitesimal expansion of link 1, will be equal in length to the actual travel of the fly-wheel at its rim past the friction-block, when the piston advances by an amount equal to the given expansion of link 1. The diagram, independently of its value as a means of estimating the relation between effort and resistance, is of use in showing clearly the direction and magnitude of the stresses to which the bed-plate is subject. As the revolution of the crank continues the dynamic frame changes. Figs. 17 to 20*a* show four positions of the engine with four dynamic frames; the bearing-points are marked in each case by dots on the circles representing sections of the pins. The changes in the direction of the stress in the connecting rod, relatively to its axis, should be observed, as well as the sudden changes which take place in the points of bearing pressure as the crank-shaft revolves. It is these sudden changes which give rise to 'knocks' in the engine when the shafts or pins and bearings or eyes do not fit. At joint *ab* four sudden changes occur—two due to changes of relative motion between the pin and eye, and two to changes in the direction of the stress in the link; at joints *bc* and *cd* there are only two changes. It must be understood that the whole dynamic frame is modified by any change in the direction or position of the resisting link. It will be found easy to construct diagrams showing the modifications—resulting from a change of this kind or from the substitution of a couple between *c* and *d* for the resisting link; this latter case corresponds to the arrangement of an engine employed to drive a long shaft with separate bearings co-axial with the crank-shaft.

19. *Wedge.*—When a wedge is employed to form a complete

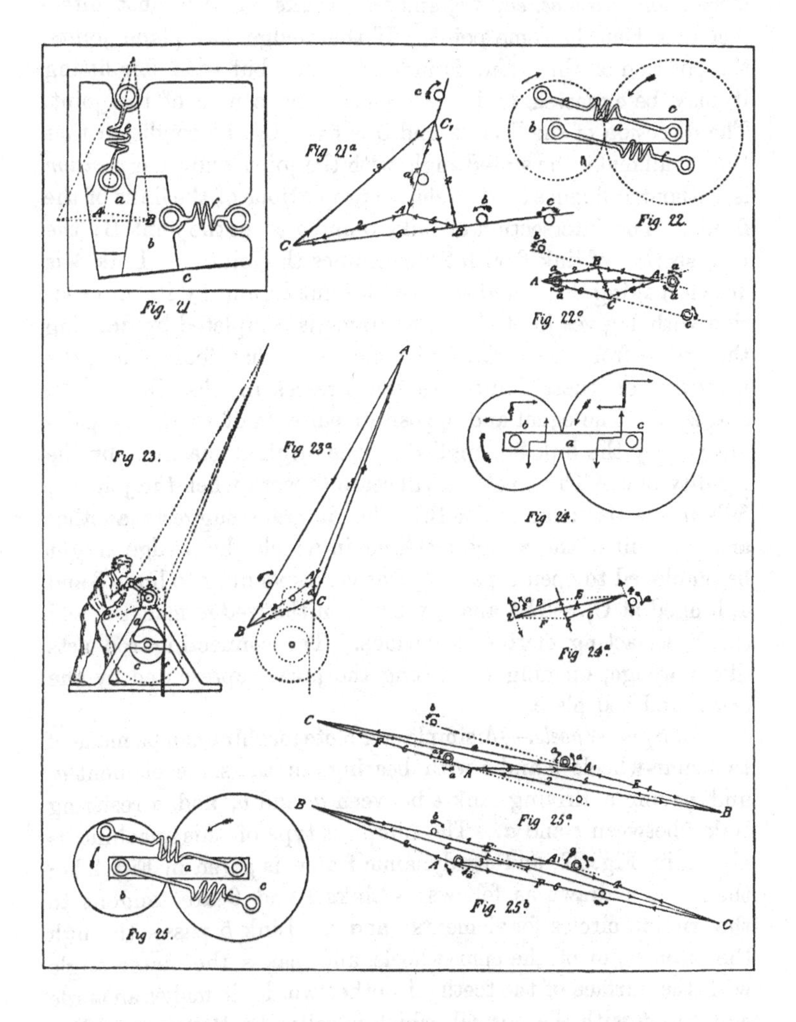

Fig. 21. Fig 21ᵃ. Fig. 22. Fig. 22ᵃ. Fig 23. Fig 23ᵃ. Fig 24. Fig 24ᵃ. Fig 25. Fig 25ᵃ. Fig. 25ᵇ.

machine we find a dynamic frame of class 1, similar to that given by the direct-acting steam machine. Fig. 21 shows a complete wedge machine. The letters on the parts indicate the elements, where e is the driving, and f the resisting element. Dotted lines on the same figure show the simple dynamic frame without friction. The dynamic frame with friction is given in Fig. 21*a*; links 1 and 6 are determined by being made tangent to the friction circles for joints *ec*, *ea*, *fb*, and *fc*. Links 2 and 5 must intersect link 1 at the same point. If the wedge has plane joints, the position of this point is indeterminate, but with fair fitting it may be expected to lie at or near the centre of the joint. The direction of the links 2 and 5 is fixed by the condition that they shall make the stated angle with the joint, and their *position* is without influence on the relative proportions of the links of the frame. The intersection of links 6 and 5 gives the joint B; the intersection of link 6 with 2 determines the joint C. Link 4 is drawn making the stated angle with joint *bc*, and by its intersection with 1 gives joint C_1. The frame is completed by drawing the link 3 from C to C_1. The element *c* has four joints, the pressures on these joints are the stresses in the links 1, 4, and 2, 6. The equal and opposite resultants of these two pairs are met by the link 3, supplied in the original machine by the rigidity of *c*. The machine will cease to work when the joint C_1 falls inside the triangle CAB. The diagram suggests another arrangement of the wedge machine in which the wedge might be employed to open a pair of jaws corresponding to links 3 and 4, hinged at C_1. The analogy between the wedge machine and the direct-acting engine is curious. The connecting rod acts like a wedge, opening or closing the jaws, represented by the crank and bed-plate.

20. *Spur Wheels.*—A simple complete machine can be made of two spur-wheels *b* and *c* with bearings in the same element *a*, and having a driving link *e* between *a* and *b*, and a resisting link *f* between *c* and *a*. The simplest type of this machine is shown in Fig. 22, and its dynamic frame is given in Fig. 22*a*; the frame is drawn as follows:—Links 1 and 6 are tangent to the friction circles for elements *e* and *e*. Link 5 passes through the pitch-point of the spur-wheels, and makes the stated angle with the surface of the teeth; in other words, it makes an angle equal to ϕ with the normal, which is called by RANKINE the line of connection. ϕ here as elsewhere signifies the angle whose tangent is μ, the coefficient of friction. The intersection of 5 with 1 and 6 gives the joints B and C; from B and C links 2

and 4 are drawn tangent to the friction circles for *ab* and *ac*, and the frame is completed by joining AA_1. Each wheel is an element having three joints, and therefore gives three half links to the frame. These three half links for wheel *b* meet at B, and represent the pull of the spring *e*, the push from the joint where the teeth meet, and the reaction from the bearing *ab*. Wheel *c* gives the three corresponding half links at C; the frame is completed by link 3, so placed as to receive the equal and opposite resultants of the second halves of the links 1 and 4, 2 and 6. This link 3 lies in the direction of the stress on the element *a*. A practical example of this machine is afforded by a man *e*, Fig. 23, turning a winch handle *b*, and lifting a weight *f* by the rope on an axle *c*, driven by a spur-wheel gearing with a pinion on the shaft of the winch handle. The man stands on the element *a*, which also supports the bearings of the spur-wheels. The dynamic frame, Fig. 23*a*, for this example is drawn precisely as for the previous typical example. As before, we know the directions of the effort in link 1 and of the resistance in link 6 (the latter is shown shifted outward to allow for the stiffness of the rope). The effort of the man need not be perpendicular to the crank, but must be exerted between elements *a* and *b*; the resisting link is the attraction between the weight and the earth, that is to say, as before, it is a link between C and A. Links 5, 2, and 4 are drawn as before, and link 3, supplied by the rigidity of element *a*, completes the frame. The friction between the man's hand and the handle is important. A friction circle is accordingly shown at the joint *bc*. A handle which the man can grasp firmly, and which turns easily round a well-oiled axis, makes the machine more efficient than when the man's hands must slip round the handle.

One of two spur-wheels may be driven by a couple, and the other resisted by a couple; this is a case frequently arising in practice, when one wheel is driven by a shaft, and the other drives a shaft, both shafts having such additional bearings as prevent the ultimate driving effort or ultimate resistance from having any effect on the bearings of the simple machine. Fig. 24*a* shows the dynamic frame for this case. There are three bars, as we have two couples; the bars are lettered as elements. Two friction circles are drawn with their centres at the centres of the bearings *ab* and *bc*. The bent arrows show the direction of rotation of pins fixed in the element *a*, relatively to *b* and *c*. Link 5 is drawn as before through the pitch-point, and making the stated angle with the surface of the teeth; the directions of the equal and parallel pressures on the pins form part of links 2 and 4, which no longer cut link 6; the other halves of links 2 and 4 are the reactions from the bearings shown as dotted lines. The diagram is completed by drawing bars to represent *a*, *b*, and *c*. The position of these bars is really immaterial, but B and C may be conveniently shown perpendicular to the direction of link 5, and these letters will now be used to signify the perpendicular distances between these links. When the couple M^D is applied between *b* and *a*, it produces two forces equal to $\frac{M_D}{B}$, the one acting to compress link 5, and the other to force the bar B against the pin in the direction shown by the full arrow 2. The first force is resisted by the tooth and wheel, as if these formed part of a link under compression, the other force by the element *a* in the direction of the dotted arrow 2. The force $\frac{M_D}{B}$, is transmitted through the tooth to act on *c*; it produces an equal force at the distance C, forcing *c* down on its pin in *a*, as shown by the full link 4. This force is resisted by an equal and opposite force, as shown by dotted link 4. Thus the couple produced in *c* is $\frac{M_D C}{B}$, and this is the resisting couple M_R, which the driving couple M_D can overcome. The forces and couples are the same as would be produced if we could construct a material frame of the link 5 and the bars ABC of the dimensions shown, and having the driving couple between B and A, with the resisting couple between A and C.

21. *Rolling Contact.*—The simple machine, made with two wheels which transmit power by a rolling contact between them, has a dynamic frame of the same character as that corresponding to spur-wheels. This arrangement is shown in Fig. 25, where the parts have the same names as were given in the case of spur-wheels. The directions of links 1 and 6, Fig. 25*a*, are known; the direction of link 5 is also known, if the machine is running with its maximum efficiency; that is to say, if there is

no more tension on element *a* than is necessary. In that case, the line of pressure at the point of contact will make the stated angle with the surface of contact. Link 2 is drawn from the intersection of 5 and 1 to the friction circle at the centre of *b*. Link 4 is drawn from the intersection of 5 and 6 to the friction circle at the centre of *c*. The intersection of 4 and 1 is then joined by link 3 to the intersection of 6 and 2. The three lines meeting at B show the positions of the three forces under which *b* is in equilibrium, and the arrows show the directions of those forces. The three lines meeting at C show the forces under which the second wheel *c* is in equilibrium, and the arrows at C show the directions of those forces. The rules given in § 8 will enable the draughtsman to determine on which side of each friction circle the link is to be tangent. Fig. 25*b* shows the manner in which the diagram becomes modified when the directions of links 6 and 1 make smaller angles than link 5 makes with the normal to the joint *bc*. Fig. 25*b* also shows the effect of rolling friction at this joint, which, however, may generally be neglected. At the point of contact the material is continually being crushed, and the material is not perfectly elastic. We have, therefore, a resisting couple analogous to that met with in the case of ropes, and the effect is to shift in a disadvantageous direction one part of link 5 by a distance equal to the arm of this couple. If excessive tension is employed in *a*, the direction of link 5 will be found by compounding that tension with the force transmitted at the periphery. The diagram where one roller is driven by a couple and the other roller resisted by a couple, is easily deduced from that for spur-wheels.

22. *Belt and Pulley.*—The complete belt and pulley machine is shown in Fig. 26, p. 306. It is composed of the pulleys *b* and *c*; the element *a* in which their bearings run, the flexible belt *d*, the driving element *e*, and the resisting element *f*. The dynamic frame without friction is given in Fig. 26*a*. Link 5 is the direction of the resultant of the tensions on the two bands, which may be considered as together forming one split link. When we assume that no more tension is used than is necessary, the ratio between the tensions on the tight and slack side of the bands is determined by the arcs round which the belts are in

contact with the pulley, and by the coefficient of friction between the belt and the pulley. Consequently, the position of link 5 may be taken as known. The intersection of the driving link 1 with 5 gives joint B; the third force acting on b is the

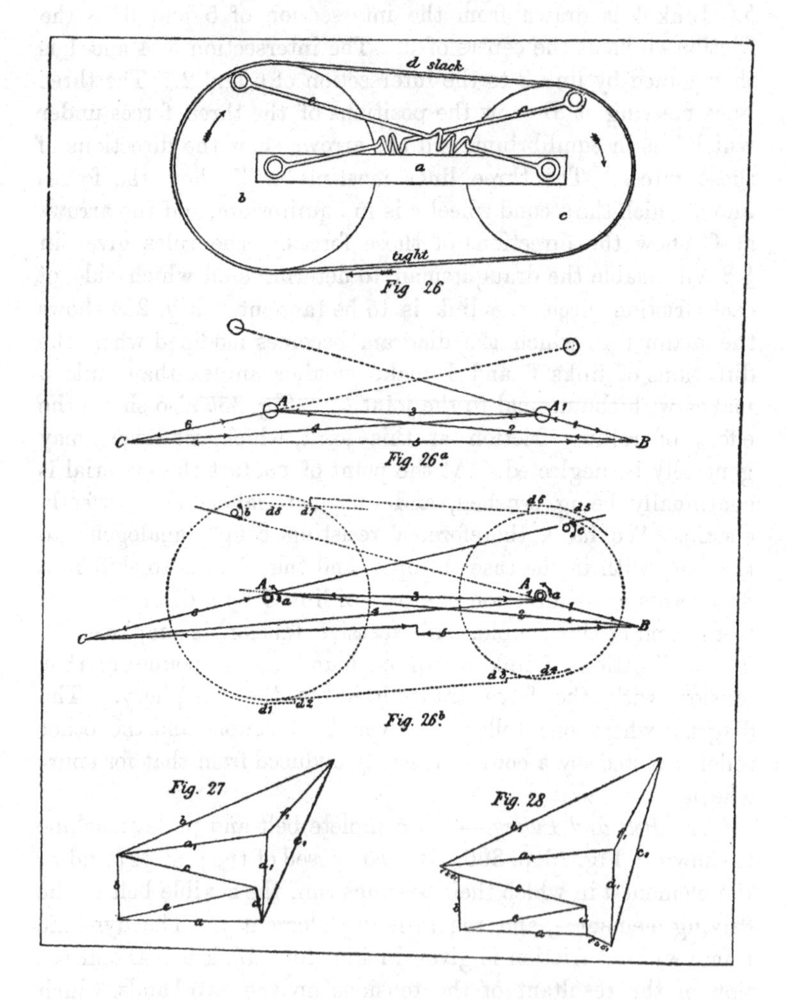

resultant of the two others, and must pass through the centre of the pulley b; this determines the direction of link 2. Similarly, C is given by the intersection of 5 and 6; link 4 is drawn from their intersection to the centre of the pulley c; the diagram is completed by drawing link 1. The forces which keep the pulley *b* in equilibrium are shown in position and direction at the joint B. Similarly, the forces which balance *c* are shown at C, but the driving and resisting links, as well as link 6, appear as if under compression. Before proceeding to draw the frame with friction, it may be well to show how the stiffness of the belt may be taken into account. In Fig. 26 the arrows show the direction of the motion of the belt. The pulley *b* at the lower side has to exert a force P in the direction of the arrow on the belt, and also a right-handed couple *m* to bend the belt; the resultant of this force and couple is an equal force shifted to the left or downwards by a distance $\frac{m}{P}$; we might, therefore, represent the actual belt by a perfectly flexible belt placed as shown by the dotted line d_1 d_2. The effect of the stiffness of the belt at the other places where it is bent and unbent, is also to shift the line of application of the force as shown by the dotted lines d_3, d_4, d_5, d_6, d_7 d_8. Thus the resultants of the forces due to the tension of the belt on each pulley will not be opposite each other; the resultant will be shifted outwards on the driving pulley, and inwards on follower, or disadvantageously in both places. The actual amount of shifting cannot be ascertained until P and P_1, the tensions on the belt, are known; but the nature of the change is easily apprehended, and is therefore included in the dynamic frame, Fig. 26*b*. This frame is drawn as for the case without friction.

23. *Compound Machines.*—It is evident that the resisting link of one complete simple machine may be used as the driving link of another simple machine. This combination gives rise to what may be called a compound machine. If the two machines have no element in common, they must be connected by two joints in the manner of which Fig. 27 gives a typical example. In this figure the lines may be considered to represent either the links of a dynamic frame, or the axes of a series of material elements, jointed without friction. The contraction of element *e* would cause an expansion of the line joining the joint *bc* with *da*. This line would indicate the position of the resisting link of the machine *abcde*, if this machine were a simple one. This same line lies between the joints b_1 a_1 and d_1 c_1 in the position

required to enable the effort produced by the first machine *abcd* to drive the second $a_1\, b_1\, c_1\, d_1\, f_1$. In fact, the whole first machine may be considered as a somewhat complex driving element, relatively to the second machine; or the whole of the second machine may be looked on as a rather complex resisting element, relatively to the first machine. The two machines may obviously be treated as entirely separate. Let the first machine drive the second by contact between the elements b and b_1 and between d and d_1. Then it is necessary that at these joints the direction of the lines of bearing pressure shall be in one straight line. This line is the direction of the resisting link for one machine, and the driving link for the next. The joints bb_1, and ee_1, between two successive machines, will be called *transmitting joints.* They do not themselves belong to either machine. Distinct elements may be introduced between the two machines as in the typical example, Fig. 28. Here the second machine is tied to the first by two links, each lettered f and e_1. These elements, which must represent forces in one straight line, may be considered as equivalent to the driving element of the second machine, and the resisting element of the first. In the example given it would be necessary to make f a tie, so that the machine, as drawn, would only work when element e was expanding. In another case, one part of the link f might be omitted as between bc and a_1b_1, and replaced by a transmitting joint between b and b_1 as above. An element placed between two machines and serving to transmit the power, will be called a *transmitting element.* We see then that the communication of power can be made from one complete machine to another, either by two joints, by two elements, or by a joint and element. We may, with perfect propriety, give the name of complete machine to any one of a series, each of which drives its successor; for we may regard the driving system simply as a more or less complex driving element, and the driven system as a more or less complex resisting element. We are not concerned with the complex play of forces which produces the driving or resisting effort, but, so far as each complete machine is concerned, only with the fact that its elements are driven or resisted in a manner which may be represented by a single driving or resisting link.

24. *Compound Machines with one common element.*—The

compound machines described in the last paragraph have no elements common to two simple machines, but we may have compound machines in which either one or two elements are common to two successive machines. The examples most commonly met with in engineering practice are those in which there is one common element, namely, the framework or support which is continuous and common to a series of successive complete machines. The common element is necessarily in equilibrium under the whole series of stresses to which it is subject, but this equilibrium is not a matter of great interest. The driving link of the first machine usually abuts at one end against the common element. If the first simple machine stood alone, one end of its resisting link would abut against the same element; when it drives a series of machines, the resisting link of each must be so placed that in each case, if that particular machine were the last of the series, the resisting link would also abut against the common element. The common element takes the place of one transmitting joint or one transmitting link in the types given in the last paragraph. A practical example will serve to show the connection between successive complete machines, having one element in common. In Fig. 29, p. 310, a horizontal engine is shown driving a train of machinery. The engine consists of the elements *abcde*. The element *d* is a fixed bed-plate; the element *c* comprises a fly-wheel and spur-wheel, which drives the pinion which is part of *g*; the spur-wheel of *g* drives a pinion which is part of *h*. A pulley, which is also part of *h*, drives a belt *l*, which, in its turn, drives a pulley *m*; a second pulley, also part of *m*, drives a second belt and pulley *n* and *o*. A piece of wood, forming part of *o*, is being turned by a tool which forms part of *d*. We have here three complete machines—1st, the engine; 2nd, the machine *ghd*, driven by two transmitting joints and *gdcg*, and driving a transmitting link *l*; 3rd, the machine *mnod*, driven by the transmitting link *l* and the joint *md*, and overcoming the useful resistance at the joint *od*. All these machines have the element *d* in common. The dynamic frame of the compound machine is shown in Fig. 29*a*, and the reciprocal figure for that frame in Fig. 29*b*. The driving element of the first machine is the steam which abuts against *d*, the bed plate or support. The resisting element is the whole series of driven machines in the last of which a resistance is overcome between an element of the machine and the bed plate *d*. The resisting element is not

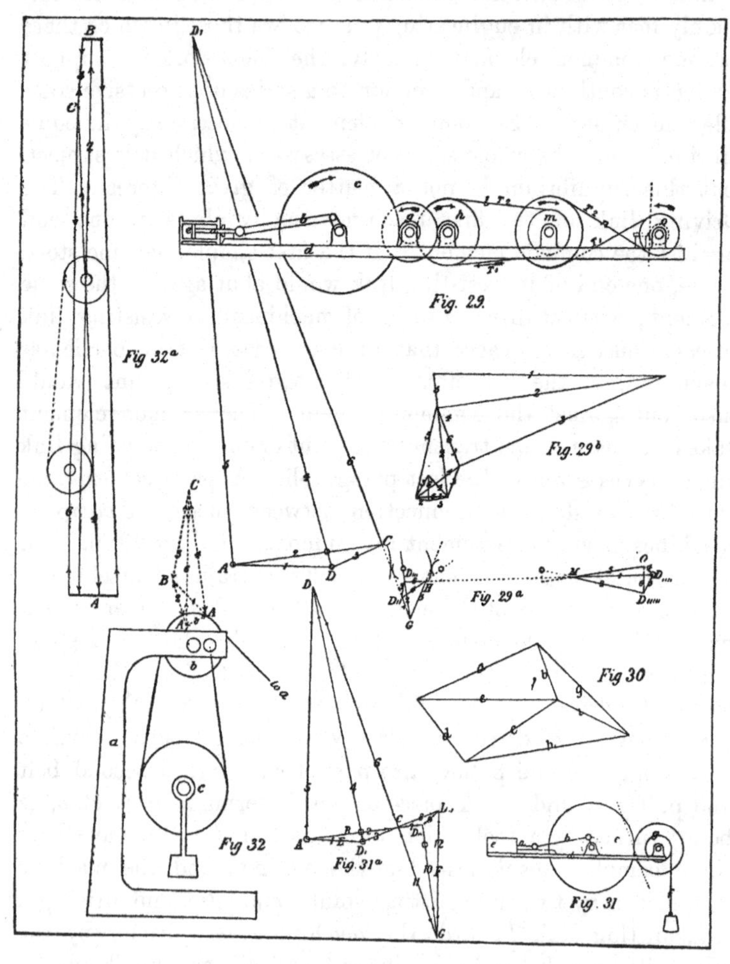

complete unless we take into account the force it exerts at both ends; the one end of the resisting link of the first machine pushes up the periphery of the fly-wheel *c*, the other end pushes down the element *d*; this is precisely the action we should have, if the first machine had been completed by a single resisting link; the circuit must in either case be completed, so that the resisting link may abut against the common element *d*, against which the first driving link also abuts. The driving link of the second machine *ghd* is in the line of the resisting link of *abcde*. The direction of this link is for both machines determined by the transmitting joint *cg*—i.e., by the form of the teeth of the wheels. The driving link of the machine *mnod* is in the line of the resultant tension due to the band *l*, which is here a transmitting element. The driving or resisting link of each successive machine, if taken by itself, would abut against the common element *d*. The first driving and last transmitting element do abut against this common element. In any machine it will be found easy to analyse the series of parts so as to divide the whole structure into a series of complete machines, each joined to its neighbour by a transmitting joint or link. Each machine can then be treated as a separate whole. We see in the example given that the dynamic frame consists of three distinct quadrilaterals with diagonals. The connecting links may be considered as double links in each case. Thus we may consider the joint C, Fig. 29*a*, as connected by one link with the joint G which it drives, while at the same time the joint D_{I} is connected by a link with D_{II}. Thus, when the reciprocal figure for the whole frame is drawn, as in Fig. 29*b*, it forms one connected whole. The stress corresponding to each connecting link is used twice as in all reciprocal figures. In Fig. 29*b* each link is numbered from 1 to 6 for each successive machine. The length of the first line 1 corresponds to the driving effort. The length of the last small line 6 corresponds to the resistance which that effort can overcome at *od*. The diagram is drawn without taking friction or stiffness into account, so that the frame shown is kinematically equivalent to the machine. It would be easy in a drawing on a large scale to show the complete effect of friction and stiffness, for we have already learnt to take these into account for each component simple machine. We see, therefore, that in any simple train of machinery there can be little difficulty in estimating the true relation between effort and resistance (neglecting weight and mass); this difficulty never exceeds that met with in analysing a simple machine, and all simple machines are of one type.

25. *Half Machines Compounded.*—In certain cases two successive machines have *two* elements in common, as well as the driving or resisting link. In this case we may consider the addition as in reality only half a machine. The typical example of this arrangement is given in Fig. 30, where links *b* and *c* are common to two complete machines—1st, *abcde*, with its resisting link; and 2nd, *bchgi*, with its driving link. It will be seen that the half machine *hig* has a certain analogy with the machines of class 2, only it is here the stiff bar which acts as a driving element by an alteration in its length; *h*, *i*, or *g* might represent the final resisting element. We are never driven to adopt this subdivision of a machine, except at one or other end of a train of machines which may not be divisible into a series of complete machines with joints or elements of transmission. Thus, if we (Fig. 31) have a steam-engine with a spur-wheel on its crank shaft, driving a single pinion from the shaft of which a weight is hanging, we cannot divide the train into two distinct complete machines, each having a quadrilateral with diagonals as its dynamic frame. We may, however, draw two distinct dynamic frames, as shown in Fig. 31*a*, where 7, 8, 9, 10, 11, and 12 are the links of a complete spur-wheel machine, similar to that of Fig. 22*a*, driven by a link 7, but link 7 is in the same line as link 2 of the well-known engine frame. Links 8 and 3, 6 and 11, are also common to the two frames. The diagram shows that we might analyse the machine in two ways, calling, for instance, the steam-engine a complete machine, and the extra spur-wheel, with its weight, a half machine; or we might call the two spur-wheels a complete machine, and the piston, steam, and connecting rod a half machine.

It is a matter of no consequence how we subdivide a train. The relative stresses in the first and last links will be the same, whether we use half machines with two common elements, or successive machines with one common element. This example shows us that the machines of class 2 may very properly be regarded as only half machines. This view is supported by the observation that when one machine of class 2 drives another of the same class, we get for the system a single complete dynamic frame, namely, the quadrilateral with two diagonals. As an example, we may take the hanging pulley and fixed pulley combined, as in Fig. 32, on which is shown the dynamic frame of the combination. When the ropes are nearly parallel, as in Fig. 32a, the bars shown by thick black lines may be considered as jointed to the six links, which would otherwise meet at the joints B, A, and A_1.

26. *Reduplication of Cords.*—When a series of fixed and hanging pulleys are employed as in the ordinary blocks and tackle, it is found that, with the usual stiff ropes, no advantage can be obtained by using more than 5 or 6 plies of rope. The reason of this is shown clearly by the dynamic frame for a compound machine of this class, Fig. 33, p. 315. In this frame the successive pulleys and plies are arranged in one plane, so that the diagram may be better followed than could be the case if the pulleys were placed so as to be co-axial. Let the driving link act between the rope a and the fixed support d, and let the force applied by the driving link be called E, and the effective radius of each sheave R. The effect of the rope a is to produce a couple m diminishing R on the driving side by a length depending on the stiffness of the rope, and inversely proportional to the diameter of the pulley and to the tension on the rope. The effect of this is shown in the diagram by a shifting of the line of action of the force towards the centre of the pulley by a distance s equal to $\frac{m}{E}$. Let F_1 be the tension on the second rope. The couple required to unbend the rope has an effect which may be represented by shifting the line of action outwards to an amount s_1, equal to $\frac{m}{F_1}$; at each pulley a similar effect is produced, and as the value of the tension in the rope diminishes at each pulley, so the value of s increases at each pulley. The effect of friction at the axle is shown by shifting the joint in the bar representing each pulley towards the driving rope by a distance $r \sin \phi$, where r is the radius of the shaft. We then have the equation

$$E\left(R - r \sin \phi - \frac{m}{E}\right) = F_1\left(R + r \sin \phi + \frac{m}{F_1}\right)$$

$$\text{or } E(R - r \sin \phi) = F_1(R + r \sin \phi) + 2m.$$

From this equation we obtain F_1, and by a similar equation we could from this obtain F_2, etc. $\Sigma F = W$, the weight which can be raised. The result is after a few turns to reduce F_n to nil, after which no more plies can be of any service. The gradually diminishing efficiency of successive pulleys is very well shown by the diagram, Fig. 33.

27. *Loaded Dynamic Frame without Friction.*—Let us now consider the effect of the mass and weight of the elements of a machine. We may feel sure that the effect of weight and inertia may be shown by means of a dynamic frame, for this frame consists of lines indicating bearing pressures at the joints, and the direction and magnitude of these bearing pressures are always determinate. In the case of a material element supported by two joints, the lines of bearing pressure will no longer be directly opposite, or in the same straight line, but will intersect in a point on the line which indicates the resultant of the forces due to the weight and inertia of the element. Let the resultant of all the forces exerted on a given element other than those exerted at the joints be called the *load* on the element. This includes the equilibrant of the force producing acceleration. Then the action of an element with two joints, as in Fig. 34, might be supplied by three forces represented by three half links 1, 2, and 3, Fig. 34, showing in position and direction the bearing pressures at the joints, and the load on the element; this mode of representing a loaded element is commonly in use where the equilibrium of arches is discussed. The load 3 is here called a half link, for in the complete self-contained machine an equal and opposite load necessarily exists in some other element. This equal and opposite load is in general supplied by the reaction of the foundations, or more strictly by the reaction due to the mass of the earth.

Where an element has more than two joints, it will be found that the arrangement or form of the joints is generally, if not always, such as to render determinate the single joint or pair of joints by which it is supported. The effect of the load in modifying the direction of the bearing pressure can for these cases be as easily taken into account as in the simple case just cited.

Let us now consider the effect of four loads L_a L_b L_c and L_d, on the four elements *abcd* of the elementary machine, Fig. 35. We may, for the present, suppose the driving and resisting element to have no weight or inertia; the effect of these elements may then be treated as equivalent to the effect of four external forces, $1a$ 1β, and $6a$ 6β. The dynamic frame for this case will be a polygon of eight sides, Fig. 35, $2a$ 2β, $3a$ 3β, $4a$ 4β, $5a$ 5β,

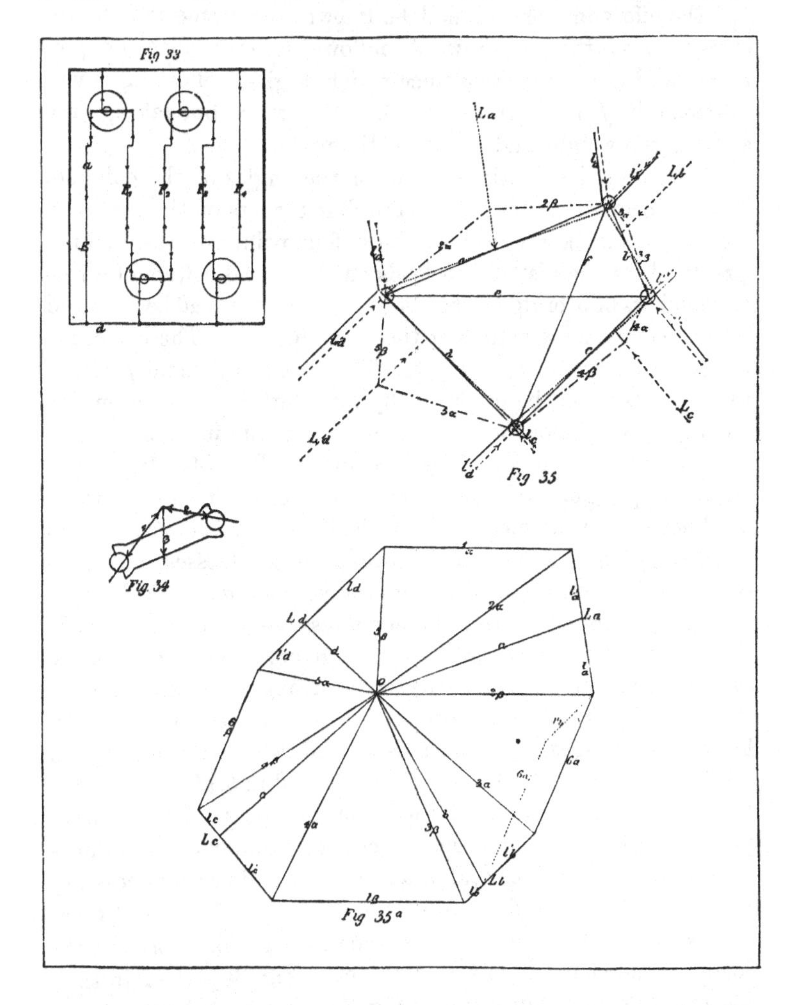

Fig 33

Fig 34

Fig 35

Fig 35a

having its angles on the lines of load, and so inclined as to be in equilibrium under these loads. The stresses in each link are the pressures at each joint. The reciprocal figure for this frame is shown in Fig. 35a. The inclinations of the links are no longer independent of the magnitudes of the forces acting in the elements *e* and *f*, as was the case when we neglected weight and inertia. The effort in *e* or in the resistance in *f* must therefore be given, as well as the loads, before the frame can be drawn. Let the effort in *e* or in link 1 be known, the frame and its reciprocal may then be drawn as follows, so as to solve the problem of finding the resistance which a given effort in *e* will overcome in *f*, neglecting the effect of friction, but taking into account the weight and inertia of the parts.

We know the position of four of the angles of the polygon, viz., the centres of the pins at the four corners of the machine. Let the loads be referred to these four points in the manner practised for the distributed loads on the actual rafters of a roof or members of a bridge; that is to say, let each load be replaced by two components acting at these four points. These components are lettered l_a l'_a, l_b l' , etc. The stress in element *f* will be the same as would be produced by the effort in *e* acting on the quadrilateral frame *abcd*, loaded at the joints in this manner. This stress in *f* is found by drawing the Fig. 35*a*, beginning with the polygon d, l_d, 1α, l'_a, a. The directions of all these are known, and the magnitude of the stresses in all except *a* and *d*. The polygon serves to determine these stresses. To find the stress in *b* we require to draw the polygon a, l_a, 6α, l'_b, b, in which there are only two unknown stresses—those in 6α and *b*, the directions of which are, however, given. We cannot draw the polygon directly with the lines arranged in the manner shown in full lines, because 6α and *b* are not contiguous. If, however, the lines are drawn as dotted, we obtain a polygon which determines the stress both in 6α and in *b*, after which the lines may be rearranged in their natural order. By a similar process we find 6β and *c*, and can complete and check the drawing by adding the lines l'_c, 1β, and l_b. It is almost unnecessary to remark that 1α and 1β are equal, and that 6α and 6β are also equal. The sides of the polygon in Fig. 35*a* represent the loads on the elements of the machine in Fig. 35, taken in their natural order. The lines *abcd* in Fig. 35*a* represent the stresses on the links *abcd*, or 1, 2, 3 and 4 of the machine in Fig. 35. Let *o* be the point where *a*, *b*, *c*, and *d* intersect (Fig. 35*a*), and let lines be drawn from *o* to the angles of the polygon; now, draw the lines $2a$ and 2β in Fig. 35, parallel to the lines of the same name in Fig. 35*a*, and so placed that they abut against the ends of link a, and intersect in the line L_a; draw $3a$ and 3β, $4a$ and 4β, $5a$ and 5β by a similar rule. The octagonal frame thus obtained, and shown by a broken dotted line, is a frame which will be in equilibrium under the four given loads and the two stresses in *e* and *f*. It is easy to see that this will be the case. The load L_a and the two stresses in $2a$ 2β of the frame form a polygon in the reciprocal figure. The same relation obtains between the other loads and the links supporting them. Moreover, the links $2a$ and 5β in the reciprocal figure give a closed polygon with the line $1a$, representing the stress in *e*. The lines abutting against the ends of *f* also give closed polygons with the stress 6, as shown in the reciprocal figure. The links $2a$, 2β, etc., of the octagonal frame do therefore represent the directions of bearing pressures at the joints on the hypothesis that there is no friction. The frame found by the method described will be called the *loaded dynamic frame without friction.*

The elements *e* and *f* have been represented as without mass; if their weight and inertia are to be taken into account, their load is to be referred to the joints in a manner similar to that indicated for the other elements. Links 1 and 6 would then be broken lines in the frame, with loads at their angles. It may be well to remind the reader that by hypothesis the machine is in equilibrium as a whole, and therefore in the reciprocal figure lines representing the loads necessarily form a closed polygon.

28. *Loaded Dynamic Frame with Friction.*—The method by which we were enabled to draw the octagonal polygon described in the last paragraph depended on our knowledge of the four points which determined the position of four angles of the polygon, or one point on each of the eight lines. When we try to ascertain the actual lines of bearing pressure, taking into account the friction of the machine in motion, we find that the conditions determining their direction are more complex, since now we do not know any fixed point in any line. The conditions are, however, only changed to this extent, that the lines of bearing pressure must make the stated angle with the joints, instead of being normal to the surfaces of those joints. By trial, an octagon is easily found fulfilling this condition as well

as the general condition of being in equilibrium under the force applied at the joints. The manner of proceeding which seems most easy is to draw, first the polygon without friction, and then to sketch a modified polygon having its sides tangent to the friction circles, or making the stated angle with the joints. The sides of this trial polygon intersect at certain points which may be called *trial points*. When the friction circles are not very large, it is easy, by the exercise of a little judgment, to draw the trial polygon so as to make these *trial points* agree very closely with the true points of intersection, even at the first attempt. Then, referring the loads to the trial points, we draw a new polygon and reciprocal figure, Figs. 36 and 36*a*, as for the frame without friction. If this second polygon has sides which make the stated angle with the joints, the problem is solved. Otherwise, a second selection of corrected trial points must be made, and a third trial polygon drawn: it will seldom if ever be necessary to make a third trial. We thus get a dynamic frame which truly represents the directions of the forces at every joint in the actual machine, and this frame will be called the *complete dynamic frame of the machine*, or the *loaded dynamic frame with friction*. The resistance which can be overcome by a given effort 1 in the driving link, is shown by the line 6 in the reciprocal figure 36*a*, which has been used as an auxiliary in drawing the loaded dynamic frame, and this resistance will be the actual resistance which could be overcome by the given effort in the given machine, under the given conditions as to speed, friction, mass, and weight.

29. *Application of the Method to an ordinary Horizontal Single-acting Steam-Engine.*—In Fig. 37, p. 321, let the lines *b* and *c* represent the centre lines of the connecting rod and crank of an engine, while the line *a* represents the direction of the motion of the piston. Let the line *f* represent the direction and position of the resistance overcome; and let this resistance be represented as in previous examples by a stress between *d* and *c*. Let the lines L_a, L_b, L_c, and L_d represent the loads on the elements *a*, *b*, *c*, and *d* of this engine, where *d* is the bed-plate, it being remembered that these loads must be such as would balance one another; in other words, L_d, the resultant reaction due to the foundation, must be the equilibrant of the three others; L_c is

the weight of the balanced fly-wheel and crank-shaft, acting directly through the centre of the main bearing; L_b is the resultant of the weight of the connecting rod, compounded with the equilibrant of the force required to give the rod the acceler-

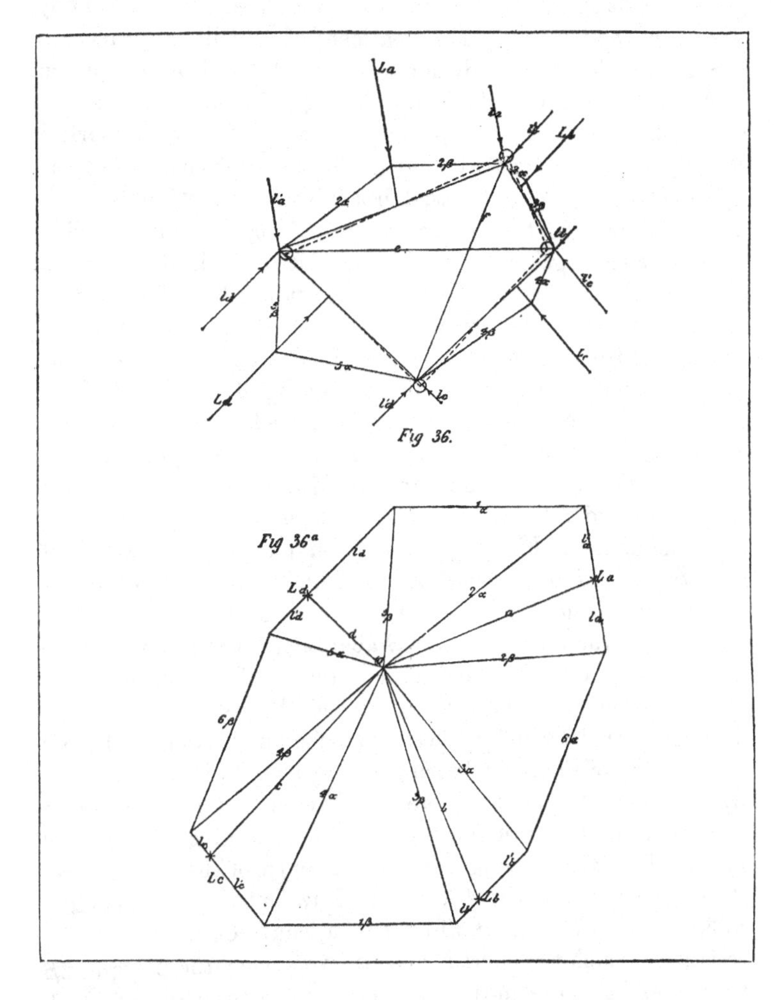

Fig 36.

Fig 36[a]

ation (in respect of rotation and translation) which it actually has at the given instant. L_a is the resultant of the weight of the piston compounded with the resistance to acceleration. The loads shown in the figure have been calculated for an actual small direct-acting engine, making 1 turn per second. To draw the dynamic frame without friction we proceed as follows:—L_b is represented by two components l_b and l'_b, acting at the joints *ab* and *bc*. L_a is wholly borne by the one joint, and the actual point of its application is a matter of indifference; L_c is wholly drawn by the joint *cd*. The treatment of L_d will be explained hereafter. The simple dynamic frame without load or friction is shown by lines 1, 2, 3, 4, 5, 6 in Fig. 37*a*, which also shows the loads referred to the proper points. Let it be remarked that l'_b is referred not to a joint in the simple frame, but to the point through which the actual bearing pressure 2β must pass. We may now begin the reciprocal figure 37*b* by drawing the effort $1a$ of the steam against the piston, the load L_a the load L_b and the directions of the reaction 5, and the resistance $6a$. The load L_b is then subdivided into its two components l_b and l', and the line 2 is drawn from the point which subdivides L_b into its two components l_b and l'_b parallel to the line *b* in Fig. 37 or 2 in Fig. 37*a*. Let the point where line 2 intersects 5 be called *o*; then the polygon $1a$, L_a, l_b, 2, 5 (Fig. 37*b*) represents the forces in equilibrium at the joint *ab*; now, returning to Fig. 37*a*, we are able to draw lines 5, $2a$ and 2β: the direction of $2a$ is given by the line of the same name in Fig. 37*b*, drawn from *o* to the intersection of L_a with L_b; the line 2β is drawn in Fig. 37*a* parallel to the line of the same name which in 37*b* joins *o* with the intersection of $6a$ and L_b; the lines $2a$ and 2β abut against the joints at the end of link 2 (Fig. 37*a*), and meet in the line L_b. The element *c* is in equilibrium under the action of the resistance $6a$, the driving force 2β which we have just found, the weight L_c and the reaction of the main bearing. The load L_c is wholly borne by the joint *cd*, and therefore the direction of the remaining component of the reaction at the bearing is given by the full line $3a$, Fig. 37*a*, passing through the centre of the joint *cd*, and the intersection of 6 with 2β. From *o* draw $3a$ in Fig. 37*b*, parallel to $3a$ in Fig. 37*a*, and it will cut off a length $6a$ measuring the resistance which the effort is able to overcome; the polygon 2β $6a$ L_c and 3β represents the four forces under the action of which *c* is in equilibrium. 3β is the resultant pressure on the joint *cd*. We may here complete the reciprocal figure by drawing the force 1β, the load L_d, and the forces 6β, $4a$, and 4β. We may also complete the loaded frame in Fig. 37a by drawing 4a and 4β parallel to the lines of the same name in Fig. 37b; a line from o parallel to link 4 of Fig. 37a will subdivide L_d in the ratio in which L_d would be subdivided if referred to the two frame joints

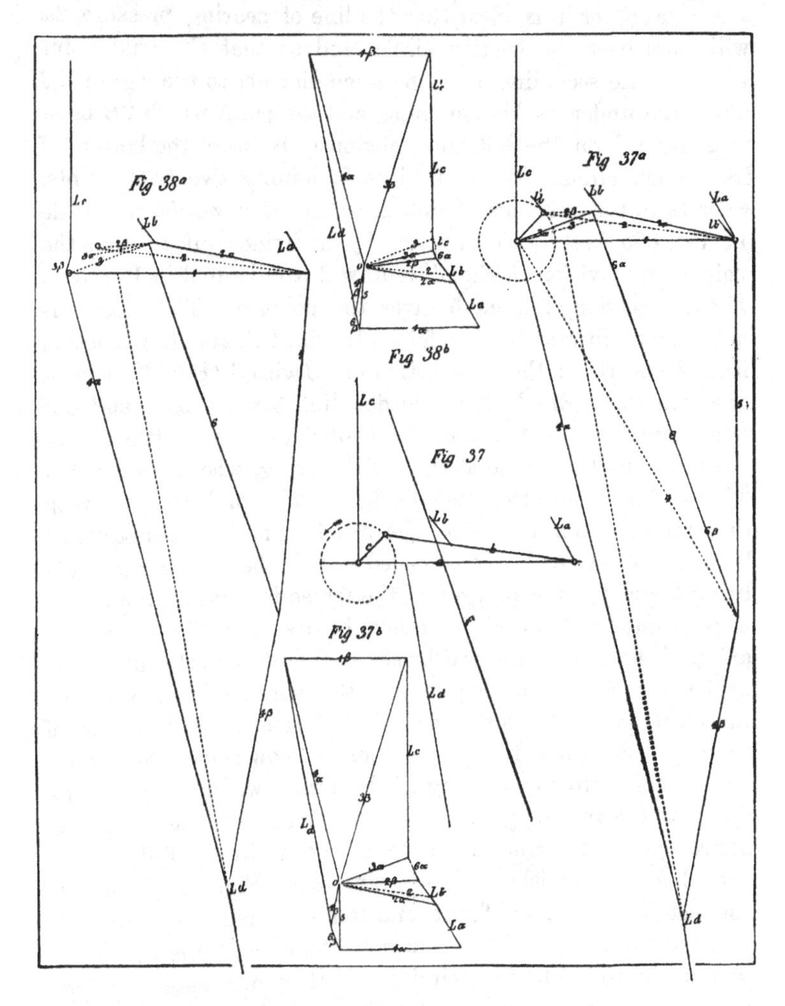

between links at the ends of link 4. The problem is rather simpler than that described in the last paragraph, inasmuch as link 5 of the dynamic frame is not loaded.

We will now consider how this figure is modified by the

introduction of friction. Beginning with link 1*a* in Fig. 38*b*, we can draw line 5 making the stated angle with the surface of the guide bars, and lines L_a L_b which are identical with the lines of the same name in Fig. 37*a*. We must, however, subdivide L_b in a new ratio, for it is clear that the line of bearing pressure 2*a* will pass over the friction circle, and so that the trial point where it intersects line 5 will be some distance to the right; 2β must pass under its friction circle, and the point which 2β must pass through on the left must obviously be near the bottom of its friction circle. Draw the line 2, joining two trial points, refer L_b to the two ends of link 2, or, in other words, subdivide L_b, Fig. 38*b*, in the ratio in which L_b subdivides link 2; from the point of subdivision in Fig. 38*b*, draw 2 parallel to 2 in Fig. 38*a*. The intersection of 2 and 5 gives the point *o*. The piece *c* is held in equilibrium by four forces L_c, 6*a*, 2β, and the reaction from the bearing: the intersection of 2β with 6 (Fig. 38*a*) gives one point through which the loaded link 3 must pass, and our trial point for the other end must obviously be a little to the left of the friction circle at the main bearing, where the tangent 3β cuts line 1, and this tangent 3β must be a little less steep than the line 3β in Fig. 37*b*. The load L_c is now to be subdivided between the two ends of the loaded link 3, the two components being l_c and l'_c; the polygon of the forces in equilibrium at the upper right-hand end of 3 can now be drawn in Fig. 38*b*, these are 2β, l_c, 6*a* and 3; the two former are known and the directions of the two latter; the polygon can therefore be drawn with the lines arranged in the order named, and thus the magnitude of 6*a* can be determined. The problem is now solved, but if we wish to complete our drawing of the frame, we must rearrange the last drawn polygon, so that the forces in the reciprocal figure come in their natural order as shown by the full lines in Fig. 38*b*, then finishing L_c we can, as in Figs. 37*a* and 37*b*, complete the reciprocal figure and frame without difficulty. If we have chosen our trial points well, the lines of the frame will be tangent to the friction circles. If they cut these or do not touch them, we must correct our choice of the trial points until the desired figure is found. As the friction circles are generally very small, a single trial is generally sufficient for a draughtsman who has mastered the theory. The result is really remarkable.

When the loads have been determined a reciprocal figure of nine lines enables us to ascertain the true relation between effort and resistance in a horizontal direct-acting steam-engine, taking into

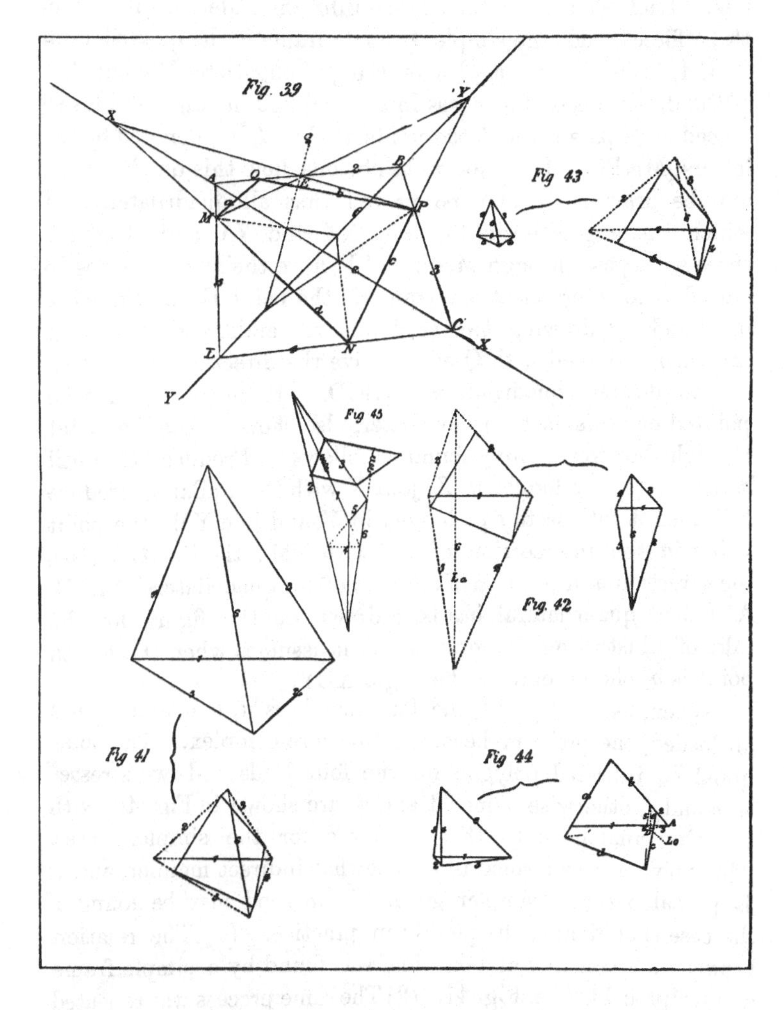

account the weight, inertia, and friction of every part of the simple train of joints and elements.

30. *Loaded Dynamic Frame when neither* e *nor* f *are attached to the Extremities of other Members of the Machine.*—This case presents some geometrical peculiarities. Let the

elements of the frame be the four lines *a*, *b*, *c*, *d*, shown as thick black lines in Fig. 39, p. 323, and let the elements *e* and *f* be joined to these at points intermediate between their extremities. Each element is then in equilibrium under the action of three forces, and the simple dynamic frame is the quadrilateral 2, 3, 4, 5, having its angles on the prolongations XX and YY of the directions of the stress in *e* and *f*, and having its sides so placed as to pass through the joints *ab*, *bc*, *cd*, *da*, denoted by the letters MNPQ. It is not quite obvious how this quadrilateral may be drawn. It may be proved that all quadrilaterals of which the angles lie in the lines XX and YY and of which three sides pass through M, N, and P, have their fourth sides so placed as to intersect at one point E; the point E can therefore be found by drawing two trial quadrilaterals, and this point can then be joined with Q and so give the direction of one side of the desired quadrilateral ABCD. Professor Tait, who pointed out this fact to the writer, also showed that the point E might be more simply found as follows:—Produce MN until it intersects *e* prolonged in X, join X with P: similarly, produce NP until it intersects *f* prolonged in Y, and join YM; the point E lies in the intersection of XP with YM; the line QE gives the direction and position of one side of the quadrilateral ABCD. A second quadrilateral has been drawn on the figure for the sake of illustrating the form which it assumes when the fourth point is *q*, chosen outside the angle XEY.

When, as in Fig. 40, the four members *a*, *b*, *c*, and *d*, are all loaded, the problem becomes still more complex. The octagonal equilibrated polygon for the four loads and two stresses in *e* and *f*, otherwise named 1 and 6, are shown in Fig. 40, with lettering analogous to that employed for the simpler cases. This polygon was formed in a somewhat indirect manner, and it is probable that a simpler geometric method may be found if the case should arise frequently in practice. (1) The relation between a stress in *e* and one in *f* was found by a simple frame and reciprocal figure, Fig. 41. (2) The same process was repeated for a stress in *e*, and a stress L_a between the elements *a* and *d*, Fig. 42. (3) The process was repeated for a stress L_b between *b* and *d*, Fig. 43. (4) The process was repeated for a stress L_c between *c* and *d*, Fig. 44. (5) By addition the stress in *e* was found which was required to overcome the given stresses due to f and to the four loads. (6) The several loads were referred to the joints. (7) Polygons of force were drawn for each point, and by these polygons the inclinations of 2β, 3β, 4β, and 5β, Fig. 40, were found, the intersections of these lines with

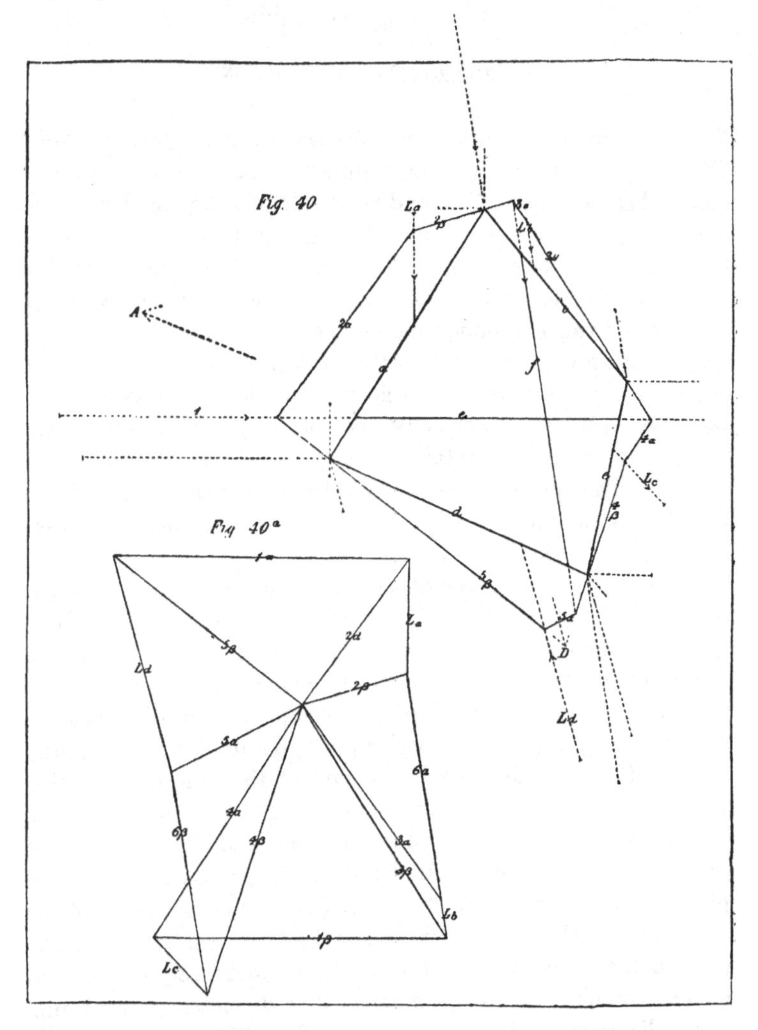

the loads (including e and f) gave the eight angles of the polygon. (8) The reciprocal figure, Fig. 40a, was drawn, by which the work was checked, and the inclinations of the sides of the polygon verified.

PART II.

THE HORIZONTAL STEAM-ENGINE.

31. IN a previous paragraph (29) the loaded dynamic frame (Fig. 38) for one position of a direct-acting horizontal steam-engine has been described, and the mode of drawing that frame explained, on the assumption that the loads L_a, L_b, L_c, L_d are known: it was stated that these loads had been calculated for one particular engine. The present paper gives the varying effort which this engine is capable of exerting at each part of its stroke with given pressures (both constant and varying) in the cylinder, and given *constant* velocities of the crank-axle. The same calculations show the efficiency of the engine at each part of the stroke, and its total efficiency under the same circumstances. This problem has, it is believed, never hitherto been solved so as to take into account all the circumstances of mass, weight, and friction.

32. No pains have been taken in the choice of the particular example. The object of the paper is not so much to draw general conclusions as to all steam engines—which, indeed, differ too much in arrangement to make this feasible—as to show how, by a method of no great complexity, full information can be obtained as to any particular engine or class of engines.

The following are the particulars of the engine:—Stroke, 16 in.; diameter of cylinder, 8 in.; length of connecting rod, 41 in.; centre of gravity of connecting rod, 20 in. from the crank pin; mass of connecting rod 34 lbs.; mass of piston and piston rod, 46 lbs.; mass of balanced crank and that part of the fly-wheel which is borne by the main bearings, 200 lbs.; the mass of the frame fixed to the earth is indefinitely large; diameter of crank-shaft in bearings, $4\frac{1}{2}$ in.; diameter of crank pin, 2·4″; diameter of pin at crosshead, 1·8″. The position of the resistance overcome (Fig. 17, Part I.) is that of a tangent to an imaginary circle of 18 in. radius, concentric with the crank

The fold-out originally positioned here is too large for reproduction in this reissue. A PDF can be downloaded from the web address given on page iv of this book, by clicking on 'Resources Available'.

FIG 46. A. Speed 1 Rev. per Second; Uniform effective pressure $p_m - p_3 = 2$ lbs. per sq. in.

A_1 unloaded no Friction; A_2 unloaded and Friction; A_3 loaded no Friction; A_4 loaded and Friction.

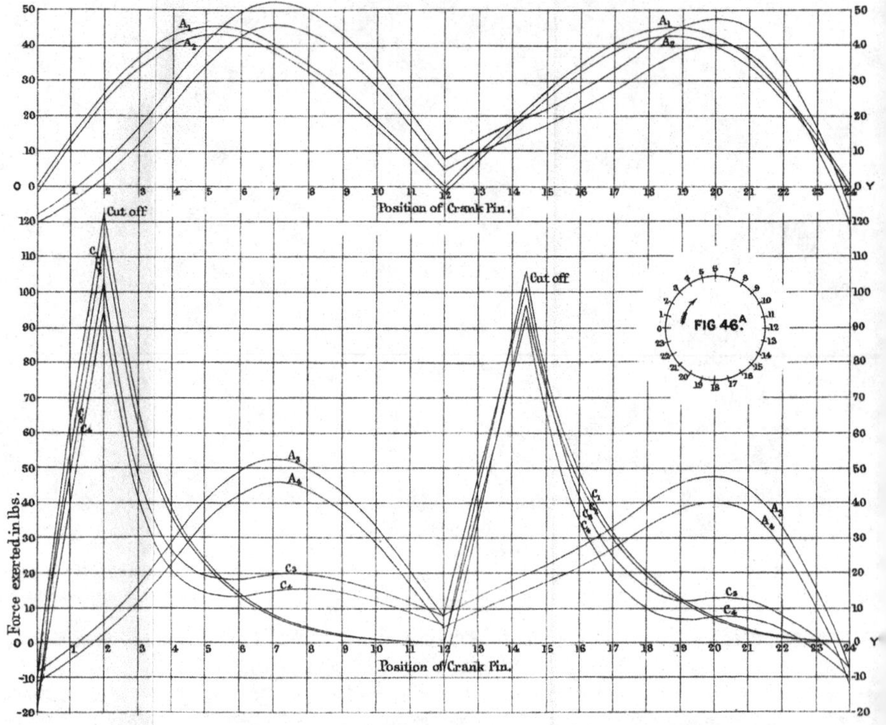

FIG 48. C. Speed 1 Rev. per Sec., $p_1 = 10$ lbs. $r = 12{\cdot}8$; p_3 0·516, giving $p_m - p_3 = 2$ lbs per sq. in.

C_1 unloaded no Friction; C_2 unloaded and Friction; C_3 loaded no Friction; C_4 loaded and Friction.

The material originally positioned here is too large for reproduction in this reissue. A PDF can be downloaded from the web address given on page iv of this book, by clicking on 'Resources Available'.

shaft, the tangent being so inclined as to cut the centre line of the engine 19·2″ from the crank-shaft centre on the side towards the piston. This line may be regarded as the direction of the resistance exerted at the teeth of a spur-wheel. The position of this line exercises a very material influence on the results obtained, which must not, therefore, as was said before, be considered as applicable to engines generally. The coefficient of friction has been taken as $= \frac{1}{12}$.

33. Eight distinct cases have been investigated. In these examples the action of the engine has been studied on four different assumptions—1st, neglecting weight, mass, and friction; 2nd, neglecting weight and mass; 3rd, neglecting friction, but taking mass and weight into account; 4th, taking mass, weight, and friction into account. A comparison of the results shows the influence of each element of the problem on the final result. These processes are obtained by what in the first part of the paper are called—1st, the simple dynamic frame; 2nd, the dynamic frame with friction; 3rd, the loaded dynamic frame; 4th, the loaded dynamic frame with friction. The examples will be called A, B, C, D, E, F, G, and H; the four assumptions will be indicated by the suffixes 1, 2, 3, 4, and it must be borne in mind that the suffix 4 corresponds to the complete solution, in which all the elements of the problem are taken into account. The results are graphically shown in a series of curves which will be called 'Effort curves.' These were drawn to a very large scale and have been reduced by a photographic process.

34. *Example A.*—Speed of engine, 1 revolution per second; effective pressure on cylinder, 2 lbs. per square inch—uniform throughout the stroke. This example has been chosen to indicate the effect of running the engine with a very low pressure and no expansion. The vertical ordinates of curve A_1, Fig. 46, indicate in pounds the total effort which the engine could exert along the assumed line of resistance, the total pressure on the cylinder being 100·5 lbs. The horizontal coordinates indicate the arc which the centre of the crank-pin has traversed, measured from the position in which the crank-pin is, at the end of its stroke, nearest to the cylinder. The coordinates of all the other curves have a similar signification. The ordinates for each curve have been calculated by separate

figures for 24 evenly spaced positions of the crank, numbered as shown in Fig. 46*a*, when the piston and connecting rod lie to the left of the crank. The portion of the curve corresponding to the forward or front stroke, between the positions 0 and 12, will be called the front branch; the portion corresponding to the backward or back stroke, between 12 and 24, will be called the back branch. In A_1 the front and back branches are symmetrical relatively to ordinate 0 or 12. The relation of the ordinates of this curve to the total pressure 100·5 lbs. on the piston depends wholly on the relative velocity of the piston and a point on the circumference of a circle 3 feet diameter concentric with the crank shaft. There are two dead points at 0 and 12, where, for an infinitely short arc, no resistance could be overcome.

In curve A_2 the front and back branches are sensibly symmetrical in the present example, although not rigidly so. Instead of a mere dead point we have now two dead arcs of different lengths, the longest being at the end of the front and beginning of the back strokes. These dead arcs together last for about 2·7 per cent. of each revolution. The negative ordinates throughout these dead arcs indicate that the engine, instead of driving, must be driven. The area enclosed by the curves above the line OY measures the work which the steam can do. The area enclosed below the line OY in A_2 measures the work which must be done (not by the steam) to pull the engine through the dead arcs. This work, in practical cases, is done by the fly-wheel, which, if large enough, might do the necessary work without allowing the speed to fluctuate sensibly —a condition assumed throughout the calculations.

The whole work done by the steam is $100{\cdot}5 \times 16 \times 12$, or 3,216 inch lbs. The area of A_1 was found to be 3,210 inch lbs., showing the error due to defective drawing to be very trifling; the area of A_2 (being the arithmetical difference between the positive and negative areas) is 2,974 inch lbs. The efficiency on assumption 2nd is therefore $\frac{2{,}974}{3{,}210} = 0{\cdot}927$.

In curve A_3 the front and back branches are no longer even approximately symmetrical to one another. There is no dead point at the end of the front stroke, and there is a long dead arc at the end of the back stroke. The maximum effort is greater

FIG 47. B. Speed 4 Rev. per Sec. Uniform effective pressure $p_a - p_b = 2$ lbs. per sq. in.
B_1 same as A_1; B_2 same as A_2; B_3 loaded no Friction; B_4 loaded and Friction.

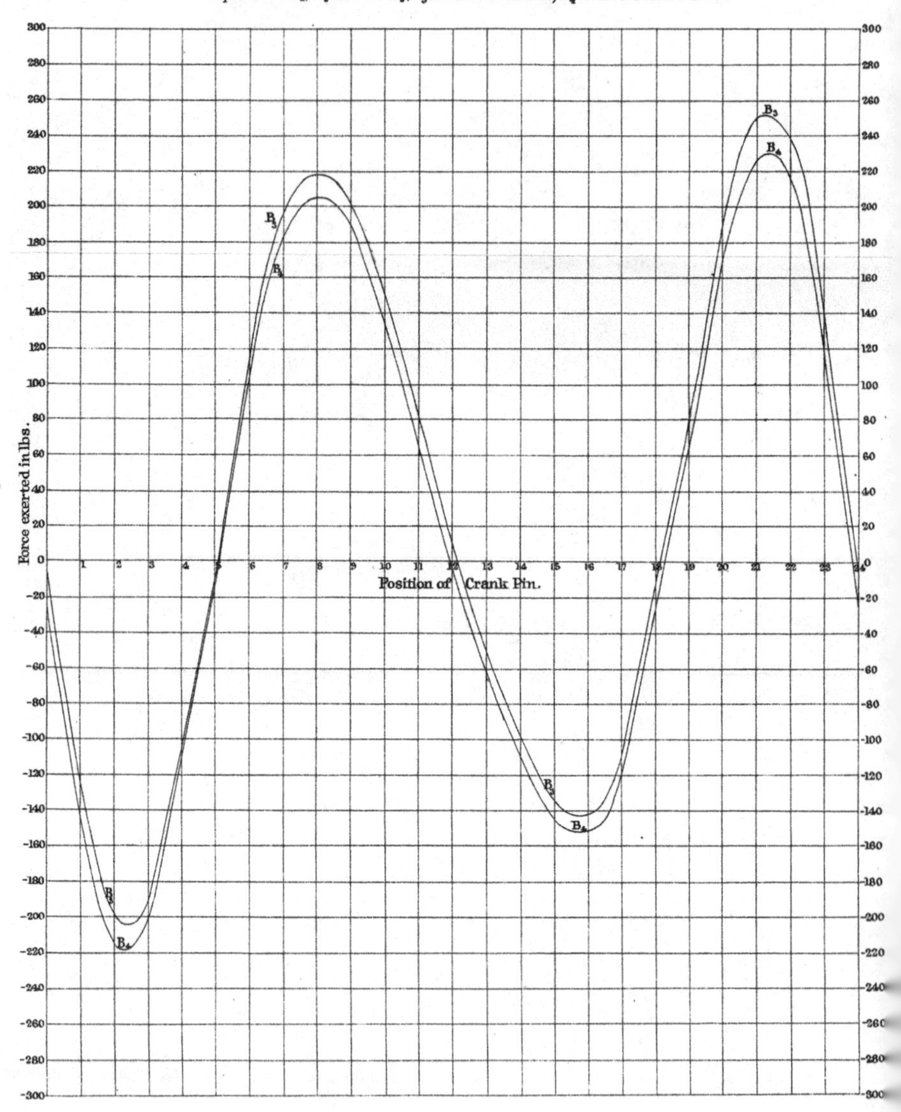

The material originally positioned here is too large for reproduction in this reissue. A PDF can be downloaded from the web address given on page iv of this book, by clicking on 'Resources Available'.

in the front than in the back stroke. These effects are due to the weight and mass of the connecting rod. A counterbalance might be so placed as to make the front and back branches resemble one another much more closely. The area of the curve A_3 (calculated as for A_2) is 3,212 inch lbs., the total error due to imperfect draughtsmanship being only 4 inch lbs.

Curve A_4 resembles A_3 in its general outline, and when these are compared with A_1 and A_2 they show the effect of the weight and mass of the parts in increasing friction. The increase in the loss due to friction is not even approximately constant throughout the stroke, but is much greater when the crank is nearly at right angles to the centre line of the engine. The loss is greatest during the back stroke, producing inequality between the useful work done during the back and front strokes. The causes of each departure from symmetry in all these curves can be followed on the diagrams of the dynamic frames, but this paper would be unduly extended if these were all to be printed. Moreover, such directions have already been given for drawing these as will enable any one to investigate the cause of each effect now described or shown on the curves. The area of A_4 is 2,602 inch lbs., so that the true efficiency of the engine running at this speed, and with this low pressure of steam, is $\frac{2,602}{3,212} = 0{\cdot}810$.

35. *Example B*, Fig. 47.—The effect of the resistance of the masses to acceleration as distinguished from the effect of the weights of the parts might have been exhibited by drawing effort curves—1st, on the assumption that although the parts resisted acceleration they had no weight; and, 2nd, on the assumption that although the parts had weight they offered no resistance to acceleration, or were moving at an infinitely slow speed; then, comparing these curves with A_4, we should have seen the effect of each element of the problem. This, however, was thought unnecessary, because the effect of the resistance of the masses to acceleration is strikingly shown by the curves B_3 and B_4, being the effort curves of the same engine, with the same pressure in the cylinder, but running four times faster. When making one revolution per second, the resistances to acceleration are smaller than the weights of the elements in motion, as may be seen in Table I. of the Appendix,

where these forces are given for each position. When the speed is increased to 4 revolutions per second, these forces are multiplied by sixteen, and their effect is then much greater than that of the weight of the parts and suffices completely to change the character of the effort curve. The negative portion of the curve lasts for nearly half the revolution of the crank, so that for nearly half of each revolution the fly-wheel would have to pull the engine round. This is true both for B_3 and B_4—for the curves without and with friction—and is simply due to the fact that during the first half of each stroke the reciprocating masses are being positively accelerated. The positive ordinates during the period when the engine is driving of course exceed those during which the engine is being driven; so that for curve B_3 the balance of positive area ought to be 3,216 inch lbs. It actually is, on the drawing, 3,256, the excess being due to small errors in drawing and computation. The area of B_4 is only 1,744 inch lbs.

The efficiency, therefore, has sunk to 0·536; almost half the power of the engine is taken up in driving itself; the pressures on the joints caused by resistance to acceleration have at this high speed greatly increased the loss due to friction. The inequality between back and front strokes is also very great, the area of the front branch being 953 inch lbs., that of the back branch only 791. The loss due to friction sinks, however, almost to nothing at one point of the front stroke, very near the place where, in Example A_4, it was a maximum. This may serve as a warning against hasty generalisations. In curve B_4 there are sensible sudden changes of efficiency at points 0 and 12, due to a sudden change in the position of the points where the elements bear on one another.

36. *Example C*, Fig. 48.—Example C was selected with the object of ascertaining how far the efficiency is affected by using the steam expansively instead of admitting it throughout the stroke. There is a very general idea that the sudden shock, as it is called, of admitting steam at a much higher pressure for a short time at the beginning of the stroke must diminish the efficiency of an engine. This is not so in the present example. An imaginary indicator diagram was selected for Example C, drawn on the supposition that the steam was

FIG 49. D. Speed 4 Rev. per Sec., P_1=10 lbs, r=12·8, p_3=0·516, giving p_2–p_3=2 lbs per sq. in.
D_1 same as C_1. D_2 same as C_2. D_3 Loaded, no Friction; D_4 Loaded and Friction.

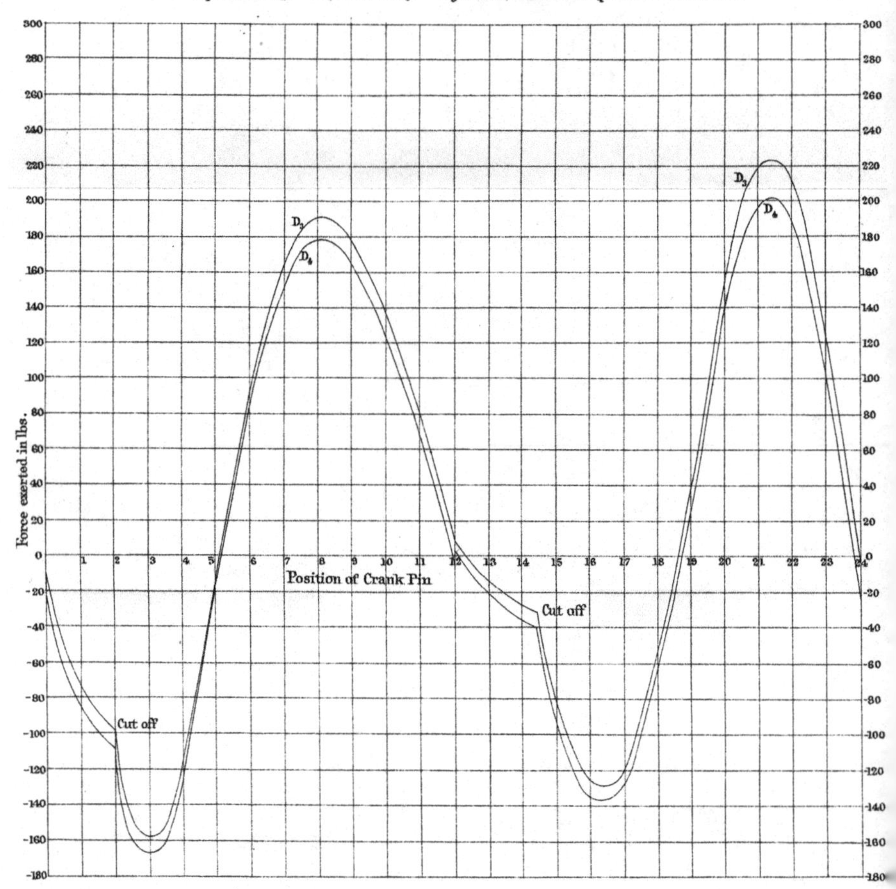

The material originally positioned here is too large for reproduction in this reissue. A PDF can be downloaded from the web address given on page iv of this book, by clicking on 'Resources Available'.

first admitted at a constant pressure of 10 lbs. per square inch; that r, the reciprocal of the fraction of the stroke during which steam enters at a constant pressure, was 12·8; that the steam expanded, according to an adiabatic curve, and was suddenly released at the end of the stroke, and that during the return stroke the back pressure was 0·516. These data give a total effective initial pressure of 476·7 lbs.; a total effective final pressure of 3·7 lbs.; and a mean effective intensity of pressure of 2 lbs. per square inch, as in Examples A and B. The work done by the steam per stroke is, therefore, as before, 3,212 inch lbs. The speed of the engine is taken, as for curve A, at 1 revolution per second. Curves C_3 and C_4 give the effects with and without friction; the area of C_3 is 3,223 inch lbs., showing a very small error of execution; the area of C_4 is 2,640; the ratio of these values gives 0·819 as the efficiency of the engine worked in this way. This efficiency is actually a little higher than that of curve A, which is repeated in this figure to allow it to be more readily compared with C. A similar result is to be observed in curve D, Fig. 49, constructed from the same data, but at the high speed of 4 revolutions per second. The area of D_3 is 3,270 inch lbs., that of D_4 1,965 inch lbs., giving an efficiency of 0·601—a value sensibly higher than that derived from curve B when the steam was admitted throughout the stroke. The errors due to imperfect drawing have generally the result of slightly increasing the effort, and the error in curve D_3 for this particular drawing (3,270 over 3,216) is nearly 1·7 per cent. The liability to error is much increased when, as here, a large portion of the area is negative. If this percentage were reckoned on the arithmetical sum of the areas, instead of on their difference, it would be insignificant. Notwithstanding this inevitable imperfection, there is every reason to expect that the errors in curves D_3 and D_4 resemble one another; and we have the less reason to suspect the accuracy of the conclusion, because we can see that since the tendency of resistance to acceleration during the beginning of each stroke is to diminish the effort, while that of a large initial pressure is to increase it, the two tendencies counteract one another without causing pressure on the main bearings or crank pin. Thus, in curves B_3 and B_4 we found that the loss due to friction at positions 5 and 18 was much reduced from this cause. In curves C_3 and C_4 we have this useful result of the mass of the moving parts without the reversal of the stresses which brings down the efficiency in B_3 and B_4. Similarly, in curves D_3 and D_4 the effect of the high initial pressure is to prevent the negative parts of the curve from falling so low as in Example B. The extremely low efficiency of A_4, B_4, and C_4 illustrates the evil effect of using a large and heavy engine running with small mean pressures of steam. We not unfrequently see large engines ordered for factories, mines, or water-works to allow for subsequent extension, or for large variations in the work required at different times. The above cases show how very serious the loss may be when an engine is habitually worked much below its power. The cases are no doubt extreme, but they show the tendency of the practice.

37. *Example E*, Figs. 50 and 51. The four cases hitherto analysed are not, strictly speaking, practical cases: they were chosen so as to bring into prominence the effects of friction and high speed, separately and combined. These effects are obviously most marked when the engine is running lightly loaded. We will now consider a more practical example. The curves E correspond to a speed of 1 revolution per second, and to an indicator diagram drawn as follows:—Initial intensity of pressure, 50 lbs. per square inch, $r = 5{\cdot}93$; back pressure, 3 lbs. per square inch; mean effective pressure, 18·095 lbs. per square inch; work done by steam, 30,640 inch lbs. per revolution.

E_1 is the effort curve, neglecting mass, weight, and friction.

E_2 is the effort curve, taking the friction into account which results from effort and resistance, but neglecting that due to weight and mass; the area of E_1 is 30,590 inch lbs., the error being 50 inch lbs. The area of E_2 is 28,400 inch lbs., making the efficiency 0·928 on this hypothesis.

E_3 is the effort curve, taking the mass and weight into account, but neglecting friction. The area is 30,700, the error being 110 inch lbs. in the drawings.

E_4 is the effort curve, taking mass, weight, and friction into account—in other words, the true effort curve. Its area is 28,030 inch lbs., giving an efficiency of 0·913, or a little less than that calculated for E_1. Now that the steam pressures are

FIG 50. E. Speed 1 Rev. per Sec., p_1=50 lbs., r=5·93, p_3=3 lbs., giving p_m−p_3=18·905 lbs per sq. in.
E_1 unloaded, no Friction; E_2 unloaded and Friction.

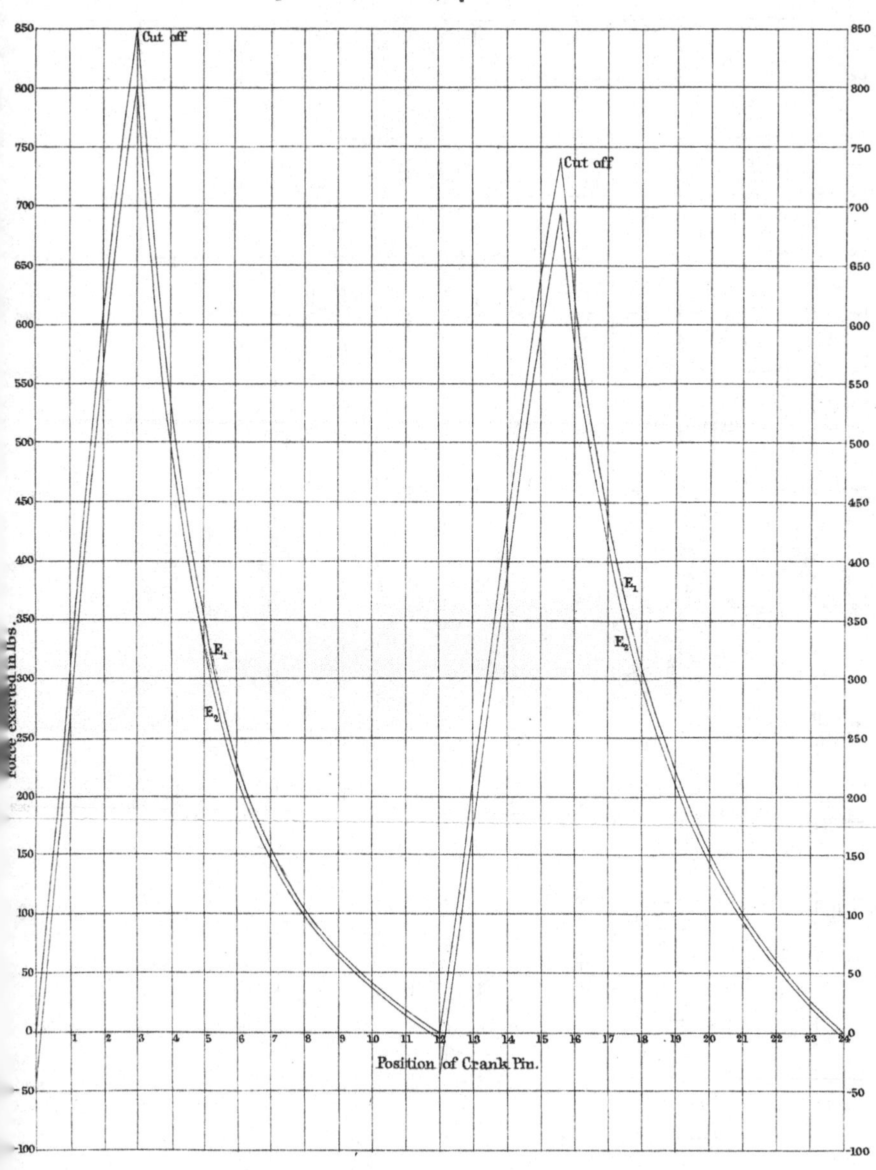

The material originally positioned here is too large for reproduction in this reissue. A PDF can be downloaded from the web address given on page iv of this book, by clicking on 'Resources Available'.

FIG 51. E. Speed 1 Rev. per Sec., p_1 = 50 lbs., r = 5·93, p_3 = 3 lbs., giving p_m − p_3 = 18·905 lbs. per sq. in.
E_3 loaded no Friction; E_4 loaded and Friction.

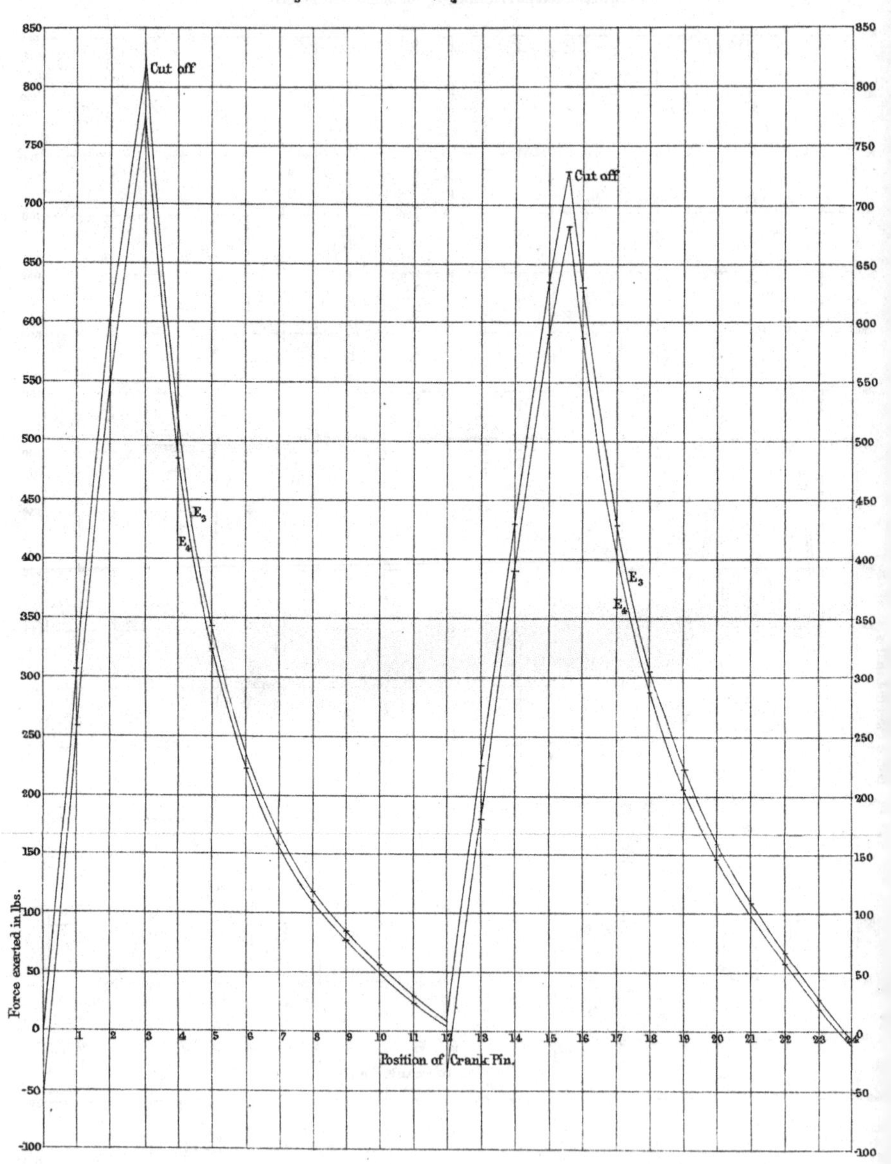

The material originally positioned here is too large for reproduction in this reissue. A PDF can be downloaded from the web address given on page iv of this book, by clicking on 'Resources Available'.

FIG 52. F. Speed 4 Rev. per Sec., p_1=50 lbs., r=5·93, p_3=3 lbs., giving p_m-p_3=18·905 lbs. per sq. in.

F_1 same as E_1, F_2 same as E_2; F_3 loaded no Friction; F_4 loaded and Friction.

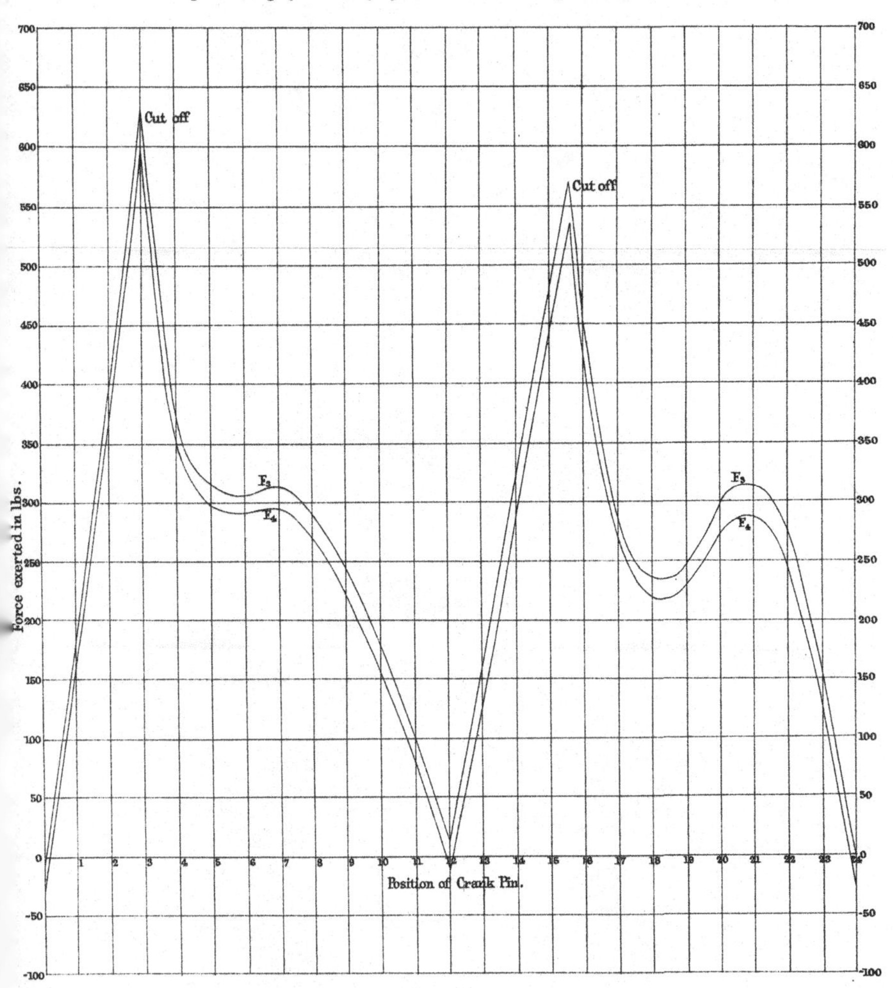

The material originally positioned here is too large for reproduction in this reissue. A PDF can be downloaded from the web address given on page iv of this book, by clicking on 'Resources Available'.

large, the loss due to the weight and mass of the parts is at this speed comparatively insignificant. The total inevitable loss due to friction in the machine alone, even excluding the accidental friction due to tightness of piston and glands, and the power required to work the valve gear, is nearly 9 per cent. of the whole indicated horse-power. This example shows the complete fallacy of the experimental method sometimes adopted with the object of testing the efficiency of an engine. The engine is run with no resistance, and the indicated H.P. observed. This is assumed approximately to represent the loss due to the engine itself when doing useful work against a large resistance; but from Example A we see that the power required to overcome the friction in the engine when a small resistance was being overcome was only 3,210–2,974 inch lbs., or 236 inch lbs.; whereas in Example E the power required is 1,190 inch lbs., or nearly five times as much. This ratio would be somewhat diminished by the constant accidental resistances in glands, etc., and by the power required for valves and pumps. It must, however, always remain very large.

In E_1 and E_2 the discontinuity of the curves at positions 0 and 12 is very marked. There must also be a slight break near positions 6 and 18, due to the change in the bearing-points on the crosshead pin; but the frictional loss at this point is so insignificant that the break in the curve is not sensible. There is very little difference in general character between curves E_1 E_2 and E_3 E_4. Mass and weight play a very small part in the general result.

38. *Example F*, Fig. 52.—Example F is important and instructive, showing the result of running the engine at 4 revolutions per second instead of 1 revolution per second, but maintaining the same indicator diagram as in Example E, the mean effective pressure being 18·905 lbs. as before. We might expect this high speed to diminish the efficiency, as in Examples B and D, but this is not the case. Curves F_1 and F_2 are identical with E_1 and E_2, and are therefore omitted; curves F_3 and F_4 are, however, very different in character from E_3 and E_4. The resistance to acceleration causes the effort to be much more uniformly distributed. It greatly diminishes the maximum pressure, and largely increases the pressure during the second

half of each stroke. The resistance to acceleration is not, however, so large relatively to the steam pressure as to produce negative ordinates, such as were found in cases B and D. The action of the reciprocating masses is, therefore, on the whole, beneficial, and we find the efficiency of the engine to be 0·917, or somewhat higher than in Example E. Much care has been taken in the computations to check this rather remarkable result, but there seems no reason to doubt its accuracy. The important conclusion to be drawn does not, however, depend on trifling differences—such as that between 0·917 and 0·913. It is that high speeds do not necessarily entail low efficiency, so that by judicious arrangement of the masses we can have small engines running rapidly with high expansion which, so far as frictional resistances are concerned, may be at least as efficient as the same engines running slowly, and necessarily more efficient than larger engines doing the same work.

39. *Examples G* and *H*, Figs. 53 and 54.—The unexpected results obtained rendered it desirable to investigate the case of infinitely slow speed, or that in which the effect of inertia was wholly disregarded. Curves G_3 and G_4 show the efforts with and without friction for the practical distribution of steam assumed in cases E and F. The total work done by the steam is 30,770 inch. lbs., the useful work 28,120, and the efficiency 0·914 —a result hardly differing sensibly from that obtained with the high speed of 4 revolutions per second.

40. *Example H.*—Lastly, curves H_3 and H_4 show the efforts with and without friction when the engine is modified so as to make the connecting rod only 28 in. long, or 3½ times the stroke. The weight of the new rod is taken as 28 lbs., and its centre of gravity 14″ from the crank end, and its radius of gyration about the crosshead 18·83 inches. The speed assumed was 1 revolution per second.[1] As calculated from the areas of the curves the whole work done by the steam is 30,740 inch lbs., the useful work 27,920 inch lbs., and the efficiency 0·908.

This result is of considerable practical value, showing that the efficiency of the engine is hardly diminished by shortening the connecting rod to this extent.

[1] The initial steam pressure, the ratio of expansion, and the back pressure were assumed to be the same as in Example E.

FIG 53. G. Speed infinitely slow; p_1=50 lbs, r=5·93, p_3=3 lbs.
p_m-p_3=18·905 lbs per Square inch.

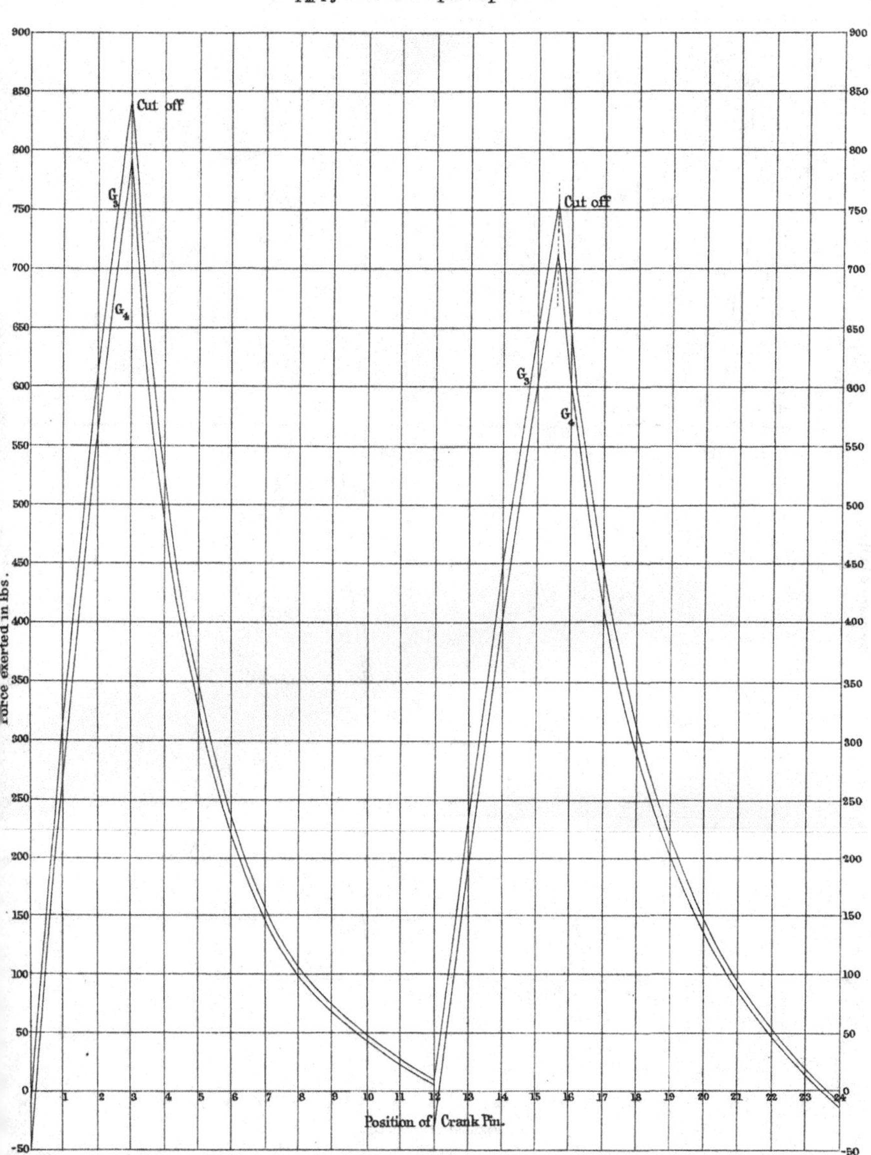

The material originally positioned here is too large for reproduction in this reissue. A PDF can be downloaded from the web address given on page iv of this book, by clicking on 'Resources Available'.

FIG 54. H. Distribution of Steam & Speed as for E; but connecting rod reduced to 3½ times Stroke.

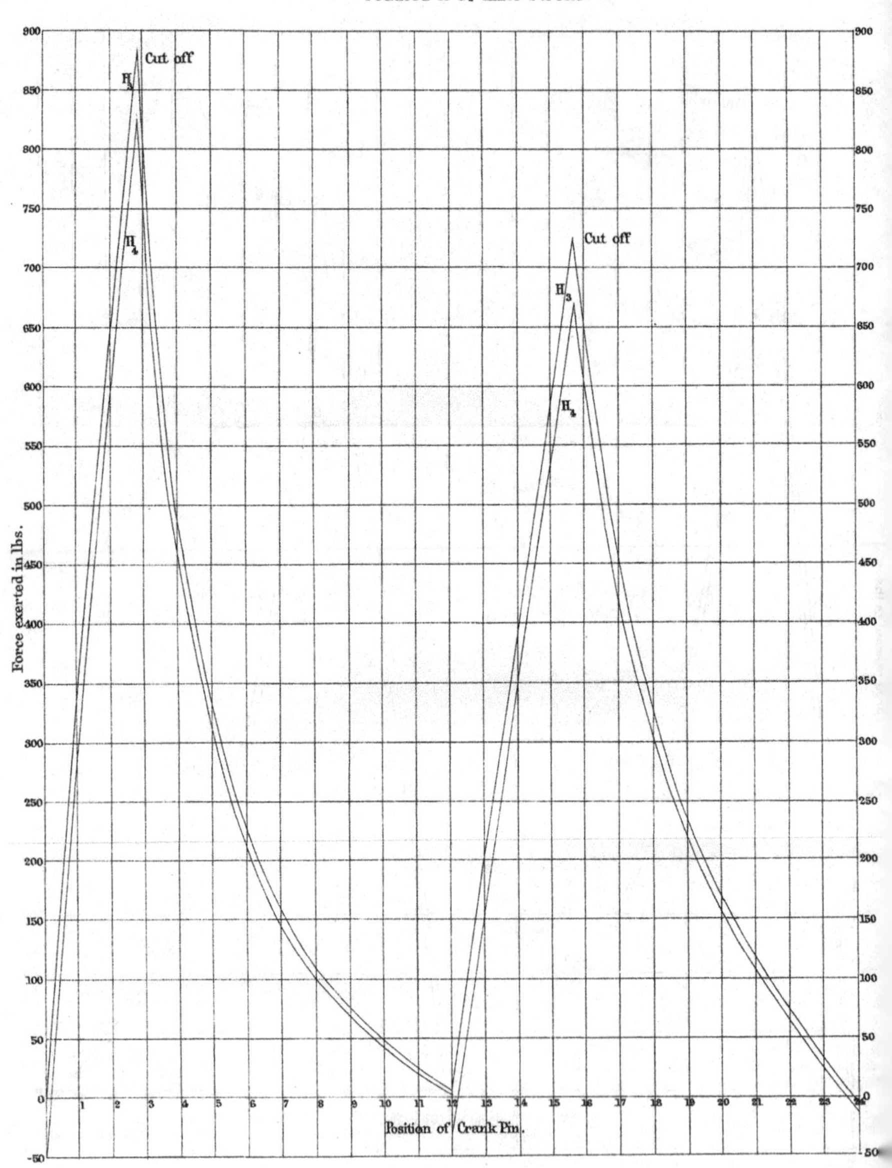

The material originally positioned here is too large for reproduction in this reissue. A PDF can be downloaded from the web address given on page iv of this book, by clicking on 'Resources Available'.

TABLE I.—*Abstract of Results.*

	Example	Pressure			Speed	Efficiency
Engine described in § 34	A_1 A_2 A_3 A_4	Low	Uniform . .	2 lbs. per square inch.	Slow, 1 revolution per second.	·927 ·810
	B_1 B_2 B_3 B_4		Uniform . .	2 lbs. per square inch.	Fast, 4 revolutions ,,	·927 ·536
	C_1 C_2 C_3 C_4		Expansion . .	Cut off $\frac{1}{12·8}$; mean effective = 2 lbs. per square inch.	Slow, 1 revolution ,	·907 ·819
	D_1 D_2 D_3 D_4		Expansion . .	Cut off $\frac{1}{12·8}$; mean effective = 2 lbs. per square inch.	Fast, 4 revolutions ,,	·907 ·601
	E_1 E_2 E_3 E_4	Practical	Expansion . .	Cut off $\frac{1}{5·93}$; mean effective pressure 18·905 lbs. per square inch.	Slow, 1 revolution ,,	·928 ·913
	F_1 F_2 F_3 F_4		Expansion . .		Fast, 4 revolutions ,,	·928 ·917
	G_1 G_2 G_3 G_4		Expansion . .		Infinitely slow, 0.	·928 ·914
Engine described in § 40.	H_3 H_4		Expansion . .	do.	Slow, 1 revolution ,,	·908

41. *General Conclusions.*—The most important conclusion to be drawn from the foregoing examples is, that the investigation of the efficiency of any given direct-acting engine is rendered comparatively easy by the new method. Next in importance may be ranked the warning not to judge hastily as to the result of any given modification in proportions or speed. The differences introduced by a change of speed are especially remarkable, and show how futile reasoning must be as to relative efforts and resistances or to stresses on the various parts of the engine, or to the efficiency of any design when the inertia of the masses is left out of the calculation.

Further conclusions may be drawn as follows:—1st, that high speeds do not necessarily involve small efficiency; 2nd, that a short connecting rod is not very disadvantageous; 3rd, that expansive working, even when carried to great lengths, does not necessarily involve a loss of efficiency.

Table I. gives an abstract of the numerical values obtained with this particular engine, but the reader must be warned against considering these results as generally applicable to other engines of different proportions.

The necessary calculations and drawings for this paper have been made by my assistant, Mr. J. A. Ewing, to whom I am much indebted both for the accuracy with which the work has been done, and for the interest he has shown in adopting the novel method of investigation.

The Appendix contains data which will allow the reader to verify the results arrived at without going through all the calculations.

This Appendix has been drawn up by Mr. Ewing.

APPENDIX TO PART II.

On the Application of Graphic Methods to the Determination of the Efficiency of Machinery.

To determine the forces required for the acceleration of the piston and connecting rod in each position of the engine, we must know the acceleration of the piston, and the angular velocity and angular acceleration of the connecting rod. If A C be the crank, and C B the connecting rod, and if A B be called x, then $\frac{d^2x}{dt^2}$ is the

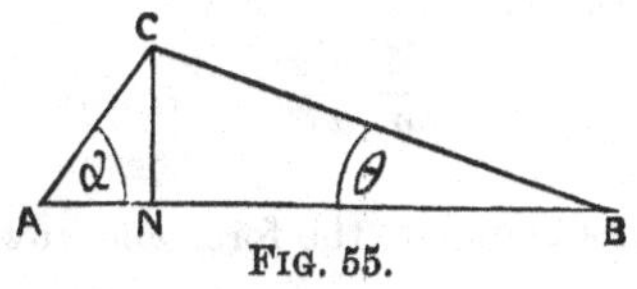

FIG. 55.

acceleration of the piston; and if the angle C B A be called θ, $\frac{d\theta}{dt}$ is the angular velocity of the connecting rod, and $\frac{d^2\theta}{dt^2}$ is its angular acceleration.

If the angle which the crank makes with A B be called α, $\frac{d\alpha}{dt}$ is the angular velocity of the crank.

Then, calling the crank radius r, and the length of the connecting rod l, we have

$$\theta = \sin^{-1}\frac{\mathrm{CN}}{\mathrm{CB}} = \sin^{-1}\left(\frac{r\sin\alpha}{l}\right),$$

$$\frac{d\theta}{dt} = \frac{r\cos\alpha}{\sqrt{l^2 - r^2\sin^2\alpha}}\cdot\frac{d\alpha}{dt}.$$

Differentiating again, and remembering that $\frac{d^2\alpha}{dt^2} = 0$, we obtain finally

$$\frac{d^2\theta}{dt^2} = -\frac{r\sin\alpha\,(l^2 - r^2)}{(l^2 - r^2\sin^2\alpha)^{\frac{3}{2}}}\left(\frac{d\alpha}{dt}\right)^2.$$

Again

$$\mathrm{AB} = \mathrm{AN} + \mathrm{BN},$$

$$x = r\cos\alpha + l\cos\theta,$$

$$\frac{dx}{dt} = -r\sin\alpha\frac{d\alpha}{dt} - l\sin\theta\frac{d\theta}{dt},$$

and $$\frac{d^2x}{dt^2} = -r\cos\alpha\left(\frac{d\alpha}{dt}\right)^2 - l\cos\theta\left(\frac{d\theta}{dt}\right)^2 - l\sin\theta\frac{d^2\theta}{dt^2}.$$

Substituting for $\frac{d\theta}{dt}$ and $\frac{d^2\theta}{dt^2}$ their values as determined above, and putting $r\sin\alpha$ for $l\sin\theta$, and $\sqrt{l^2 - r^2\sin^2\alpha}$ for $l\cos\theta$, we obtain finally

$$\frac{d^2x}{dt^2} = -r\left(\frac{d\alpha}{dt}\right)^2\left\{\cos\alpha + \frac{r l^2\cos 2\alpha + r^3\sin^4\alpha}{(l^2 - r^2\sin^2\alpha)^{\frac{3}{2}}}\right\}.$$

The forces required to accelerate the piston and the connecting rod may now be calculated as follows :—

For the piston.—If M be the mass of the piston and piston rod in lbs., the force in lbs. is

$$\frac{M}{g}\frac{d^2x}{dt^2}.$$

When this quantity is negative, the force acts towards the centre of the crank shaft.

For the connecting rod.—The motion may be looked at as a translation of the whole rod in the direction of motion of the piston, combined with a rotation of the rod about the crosshead. Hence the force producing acceleration is the resultant of three components :—F_1, the force required for the linear acceleration in the direction of motion of the piston ; F_2, the force required for rotation about the crosshead at the angular velocity which the rod has at

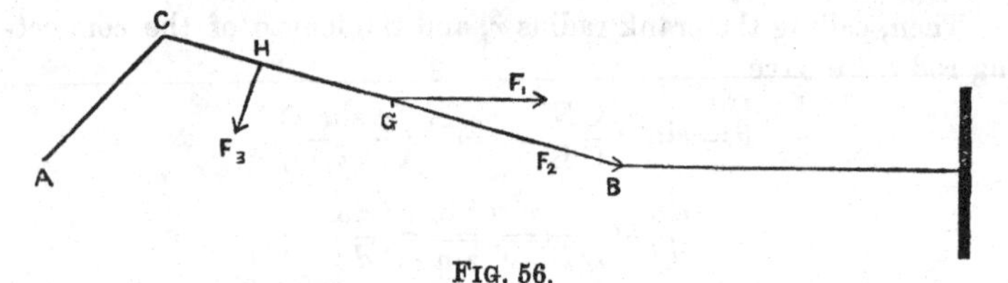

FIG. 56.

the instant under consideration. This acts towards the centre of rotation, and is equal and opposite to the so-called 'centrifugal force ;' and, lastly, F_3, the force required to give the rod the angular acceleration which it has at the given instant.

Let M′ be the mass of the connecting rod in lbs., and G its centre of mass, distant l_0 from the crosshead B ; also let k be its radius of gyration about B. Then the first component mentioned above, or F_1, is

$$\frac{M'}{g}\frac{d^2x}{dt^2},$$

and acts through G parallel to the path of the piston.

The second component, F_2, is

$$\frac{M'l_0}{g}\left(\frac{d\theta}{d}\right)^2,$$

(*Note.*—In Figs. 55 and 56 the engine has been represented as seen from behind, if the engine in the previous figures be considered as viewed from the front.)

and acts along the line of the rod towards B. The third component, F_3, is

$$\frac{M' l_0}{g} \cdot \frac{d^2\theta}{dt^2},$$

and acts at right angles to the rod through the centre of the percussion H, which is at a distance $\frac{k^2}{l_0}$ from B.

These three forces may be most conveniently compounded by shifting F_3 to G, and introducing a couple whose moment is $F_3 \cdot HG$, or

$$\frac{M'(k^2 - l_0^2)}{g} \frac{d^2\theta}{dt^2}, \text{ or } \frac{I}{g} \frac{d^2\theta}{dt^2},$$

where I is the moment of inertia of the rod about G.

Forces equal and opposite to those components form the several parts of the whole resistance to acceleration, and when these are combined with the weight, the resultant is the whole load on the element. (See Part I., § 27.) This composition is effected graphically, and a single force is obtained acting through G, which has then to be shifted parallel to itself to such a distance as to give rise to the moment $\frac{I}{g} \frac{d^2\theta}{dt^2}$. This process gives a single force of determinate magnitude and position, as the load on the element in each position of the engine.

The following tables show the component parts of the acceleration and force in the cases which have been actually examined. In Table II. the connecting rod is 41″ long, and its mass is 34 lbs.; l_0 is 20″, and $\frac{k^2}{l_0}$ is 34·08 inches. In Table III. the connecting rod is 28″ long, and its mass is 28 lbs.; l_0 is 14″, and $\frac{k^2}{l_0}$ is 25·32 inches. In both cases the crank radius is 8″, the mass of the piston and piston rod 46 lbs., and the speed is 1 revolution per second, whence

$$\frac{d\alpha}{dt} = 2\pi.$$

The positions of the crank in column 1 are numbered thus:—Positions 0 and 24 are the same, and correspond to $\alpha = 0$. Position 12 corresponds to $\alpha = 180°$. The movement in Fig. 56 is contrary to that of the hands of a watch, and the interval between two successive positions is 15°. Looked at from the other side, as in earlier figures of the engine, the movement would be in the direction of the hands of a watch.

TABLE II.

Connecting rod, 41″. Crank, 8″. Mass of connecting rod, 34 lbs. Mass of piston and piston rod, 46 lbs.
Speed, 1 revolution per second. $l_o = 21''$. I = 9,576 (inches and lbs.)

	Piston, mass = M		Connecting rod, mass = M′					
Position of Crank	$\frac{d^2x}{dt^2}$ feet and seconds	$\frac{M}{g}\frac{d^2x}{dt^2}$ lbs.	$\frac{d\theta}{dt}$ radians and seconds	$\frac{d^2\theta}{dt^2}$ radians and seconds	F_1 or $\frac{M'}{g}\frac{d^2x}{dt^2}$ lbs.	F_2 or $\frac{M'l_o}{g}\left(\frac{d\theta}{dt}\right)^2$ lbs.	F_3 or $\frac{M'l_o}{g}\frac{d^2\theta}{dt^2}$ lbs.	Moment $\frac{I}{g}\frac{d^2\theta}{dt^2}$ lbs. and inches
0	−31·45	−44·88	1·226	0	−33·21	2·78	0	0
1	−29·94	−42·77	1·186	−1·925	−31·61	2·60	− 3·56	− 47·7
2	−25·50	−36·46	1·067	−3·759	−26·93	2·10	− 6·95	− 93·2
3	−18·67	−26·66	0·876	−5·385	−19·71	1·42	− 9·95	−133·5
4	−10·60	−15·13	0·622	−6·702	−10·69	0·71	−12·38	−166·1
5	− 2·29	− 3·27	0·323	−7·557	− 2·42	0·19	−13·97	−187·3
6	5·24	7·48	0	−7·854	5·53	0	−14·51	−194·6
7	11·86	16·36	−0·323	−7·557	12·52	0·19	−13·97	−187·3
8	15·72	22·46	−0·622	−6·702	16·60	0·71	−12·38	−166·1
9	18·56	26·52	−0·876	−5·385	19·60	1·42	− 9·95	−133·5
10	20·14	28·76	−1·067	−3·759	21·27	2·10	− 6·95	− 93·2
11	20·91	29·87	−1·186	−1·925	22·08	2·60	− 3·56	− 47·7
12	21·19	30·27	−1·226	0	22·37	2·78	0	0
13	20·91	29·87	−1·186	1·925	22·08	2·60	3·56	47·7
14	20·14	28·76	−1·067	3·759	21·27	2·10	6·95	93·2
15	18·56	26·52	−0·876	5·385	19·60	1·42	9·95	133·5
16	15·72	22·46	−0·622	6·702	16·60	0·71	12·38	166·1
17	11·86	16·36	−0·323	7·557	12·52	0·19	13·97	187·3
18	5·24	7·48	0	7·854	5·53	0	14·51	194·6
19	− 2·29	− 3·27	0·323	7·557	− 2·42	0·19	13·97	187·3
20	−10·60	−15·13	0·622	6·702	−10·69	0·71	12·38	166·1
21	−18·67	−26·66	0·876	5·385	−19·71	1·42	9·95	133·5
22	−25·50	−36·46	1·067	3·759	−26·93	2·10	6·95	93·2
23	−29·94	−42·77	1·186	1·925	−31·61	2·60	3·56	47·7
24	−31·45	−44·88	1·226	0	−33·21	2·78	0	0

Connecting rod, 28″. Crank, 8″. Mass of connecting rod, 28 lbs. Mass of piston and piston rod = 46 lbs.
Speed, 1 revolution per second. $l_0 = 14''$.

	Piston, mass = M		Connecting rod, mass = M′					
Position of Crank	$\frac{d^2x}{dt^2}$ feet and seconds	$\frac{M}{g}\frac{d^2x}{dt^2}$ lbs.	$\frac{d\theta}{dt}$ radians and seconds	$\frac{d^2\theta}{dt^2}$ radians and seconds	F_1 or $\frac{M'}{g}\frac{d^2x}{dt^2}$ lbs.	F_2 or $\frac{M'l_0}{g}\left(\frac{d\theta}{dt}\right)^2$ lbs.	F_3 or $\frac{M'l_0}{g}\frac{d^2\theta}{dt^2}$ lbs.	Moment $\frac{I}{g}\frac{d^2\theta}{dt^2}$ lbs. and inches
0	−33·76	−48·23	1·795	0	−29·36	3·27	0	0
1	−31·93	−45·61	1·739	− 2·703	−27·77	3·07	− 2·75	− 31·1
2	−26·70	−38·14	1·571	− 5·342	−23·22	2·50	− 5·42	− 61·3
3	−18·81	−26·87	1·296	− 7·797	−16·36	1·70	− 7·91	− 87·5
4	− 9·41	−13·44	0·926	− 9·863	− 8·18	0·87	−10·01	−113·3
5	− 0·08	− 0·11	0·483	−11·269	− 0·07	0·24	−11·43	−129·4
6	7·85	11·21	0	−11·770	6·83	0	−11·94	−135·2
7	13·54	19·34	−0·483	−11·269	11·77	0·24	−11·43	−129·4
8	16·91	24·16	−0·926	− 9·863	14·70	0·87	−10·01	−113·3
9	18·41	26·30	−1·296	− 7·797	16·01	1·70	− 7·91	− 87·5
10	18·88	26·97	−1·571	− 5·342	16·42	2·50	− 5·42	− 61·3
11	18·91	27·01	−1·739	− 2·703	16·44	3·07	− 2·75	− 31·1
12	18·80	26·86	−1·795	0	16·35	3·27	0	0
13	18·91	27·01	−1·739	2·703	16·44	3·07	2·75	31·1
14	18·88	26·97	−1·571	5·342	16·42	2·50	5·42	61·3
15	18·41	26·30	−1·296	7·797	16·01	1·70	7·91	87·5
16	16·91	24·16	−0·926	9·863	14·70	0·87	10·01	113·3
17	13·54	19·34	−0·483	11·269	11·77	0·24	11·43	129·4
18	7·85	11·21	0	11·770	6·83	0	11·94	135·2
19	− 0·08	− 0·11	0·483	11·269	− 0·07	0·24	11·43	129·4
20	− 9·41	−13 44	0 926	9·863	− 8·18	0·87	10·01	113·3
21	−18·81	−26·87	1·296	7·797	−16·36	1·70	7·91	87·5
22	−26 70	−38·14	1·571	5·342	−23·22	2 50	5·42	61·3
23	−31·93	−45·61	1·739	2·703	−27·77	3·07	2·75	31·1
24	−33·76	−48·23	1·795	0	−29·36	3·27	0	0

The engine of Table II. was also examined when running at a speed of 4 revolutions per second. This, of course, had the effect of multiplying $\frac{d\theta}{dt}$ by four, and $\frac{d^2\theta}{dt^2}$ and $\frac{d^2x}{dt^2}$ by sixteen. The Table for this set of circumstances may, therefore, be deduced from Table II. by multiplying column 4 by four, and columns 2, 3, 5, 6, 7, 8, and 9 by sixteen.

ABSTRACTS OF
FLEEMING JENKIN'S SCIENTIFIC PAPERS

WITH A LIST OF HIS BRITISH PATENTS

ABSTRACTS OF SCIENTIFIC PAPERS.

I. *On Gutta-Percha as an Insulator at various Temperatures.* 'British Association Report for 1859,' p. 248.

II. *On the Insulating Properties of Gutta-Percha.* 'Proceedings of the Royal Society,' 1860, vol. x. p. 409.

These papers contain abstracts of experiments made at the works of Messrs. R. S. Newall and Co., to determine in absolute measure the insulation resistance of the gutta-percha coating of submarine cables, and the effect of temperature on that resistance. The author draws attention to the phenomenon of 'polarisation' or the apparent increase which the resistance of the insulator undergoes for some time after the application of the testing battery. Tables are given showing the specific resistance of the gutta-percha used in the Red Sea Cable after one and after five minutes' electrification; also of the specific resistance of pure gutta-percha, at temperatures ranging from 50° to 80° Fahrenheit.

[An account of the same experiments was printed as an Appendix to the Report of the Committee on Submarine Cables, 1859, in connection with evidence submitted to the Committee by Professor Jenkin.]

III. *On Permanent Thermoelectric Currents in Circuits of one Metal.* 'British Association Report for 1861,' p. 39; 'Chemical News,' October 26, 1861, vol. iv. p. 222.

After referring to experiments by Seebeck and Magnus, on the production of transient thermoelectric currents in circuits of one metal by bringing together cold and hot ends of a wire, the author announces the discovery that he had obtained permanent currents by looping the two ends of a wire together and heating one of the two loops, and that these currents were usually much greater when there was a loose contact between the wires than when the loops were tightly drawn together. He first observed that if one loop was heated when both were held tightly together, and then the loops were separated, a transient current was produced in the same direction as the current produced when the hot and cold ends were suddenly joined. This led him to conclude that the currents in question were due to the nearness of the ends, and that by maintaining a state of loose contact a permanent current might be established—a conclusion which was verified by the experiments. Trial was made of the comparative amounts of the effect in iron, copper, and platinum. Experiments with joints between two metals showed that the thermoelectric difference at a joint was often greatly increased by substituting loose contact for tight contact, and even the direction of the current was sometimes reversed.

IV. *On the True and False Discharge of a Coiled Electric Cable.* By Professor W. Thomson and Mr. Fleeming Jenkin; 'Philosophical Magazine,' September 1861.

Mr. Jenkin had communicated to Professor Thomson, in 1859, experimental results which showed that when a battery had been applied to the near end of a coiled cable whose distant end was connected to earth, if the near end were suddenly disconnected from the battery and put to earth through a galvanometer, the deflection of the galvanometer showed a 'false discharge,' that is, a current still entering the cable from earth, in the same direction as the current formerly supplied by the battery. In communicating this fact to the British Association in 1859, Professor Thomson had explained it as a result of the electromagnetic induction between different portions of the coil, and had anticipated that no such false discharge would be observed in a straight cable. The paper describes and discusses Jenkin's experiments which, on being repeated, showed that when the near end was put to earth the very first deflection of the galvanometer there showed a back-flow or true discharge from the cable, but this was quickly followed by a much larger opposite deflection, forming the false discharge. The theoretical conclusion that the false discharge would not be observed in submerged cables was verified by Mr. Jenkin in experiments on the Bona Cable; and still more conclusively by Mr. Webb, whose letter on the subject (to the *Engineer* of August 26, 1859) is appended to the paper.

[This paper is reprinted as Article LXXXIII. in vol. ii. of Sir W. Thomson's Mathematical and Physical Papers.]

V. *On the Retardation of Signals through long Submarine Cables.* 'British Association Report,' 1859, p. 251.

[Abstract of a part of Art. VI. below.]

VI. *Experimental Researches on the Transmission of Electric Signals through long Submarine Cables.* Part I. '*Laws of Transmission through various Lengths of one Cable;*' 'Philosophical Transactions of the Royal Society,' 1862 ; Abstract in 'Proceedings of the Royal Society,' June 19, 1862, vol. xii. p. 198.

In 1855 Sir W. Thomson had published the mathematical theory of the transmission of electric signals, and had exhibited the retarding effect of electrostatic capacity and resistance by a curve showing the gradual rise—with respect to time—of the current in the remote instrument when the end operated on was put and kept in connection with the battery. By experiments on the Red Sea Cable while stored in Messrs. Newall's works in 1859, Mr. Jenkin verified the form of this 'curve of arrival,' using as the receiving instrument the marine mirror galvanometer of Thomson, whose quickly vibrating needle allowed it to respond accurately to the variations of current. 'When the key was pressed down the spot of light remained apparently motionless for a short but sensible time, then shot along the scale, moving rapidly at first but gradually losing speed, until at last it moved very slowly to a maximum deviation, at which it remained quite still : these movements truly showed the gradual change of the received current from nothing to a maximum, a change requiring 50 seconds for its completion.' This was with a length of 2,168 knots, and experiments with other lengths of cable verified Thomson's law that the retardation varies as the square of the length. The results also demonstrated that, in accordance with the theory, the retardation is independent of the electromotive force employed to transmit signals, and that if after contact with the battery the sending end of the cable be put to earth, the rate of decrease of current at the remote end is the same after any interval as the rate of increase in the arrival curve after an equal interval from the beginning of contact with the battery. The experiments were further directed to supply practical data with regard to the possible speed of signalling through cables, a subject then attracting attention especially as bearing on the scheme for a new Atlantic line. Observations were made of the amplitude of fluctuation in the received current produced by the sending of dots and dashes (short and longer applications of the sending battery), the amplitudes being expressed as percentages of the whole strength which the received current would reach if the sending battery were kept in contact with the line for a time long enough to allow the current to become sensibly steady. A limit to the possible speed of signalling was demonstrated by sending dots so rapidly that no variation in the received current could be detected, notwithstanding the sensibility of the receiving instrument. When dots and dashes were sent alternately, and sufficiently slowly to be separately observed, the end of the dot curve was found to go lower than the end of the dash curve, a fact which limited very much the speed of sending possible with an instrument of the Morse type. This was when the interval of connection to earth was the same after both dot and dash, and it was pointed out that this defect in the signals could be remedied by making the interval of earth contact after the dash longer than that after the dot, in a definite ratio. Further, to prevent the received current from sinking during the longer intervals required for spaces, it was suggested that each of these should be produced by a number of very rapid short applications of the battery, with short earth contacts between. To carry out these ideas automatic sending became necessary, and Mr. Jenkin showed that the suggested system was practicable by an experiment in which, in place of an ordinary sending key, a strip of paper was used as sender, perforated with long and short holes which allowed electric contacts to be made for the desired lengths of time. By this means common Morse signals were successfully transmitted through the cable with much greater rapidity than would have been possible in the ordinary mode of sending. [This ingenious system of automatic sending was developed in much detail and formed the chief subject of a joint patent of Sir W. Thomson and Mr. Jenkin in 1860. The system did not, however, come into use. It was rendered less necessary by the adoption on all long submarine cables of Sir W. Thomson's mirror galvanometer, and subsequently of his Siphon Recorder, with the result that any sensible fluctuation in the received current could be followed and interpreted as a signal, irrespective of its amplitude and of its position on the curve.] A mathematical investigation by Sir W. Thomson of the relation which the amplitude of fluctuation in the received current bears to the period of the sent currents, when these are sent at regular intervals with earth contacts between, is given as an appendix to the paper. The experiments included observations of the speeds of sending of uniformly spaced short currents on various lengths of cable, at which the amplitude of the signals became (1) so small as to escape detection and (2) just capable of being received without confusion. The results agreed very exactly with the deductions from Sir W. Thomson's theory.

VII. *On the Construction of Submarine Telegraph Cables.* 'Proceedings of the Institution of Mechanical Engineers,' July 1862, p. 211.

After naming the essential features of a submarine cable, the author points out that 'for every given ratio between the cost of the materials of the insulator and of the conductor there exists a corresponding ratio between the diameters of the conductor and insulator which will give the maximum efficiency at a minimum cost; and practically the thickness of the gutta-percha is almost always in excess of this theoretical thickness; and if a constant ratio is maintained between the diameters of the conductor and insulator the number of words per minute which can be sent through a given length of core is simply proportional to the quantity of the materials used.' The process of manufacture is described; namely, the twisting of several copper wires into a conducting strand, the filling of its interstices with Chatterton's compound, the coating of it with successive layers of gutta-percha, the serving of the core with hemp or jute yarn; the sheathing with iron or steel wires—sometimes separately covered with hemp and laid on spirally but without twist in the individual wires; and the final basting of the cable with Clark's bituminous compound. The relative merits of gutta-percha and india-rubber as insulators for submarine lines are discussed; the most important point in favour of india-rubber being its lower specific inductive capacity. A number of actual cables are described, and exhibited by diagrams, and certain novel proposed forms are mentioned. Reference is made to the tendency which the spiral lay in the sheathing has to make the cable, if laid too slack, form bights which are drawn into kinks if the cable is picked up. Tables are given with particulars of the dimensions and weights of the principal submarine cables working at the date of the paper, the depth of water in which they were laid, and the time during which they had been in use. In a discussion following the paper, the author referred to the failure of cables through faults in the insulated covering, or through the destruction of the sheathing: in either case the failure was, in general, complete only when the copper conductor parted. The electrical currents did not initiate faults in the insulator, but where small faults existed previously they were developed by the currents. The faults would be partially sealed for a time by the use of positive currents, but this was done at the expense of the copper, which was gradually eaten away until at last it was severed, and signals then failed. This he believed had been what had taken place in almost all failures where the cable itself had not been broken.

VIII. *Report on Electrical Instruments.* 'Jurors' Report on the International Exhibition, London, 1862.'

The Report describes and discusses the exhibits under the following heads :—I. Introduction. II. The Construction of Telegraphic Lines. III. Instruments employed in the Transmission of Messages. IV. Philosophical Instruments, or Instruments used in Experimental Research. V. Practical Applications of Electricity other than Telegraphic. Chapter II. contains comprehensive statistics of submarine cables laid up to the date of the Report.

The following sentences may be quoted from the Introductory chapter for the sake of their historical interest :—

'The electrical instruments now exhibited are numerous and excellent, whereas, in 1851, the Jury Report states that they were but few in number. . . . The past eleven years have not been marked by any great discovery in electrical science, nor by any very important novelty in the practical application of its principles. We have to register no such marvellous invention as the electric telegraph nor any new motive power superseding steam. On the contrary, it must be acknowledged that many sanguine anticipations in this direction remain unfulfilled. We have on the other hand to record a great extension of the telegraphic system, and especially the introduction of submarine cables. . . . The absence of useless though specious inventions is illustrated by the fact that electromotors, or machines for the production of motive power by the voltaic current, are few in number and quite unimportant. The researches of Dr. Joule have shown that with the present prices of materials it would be utterly vain to expect that the power to be obtained from the conversion of zinc into its sulphate should compete economically with that resulting from the combustion of coal. Let a battery be invented in which a cheap material only is consumed, and it will then be time to consider which is the best arrangement for converting the voltaic current into mechanical effect.

'One by one the causes of failure [in submarine cables] have been discovered and eliminated, with such success that the cables lately laid have thus far been uniformly successful. The electrical tests have especially been brought to great perfection, and indeed these researches into the electrical properties of the materials used as conductors or insulators, and into the phenomena accompanying the transmission of signals, have been prosecuted with such diligence and success, that they may almost be said to form a new branch of electrical science. The whole subject has been rescued from empiricism, and made the matter of accurate measurement and calculation. . . .

'The measurement of resistance has perhaps hitherto attracted more attention than that of the other electrical quantities; and resistance coils, based on different arbitrary units, are sent by exhibitors from France, Switzerland, Germany, and Italy, as well as by English makers. No two of those sent even from the same country are alike; a state of things attended with much the same inconvenience to electricians as would be felt by engineers if every man chose the length of his own footrule. Except those founded on the so-called absolute measure, none of the units have been chosen with any reference to the other electrical quantities, and still less with any regard to the units of force and work.' After referring to 'the beautiful and coherent system of Weber and Thomson for the expression of these quantities in absolute units, chosen with reference to their relations with each other and with the units of force and work which must henceforth be looked on as the connecting link between all physical measurements,' the author continues—'This admirable system, which cannot fail of ultimate adoption, is not yet so generally known as to have produced many instruments intended specially for its illustration or practical application. . . .'

'Next may be mentioned the instruments used for the automatic regulation of the carbon electrodes of *electric lamps.* No less than eight distinct contrivances are shown for this purpose: some more suitable for optical experiments, some for signals, and some for lighthouses or similar practical applications. Faults might still be found with each instrument, but it is impossible to watch the quiet steady flame of Mr. Holmes' light, or to see M. Serrin's lamp extinguished and relighted at a distance twenty times a minute by the simple interruption and re-establishment of the electric current, without feeling that on future occasions electric lamps will be classed rather among the practical applications of electricity than among philosophical instruments. . . . The magneto-electric light is a novelty of great importance, in which steam power is substituted for the voltaic battery as the generator of the powerful current required. It is probable that this invention will enable the electric light to be extensively used in lighthouses and elsewhere.'

['Rapport sur les Appareils Électriques, Exposition de Londres, 1862;' 'Annales Télégraph.' vii. 1864, and viii. 1865.]

The next group of papers, IX. to XVI., relate to the work done by

Professor Jenkin as a member of the British Association Committee on Standards of Electrical Resistance, 1862–1869. Besides taking an active share in the classical experiments of the Committee, which established the absolute system of electrical measurements as a practical method and fitted it for general adoption, Professor Jenkin, as the Committee's Secretary, drafted the Reports, and afterwards edited a volume in which they were republished,[1] along with Articles XIII. and XX. below. The following articles are parts of the Reports with which his name is specially connected.

IX. *Appendix H to the First Report of the Committee on Electrical Standards.* '*Description of the Electrical Apparatus arranged by Mr. Fleeming Jenkin for the Production of exact Copies of the Standard of Resistance.*' 'British Association Report for 1862;' Reprint, p. 35.

The apparatus is a modification of Wheatstone's Bridge in which the two arms forming the 'ratio' of the bridge are made adjustable by connecting them through a short length of wire furnished with a sliding contact. The two portions into which the contact piece divides this resistance of the wire supplement, respectively, the resistance coils of the two arms of the bridge which form the 'ratio.' Provision is also made for interchanging the known and unknown resistance in the other two arms of the bridge, so as to eliminate error due to inexactness in the ratio—a process analogous to double weighing in the use of a balance. The apparatus includes a key which closes first the battery circuit and then the galvanometer circuit by a single motion of the finger. [The arrangement of bridge described in this paper is discussed in Maxwell's 'Treatise on Electricity,' vol. i. § 350.]

X. *Appendix C to the Second Report of the Committee on Electrical Standards.* '*On the Elementary Relations between Electrical Measurements.*' By Professor J. Clerk Maxwell and Mr. Fleeming Jenkin. 'British Association Report for 1863,' p. 130; Reprint, p. 59; 'Philosophical Magazine,' 1865, vol. xxix.

This important paper forms a concise and comprehensive treatise on Absolute Magnetic and Electric Units and the theory and methods of electrical measurements. Its scope is best shown by the following Table of Contents:—

[1] *Reports of the Committee on Electrical Standards appointed by the British Association for the Advancement of Science.* Edited by Professor Fleeming Jenkin, F.R.S. E. & F. N. Spon: London and New York, 1873.

PART I.—INTRODUCTORY.

Objects of treatise.
Derivation of units from fundamental standards.
Standard mechanical units.
Dimensions of derived units.

PART II.—THE MEASUREMENT OF MAGNETIC PHENOMENA.

Magnets and magnetic poles.
Magnetic field.
Magnetic moment.
Intensity of magnetisation.
Coefficient of magnetic induction.
Magnetic potentials and equipotential surfaces.
Lines of magnetic force.
Relation between lines of force and equipotential surfaces.

PART III.—MEASUREMENT OF ELECTRIC PHENOMENA BY THEIR ELECTROMAGNETIC EFFECTS.

Preliminary.
Meaning of the words 'electric quantity.'
Meaning of the words 'electric current.'
Meaning of the words 'electromotive force.'
Meaning of the words 'electric resistance.'
Measurement of electric currents by their action on a magnetic needle.
Measurement of electric currents by their mutual action on one another.
Weber's Electro-dynamometer.
Comparison of the electromagnetic and electrochemical action of currents.
Magnetic field near a current.
Mechanical action of a magnetic field on a closed conductor conveying a current.
General law of the mechanical action between electric currents and other electric currents or magnets.
Electromagnetic measurement of electric quantity.
Electric capacity of a conductor.
Direct measurement of electromotive force.
Indirect measurements of electromotive force.
Measurement of electric resistance.
Electric resistance in electromagnetic units is measured by an absolute velocity.
Magneto-electric induction.
On material standards for the measurement of electric magnitudes.

PART IV.—MEASUREMENT OF ELECTRIC PHENOMENA BY STATICAL EFFECTS.

Electrostatic measure of electric quantity.
Electrostatic system of units.
Ratio between electrostatic and electromagnetic measures of quantity.
Electrostatic measure of currents.
Electrostatic measure of electromotive force.
Electrostatic measure of resistance.
Electric resistance in electrostatic units is measured by the reciprocal of an absolute velocity.
Electrostatic measure of the capacity of a conductor.
Absolute condenser—practical measurement of quantity.
Practical measurement of currents.
Practical measurement of electromotive force.
Comparison of electromotive forces by their statical effects.
Practical measurement of electric resistance.
Experimental determination of the ratio v between electromagnetic and electrostatic measures of quantity.

PART V.—ELECTRICAL MEASUREMENTS DERIVED FROM THE FIVE ELEMENTARY MEASUREMENTS.

XI. *Appendix D to the Second Report of the Committee on Electrical Standards. 'Description of an Experimental Measurement of Electrical Resistance, made at King's College.'* By Professor J. Clerk Maxwell and Messrs. Balfour Stewart and Fleeming Jenkin. Part II. '*Description of the Apparatus,*' by Mr. Jenkin. 'British Association Report for 1863,' p. 163 ; Reprint, p. 96.

This paper describes and illustrates the apparatus constructed by the Committee to carry out Sir W. Thomson's method of finding in absolute measure the resistance of a coil of known dimensions by spinning it with known velocity about a vertical axis and observing how a magnet suspended at the centre of the coil is deflected by the currents which the earth's horizontal magnetic field induces in the revolving wire. The B.A. unit of resistance was determined by means of this apparatus. The paper includes a notice of a modified Wheatstone Bridge in which a group of coils arranged in multiple arc form a means of adjusting the resistance in one arm of the bridge. It is added that the idea of using large coils combined with small ones in multiple arc to obtain extremely minute differences of resistance had been suggested to the writer by Sir W. Thomson.

XII. *Appendix A to the Third Report of the Committee on Electrical Standards. 'Description of a further Experimental Measurement of Electrical Resistance, made at King's College.'* By Professor J. C. Maxwell and Mr. Fleeming Jenkin, with the assistance of Mr. Charles Hockin. 'British Association Report for 1864;' Reprint, p. 115.

This paper gives the results of a second group of experiments by which the B.A. unit of resistance was again determined by the same method as had been used the year before. The Committee made this and the former set of determinations (Art. XI.) the basis of their Standard of Resistance, of which they issued copies during the following year.

XIII. *Report [to the Royal Society] on the New Unit of Electrical Resistance proposed and issued by the Committee on Electrical Standards appointed in* 1861 *by the British Association.* 'Proceedings of the Royal Society,' 1865, vol. xiv. ; 'Philosophical Magazine,' 1865, vol. xxix. ; 'Reprint of Reports on Electrical Standards,' p. 191 ; 'Annal. Phys. Chem.' 1865.

In this Report Professor Jenkin gives an historical account of the early and arbitrary units of resistance suggested or used before the matter was taken up, at the suggestion of Sir W. Thomson, by a Committee of the British Association. The report goes on to mention the system of absolute electromagnetic units, first distinctly proposed by W. Weber in 1851, and accepted and extended by W. Thomson immediately on its appearance, which the Committee adopted as the basis of their practical electrical units. The numerical relation of the practical to the absolute electromagnetic unit of resistance is defined, and the experiments made to determine the unit are shortly referred to. The question of permanence of the standard is discussed in connection with Matthiessen's researches on the electrical permanency of metals and alloys, and with the system adopted by Werner Siemens of defining a standard of resistance by the length and section of a column of mercury. A table is given showing the relative values of the B.A. and other units, and another table shows the degree of agreement in the several determinations made by the Committee.

XIV. *Reply to Dr. Werner Siemens's Paper 'On the Question of the Unit of Electrical Resistance.'* 'Philosophical Magazine,' September 1866.

In the 'Philosophical Magazine' for May 1866, Dr. Werner Siemens had criticised the action of the British Association Committee in selecting a unit of resistance based on the absolute electromagnetic system, and had advocated as unit the resistance at 0° C. of a column of mercury 1 metre long and 1 square millimetre in cross section—a unit actually employed by him before the Committee's labours began. He had also complained that the Committee had not done justice to his work. Professor Jenkin replies that the scientific point in dispute involves two distinct questions, namely (1) What is the best unit of electrical resistance? and (2) What is the best method of making and reproducing any unit? He contends that there is no reason to adopt Dr. Siemens's definition more than any other; and that, on the other hand, the absolute system of units is of great convenience in dealing with the relations of electrical magnitudes to one another and to other quantities in dynamics. He then points out that as regards reproduction it had not been shown that a mercury unit could be reproduced with greater accuracy than the B.A. unit, and that as regards permanency the method practised by the Committee gave all the guarantees that Dr. Siemens's plan could give, and more. The last part of the paper refers at length to the historical and personal question raised by Dr. Siemens, and in it Mr. Jenkin, while acknowledging fully the independent invention by Dr. Siemens of many methods in electrical measurement and his successful reduction of them to practice, claims 'for Professor Thomson the honour of having been the first to insist on a measurement of the conducting-power of the copper in submarine cables, and to express the quality of the insulation in terms of resistance.'

XV. *On a Modification of Siemens's Resistance-Measurer. Appendix II. to the Fifth Report of the Committee on Electrical Standards.* 'British Association Report for 1867,' p. 481; 'Reprint of Reports on Electrical Standards,' p. 144.

The Resistance-Measurer of C. W. Siemens (described in Appendix I. to the same Report) was a species of differential galvanometer with two coils, one on either side of the magnet, and capable of being moved so that one receded from the magnet while the other approached it. In Jenkin's modification the two coils are set at right angles to each other, the magnet being at the centre of both, and the coils are capable of being turned together on a horizontal plane. A current is divided through them, and through the known and unknown resistance respectively, and the coils are turned until their electromagnetic effects on the needle neutralise one another.

XVI. *Experiments on Capacity. Appendix IV. to the Fifth Report of the Committee on Electrical Standards.* 'British Association Report for 1867,' p. 483; 'Reprint of Reports on Electrical Standards,' p. 146.

This paper is interesting as an account of the earliest attempt to form a standard of electrostatic capacity, based on the electromagnetic system of units. Practical electricians had previously constructed condensers of arbitrary capacity, equal to that of a knot of some submarine cable, out of sheets of tinfoil separated by paraffined paper, gutta-percha, or mica. Mr. Jenkin adopted Mr. Latimer Clark's construction, in which mica formed the insulator, and made a condenser whose capacity he adjusted so that it should be approximately equal to 10^{-14} absolute electromagnetic metre-gramme-second units. He points out that a tenfold multiple of this standard [or what is now known as a Microfarad] would be a convenient unit in submarine telegraph work. The capacity of the condenser was measured by comparing the discharge from it through a ballistic mirror galvanometer, whose needle was weighted for the purpose, with the steady deflection produced when the battery used to charge the condenser was placed in circuit with the galvanometer through a known large resistance. Reference is made to the absorption and residual discharge phenomena exhibited by mica in common with other solid dielectrics, and to their influence in causing the capacity of a mica condenser to be somewhat indefinite. The experiments are recorded in full.

XVII. *On the Retardation of Electrical Signals on Land Lines.* 'British Association Report for 1864'; 'Philosophical Magazine,' June, 1865.

The purpose of this paper is to examine the application of Sir W. Thomson's theory of the transmission of electric signals to land telegraphs, in the light of certain experiments published by M. Guillemin in 1860. The author gives a short statement of the results of Thomson's theory, and quotes, on the authority of Mr. Charles Hockin, the following series as a convenient expression to be used in drawing the 'curve of arrival:'—

$$x = C\left\{1 - 2\left[\left(\frac{3}{4}\right)^{\frac{t}{a}} - \left(\frac{3}{4}\right)^{\frac{4t}{a}} + \left(\frac{3}{4}\right)^{\frac{9t}{a}} - \left(\frac{3}{4}\right)^{\frac{16t}{a}} + \left(\frac{3}{4}\right)^{\frac{25t}{a}} - \text{etc.}\right]\right\}$$

Where

x is the current at the receiving end, after any interval t from the time of application of the battery.

C is the current at the receiving end after an indefinitely long application of the battery.

$$a = \frac{kcl^2}{\pi^2} \log_e \left(\frac{4}{3}\right),$$

k being the resistance of the conductor, per unit of length; c the capacity per unit of length, both in absolute electrostatic measures and l the length. The series quoted above applies to the case of perfect insulation : another expression is given, applicable to wire whose insulation is imperfect.

Guillemin's experiments had apparently been made without regard to the theory of Thomson, but with reference to an equation for the establishment of the electric current given by Ohm as early as 1827, where an unknown coefficient appears instead of the quantity c, the place of which in the theory was pointed out by Thomson. The results of Guillemin's experiments were however, as was to be expected, in accordance with the completed theory. His method of determining the curve of arrival was to cause a periodic variation of potential to occur at the sending end, by means of a revolving cylinder which connected the line there alternately to battery and to earth at regular short intervals. The circuit was a loop line with its ends near together, and the same mechanism which sent the currents was employed to connect the receiving end with a galvanometer for an excessively short interval once for every current sent into the line. The successive impulses thus given to the galvanometer needle produced a steady deflection which served to measure the potential at the receiving end at the particular epoch in the curve of arrival when momentary connection with the galvanometer was made. Professor Jenkin proceeds, using the data supplied by Guillemin's results, and assuming a value for the resistance of the line, to apply Thomson's formula to calculate the capacity c, and finds for it a value nearly twice as great as the calculated capacity of a perfectly insulated wire suspended at a uniform distance from a conducting plane, namely, $\dfrac{1}{2\log\dfrac{4h}{d}}$, where h is the height and d the diameter of the wire. He ascribes the difference partly to the approach of the posts to the wire, partly to the capacity of the insulators, and possibly to the effects of polarisation at the points of support. He adds that an additional retardation (which has the effect of making the capacity as calculated for M. Guillemin's results appear larger than its actual value) is produced by the electrodynamic induction from wire to wire. He concludes with an estimate of the rate at which signals can be sent by instruments such as the Wheatstone automatic transmitter, where a limit of speed is set by the retardation discussed in the paper.

XVIII. *Lectures on the Construction of Telegraph Lines.* Royal Engineers Institute, Chatham, 1863.

XIX. *Lectures on the Maintenance of Efficiency of Telegraphic Lines.* Royal Engineers Institute, Chatham, 1863.

XX. *Cantor Lectures on Submarine Telegraphy*, delivered before the Royal Society of Arts (January to February 1866). 'Journal of the Society of Arts,' 1866; 'Reprints of Reports on Electrical Standards,' p. 200.

Lecture I. is on the insulated conductor and its properties, and deals with the mechanical and electrical qualities of copper, gutta-percha, india-rubber, and Hooper's material; the question of permanency and of absorption of water; the manufacture of the core, and its mechanical properties when completed.

Lecture II., on shallow and deep water cables, describes the serving and sheathing of the core, and treats of the strength of the sheathing; the mechanical properties of the completed cable; the maintenance of cables in shallow seas; returns from cables; statistics of deep and shallow water cables; and concludes with abstracts of a number of specifications.

Lecture III., on laying and repairing cables, gives an account of the stowage on board ship, the use of tanks, the precautions against fouling during the running out, the brakes and other paying-out gear; the theory of submersion, as sketched by Sir W. Thomson and independently elaborated by Messrs. Brook and Longridge, with applications of the theory; the lifting and repairing of cables in shallow and deep water.

Lectures IV. and V., on electrical tests, define the terms used, and describe the ordinary tests of the conductor and insulator during and after manufacture; the tests at sea; the tests of short lengths by the electrometer; the tests of joints; capacity tests; tests for faults, namely the loop test, for a cable in the factory or on board ship, and the potential test for a submerged cable, both of which, although first used by other electricians, had been independently devised by the lecturer.

XXI. *Lectures on Electrical Measurements*, delivered at the Royal Engineer Establishment, Chatham (February 1867). Written by Captain R. H. Stotherd, R.E., and revised by the lecturer.

In these three lectures an account is given of the theory of electrical units and the relation of electrical quantities to one another; the principles and practice of measurements of resistance, permanent currents, transient currents, and currents of periodically varying magnitude; of capacity and specific inductive capacity; of potential; of the division of potential by resistance slides, and the combination of two slides to give minute subdivision; of the measurement of potential by electrometers. The lectures conclude with a detailed account of the tests employed during the laying of the 1866 Atlantic Cable, with diagrams showing the electrical arrangements on shore and on the ship.

XXII. *On the Submersion and Recovery of Submarine Cables.* 'Proceedings of the Royal Institution,' May 21, 1869 (Abstract).

This lecture describes the general construction of deep-sea cables, with special reference to the French Atlantic Cable of 1869, particulars of which are stated numerically in full. An account is given of the arrangements for coiling on board the 'Great Eastern,' and for paying out the cable. It is shown how the strains to be expected during the laying are calculated. The cable, while sinking, meets with a resistance in a direction normal to its length, equal (for the Atlantic Cable) to $0{\cdot}154\,v^2$ per foot of length, v being the velocity of sinking, measured in a direction normal to the length of the cable. Its weight being 0·2575 lbs. per foot the velocity of settling is given by the equation

$$0{\cdot}154v^2 = 0{\cdot}2575.$$

The result of this resistance is that the cable lies in a straight line, the inclination of which depends on the velocity of the ship and on v. A formula for the inclination is given, and it is mentioned that the rough Atlantic cable, when the ship was going at the speed of six knots per hour, lay at an angle of $6\frac{3}{4}°$, so that the inclined plane was seventeen miles long, and each piece of the cable took nearly three hours to reach the bottom. The stress on the cable at the top is less than the weight of a length hanging plumb from the surface to the bottom, by the resistance which the rope meets in slipping down along the plane —a quantity depending on the amount of slack paid out. A formula and coefficient, applicable to the Atlantic Cable, are given for this resistance, and the stresses are calculated in particular examples. The process of grappling is described. Figures are given for the stresses which occur during the recovery of cables laid with assumed percentages of slack; and the conclusion is stated that the strength of the cable is from three to four times greater than the strain which in fair weather need come on the cable when being picked up from a depth of two miles.

XXIII. *On the Construction and Submersion of Submarine Telegraph Cables.* 'Journal of the Society of Telegraph Engineers,' 1872, vol. i. p. 114.

This paper, which appears in the section of the Journal devoted to 'Abstracts and Extracts,' and is marked 'Communicated by G. E. Preece,' is substantially an abstract of (1) the first and second of the Cantor Lectures (Art. XX. above), and (2) the Royal Institution Lecture (Art. XXII.).

XXIV. *On a Method of Testing Short Lengths of Highly Insulated Wire in Submarine Cables.* 'Journal of the Society of Telegraph Engineers,' 1873, vol. ii. p. 169.

The paper describes an electrometer method of testing insulation, introduced by Sir W. Thomson and Professor Jenkin during the manufacture of the Western and Brazilian Company's cables, and since then generally adopted in shore tests of submarine Cable core. The method consists in putting the cable, charged to the full potential of the battery, in connection with one pair of quadrants of a Thomson's electrometer, while the battery itself is applied to the other pair. The loss of potential which the cable suffers by leakage is then seen by the gradually increasing deflection of the electrometer needle; and when this becomes excessive the potential of the other pair of quadrants is lowered, by the use of a resistance slide which allows any convenient proportion of the battery potential to be applied, instead of the whole potential. This brings back the needle of the electrometer, and observations of the rate of leakage can then be continued. The method has many advantages, in expedition, accuracy, and convenience, over the direct process of testing by means of a sensitive galvanometer; and it has the further merit of being equally applicable to cores or cables of all lengths. Several cables can be tested at once, since it is only necessary to connect each to the electrometer a few seconds before its potential is to be observed, after which it can be disconnected and allowed to go on losing charge while readings of other cables are being taken. The formula for the reduction of the observations is given in the paper; the electrical connections are exhibited by a diagram; and an example of an actual test is quoted in full.

XXV. *Note on the Electrification of an Island.* 'Nature,' May 5, 1870, vol. ii. p. 12 ; 'American Journal of Science,' 1870.

This note describes a curious observation made by Mr. Gott, the superintendent of the French Cable Company's station on the Island of St. Pierre Miquelon, that the Morse signals sent from another station on the island could be read on the Siphon Recorder of the French Company, when that instrument was placed between an earth plate at the station and another earth three miles out at sea. A sample of recorder slip with a Morse message stolen in this way is reproduced in the paper. The author explains the result by describing the potential of the ground in the neighbourhood of both stations as alternately raised and lowered by the battery used to send Morse signals, so that the island discharged itself through the short insulated line which connected the French Company's station with their distant sea 'earth.' He adds that the action could be avoided by the exclusive use of sea 'earths,' according to the plan introduced by Mr. C. F. Varley to eliminate natural earth-currents.

XXVI. *On the Practical Application of Reciprocal Figures to the Calculation of Strains on Framework.* 'Transactions of the Royal Society of Edinburgh,' 1869, vol. xxv. p. 441.

The scope of this paper is best shown by quoting the introductory paragraphs :—

'The theory of reciprocal figures used as diagrams of forces was first completely stated by Professor J. Clerk Maxwell, in a paper published in the "Philosophical Magazine," April 1864. The following definition of reciprocal plane figures, and their application to statics, are there given as follows :—

'"Two plane figures are reciprocal when they consist of an equal number of lines, so that corresponding lines in the two figures are parallel, and corresponding lines which converge to a point in one figure form a closed polygon in the other."

'"If forces represented in magnitude by two lines of a figure be made to act between the extremities of the corresponding lines of the reciprocal figure, then the points of the reciprocal figure will all be in equilibrium under the action of these forces."

'The demonstration of this statement is given. The conditions under which stresses are determinate, and some examples of reciprocal figures, are also given in the paper, which leaves nothing to be desired by the mathematician.

'Few engineers would, however, suspect that the two paragraphs quoted put at their disposal a remarkably simple and accurate method of calculating the stresses in framework; and the author's attention was drawn to the method chiefly by the circumstance that it was independently discovered by a practical draughtsman, Mr. Taylor, working in the office of the well-known contractor, Mr. J. B. Cochrane. The object of the present paper is to explain how the principles above enunciated are to be applied to the calculation of the stresses in roofs and bridges of the usual forms.'

The construction of reciprocal figures for various typical forms of bridge and roof trusses, under vertical and horizontal loads, is then explained fully and illustrated by a large number of diagrams. [In this connection reference may also be made to §§ 52–56 of Professor Jenkin's article on BRIDGES in the 'Encyclopædia Britannica,' where the use of Reciprocal Figures as a practical method in Graphic Statics is again explained, with the advantage of the simplified system of lettering introduced by Mr. R. H. Bow.]

After discussing a great variety of examples, the author points out the advantages which the method of reciprocal figures has over graphic processes formerly in use and over algebraic methods of calculating stresses, and concludes by repeating that the merit of discovering the method is entirely due to Professor Maxwell and Mr. Taylor, the object of his paper having been to put the theory in a form intelligible to the engineer.

XXVII. *On Braced Arches and Suspension Bridges.* 'Transactions of the Royal Scottish Society of Arts,' 1870, vol. viii.

The author explains the difference between an ordinary beam or bridge-frame, in which the reactions of the supports are vertical, and a braced arch or braced suspension bridge, in which the forces at the supports are indeterminate until account is taken of the stiffness of the rib and yield of the abutments. He had perceived in 1861 that the true form of a stiff rib or stiff chain would be that in which two members would be braced together like the top and bottom members of a girder; that in the arch both members might be in compression, and in the chain both might be in extension, whereas in a girder one is compressed and one extended. He had found, however, that he was unable to calculate, except on unproved assumptions, the distribution of stresses, being unable to find the force at the supports. He drew Professor Clerk Maxwell's attention to the problem, and Maxwell published ('Phil. Mag.,' May 1864) a method by which all stresses in framed structures could be positively determined. Professor Jenkin then gives an abstract of the method, which is based on the assumption that no horizontal force is applied at the abutments except what results from the load on the bridge. The calculations referring to a practical example are detailed in full in an appendix to the paper. They show that, for a uniform load, the chain of a braced suspension bridge, the figure of which is described in the paper, should be tapered at the middle to nearly one-tenth of its cross section at the piers ; and the bottom member, which like the chain is in tension throughout, should have its section increased in the ratio of 8 to 1 from the piers to the centre, where its section should be nearly equal to that of the chain at the piers. A similar design applies to the braced arch : and this distribution of material, slightly modified to suit practical requirements, was patented by the author. A comparison is made between the weight of a braced arch or suspension bridge built on this plan and the weights of bridges of other types, to illustrate the advantage of the proposed design : and it is pointed out that the braced arch has the further merit that none of its parts are exposed to any considerable tension, a point of special importance when the material is cast-iron or timber.

The calculations proved the possibility of constructing a suspension bridge which should be as stiff as a girder without any compressive members except some light diagonals, and showed that a braced arch might be made stable under unsymmetrical loads, both arch and suspension bridge being much lighter than the equivalent girders.

For an abstract of the theory of braced arches, identical in substance with that given in the above paper, reference may be made to § 56 of the article Bridges in the 'Encyclopædia Britannica.'

XXVIII. *On the Application of Graphic Methods to the Determination of the Efficiency of Machinery.* 'Transactions of the Royal Society of Edinburgh,' 1877, vol. xxviii. p. 1. *Ditto*, Part II., '*The Horizontal Steam-Engine.*' 'Transactions of the Royal Society of Edinburgh,' 1878, vol. xxviii. p. 703.

This very important paper, the scope of which is much wider than the title indicates, has been reprinted in full (pp. 271–341). It describes a novel process of performing what may be termed a dynamical analysis of machines, which allows the method of Reciprocal Figures (explained in an earlier paper) to be applied in determining the forces that act throughout the mechanism. The investigation of frictional efficiency is one use to which this analysis may readily be applied. The Keith Gold Medal for the period 1877–9 was awarded to Professor Jenkin for this paper by the Royal Society of Edinburgh.

XXIX. *Sur l'Application de la Méthode Graphique à la Détermination de l'effet utile des Machines.* 'Association Française pour l'avancement des Sciences, Congrès de Rheims,' August 1880.

[An abstract of the preceding paper.]

XXX. *On a Constant Flow Valve for Water and Gas.* 'Transactions of the Royal Scottish Society of Arts,' March 1876.

The paper describes a regulator, invented and patented by Professor Jenkin, by which the flow of fluid past an orifice is maintained constant notwithstanding great variations of pressure on either side of the orifice. An inlet pipe discharges into a chamber through an equilibrium throttle valve. Above this is a partition which divides the chamber into two parts. Part of the partition is constituted by a piston, which is free to rise and fall, and is connected to the equilibrium throttle valve below. The fluid passes into the upper part of the chamber through an adjustable hole in the partition, and thence to the delivery pipe. When the pressure of supply is excessive the piston rises and throttles the entering fluid ; when the pressure of supply is reduced the piston rises and opens the throttle valve. The effect is to maintain a constant difference of pressure on the two sides of the piston, and so to maintain a constant flow through the hole in the partition, notwithstanding variations in the pressure against which the fluid is delivered or in that under which it is supplied. Several forms of the regulator are figured in the paper for the supply of gas and water. The drawing of a large form, for use on water mains, is reproduced by Professor Unwin in the article Hydromechanics in the 'Encyclopædia Britannica.'

XXXI. *On Friction between Surfaces moving at Low Speeds.* By Fleeming Jenkin and J. A. Ewing. 'Philosophical Transactions of the Royal Society,' 1877, vol. 167, Part II., p. 509. bstract, 'Pro c. I. S.,' No. 179, 1877.)

This paper describes an experimental investigation of the influence of velocity on friction at very low speeds, undertaken with the view of seeing whether there is continuity between the friction of rest and the friction of motion, when solid bodies rub against one another. Under certain conditions the coefficient of friction between surfaces at rest exceeds the coefficient between the same surfaces in relative motion, and the authors looked for evidence of a possible continuity between the two values in a rise of friction when the speed became reduced. The friction of a steel axle in bearings of steel, brass, wood and agate, was examined at speeds ranging from about 0·01 ft. per second as an upper limit, down to 0·0002 ft. per second. The axle carried a massive disc, which, once set spinning, was brought to rest by the friction which it was the object of the experiments to measure. The rate of retardation of the disc was determined by means of a pendulum, which swung over against its circumference and traced a sinuous line on a strip of paper fixed to the disc. To avoid friction between the pendulum and the paper, the former was provided with a frictionless marking pointer on the plan invented by Sir W. Thomson for use in his Siphon Recorder for submarine telegraph signals. A fine siphon of capillary glass tube, fixed to the pendulum, and supplied with ink from an insulated cistern, was kept electrified by a small induction machine, and its point was thereby maintained in a continuous state of oscillation towards and from the paper, on which it deposited a particle of ink at every contact.

When the rubbing surface of the axle and journals was dry no change could be detected in the coefficient of friction throughout the range of velocities experimented on, but when the journals were of wood, and lubricated with water or oil, an increase of friction at the lower limit of velocity was found, amounting to about 20 per cent. of the value at the higher limit of velocity. Out of all the sets of circumstances examined a distinct increase of friction with reduced velocity was observed in those, and only in those, cases in which the friction of rest was notably greater than the friction of motion—a result which made the hypothesis probable that there is continuity between the two kinds of friction. The method of experiment and the numerical results obtained are detailed fully in the paper.

XXXII. *Remarks on the Phonograph.* By Fleeming Jenkin and J. A. Ewing. 'Proceedings of the Royal Society of Edinburgh,' 1878, vol. ix. p. 579.

XXXIII. *On the Wave Forms of Articulate Sounds.* By Fleeming Jenkin and J. A. Ewing. (1) 'Proc. R. S. E.,' 1878, vol. ix. p. 582. (2) 'Proc. R. S. E.,' 1878, vol. ix. p. 723.

XXXIV. *The Phonograph and Vowel Theories.* By Fleeming Jenkin and J. A. Ewing. (1) 'Nature,' 1878, vol. xvii. p. 423. (2) 'Nature,' 1878, vol. xviii. p. 167.

XXXV. *The Phonograph and Vowel Sounds.* By Fleeming Jenkin and J. A. Ewing. (1) *The Vowel Sound ō*, 'Nature,' 1878, vol. xviii. p. 340. (2) 'Nature,' vol. xviii. p. 394. (3) 'Nature,' vol. xviii. p. 454.

XXXVI. *On the Harmonic Analysis of Certain Vowel Sounds.* By Fleeming Jenkin and J. A. Ewing. 'Transactions of the Royal Society of Edinburgh,' 1878, vol. xxviii. p. 745.

This group of papers (XXXII.–XXXVI.) refers to an experimental investigation of the wave-forms of articulate sounds, undertaken with the aid of the phonograph, which had then been newly invented by Mr. T. A. Edison. In experimenting with the phonograph the authors had observed that the quality of vowel sounds spoken by the instrument did not vary so much as Helmholtz's theory led them to expect, when the barrel of the phonograph was turned faster or slower than its normal rate. This led them to undertake an extensive analytical examination of records embossed on the tinfoil of the phonograph, in the hope of finding that common feature in the wave-forms corresponding to any one vowel sound, by virtue of which the vowel is recognised when spoken or sung by different voices or at different parts of the scale. The phonograph records were magnified for the purpose of analysis by a system of light levers terminating in a pen formed of an electrified capillary tube which worked like the siphon of Sir W. Thomson's recorder, and traced a magnified version of the wave forms on a moving strip of paper. This magnified transcript was taken without injuring the original tinfoil records, which were still available for reproducing the spoken sounds. The wave-forms traced on paper were then subjected to harmonic analysis to determine the amplitudes of the constituent simple tones. By this means tables were constructed showing the relative amplitude of the prime tone and the first five overtones in the vowel sounds ō, ā, a°, and ū, sung at various pitches by several voices. The results were compared with Helmholtz's theory of vowel sounds as depending on the reinforcement of certain overtones by the oral cavity. They showed that the quality of a vowel sound does not depend either on the absolute pitch of reinforcement of the constituent tones alone, or on the simple grouping of relative partial tones independently of their absolute pitch. The authors' conclusions are stated at length in the last paper, where the method of experiment and the numerical results arrived at are also fully described and a number of the magnified traces of sound waves are reproduced by photo-lithography. The other papers are for the most part preliminary notices of the last paper, but contain some incidental observations regarding the reproduction of other sounds than vowels by means of the phonograph. Amongst these is the curious observation that any single element of spoken sound, whether vowel or consonant, is phonetically reversible, as nearly as the phonograph allows the character of the sound to be determined ('Nature,' vol. xvii. p. 423).

XXXVII. *Nest Gearing.* 'Report of the British Association for 1883,' p. 387.

Under this name Jenkin describes a number of forms of friction gear of his invention, for the transmission of power by rolling contact. The chief peculiarity of 'nest gears' is that the rollers composing them are so grouped that their mutual pressures are balanced, to avoid producing thrust against the bearings. This invention was the subject of several patents, and was employed in some of the early forms of telpher locomotives, which are illustrated in the paper on 'Telpherage,' p. 262.

XXXVIII. *Gas and Caloric Engines*, being one of the series of Lectures on Heat and its Mechanical Applications delivered at the Institution of Civil Engineers, February 21, 1884.

The lecturer discusses the general theory of hot-air and explosive-gas engines, and gives particulars of the results obtained by various makers. The gas-engines of Otto and Clerk are described in some detail, and experiments on efficiency and on the nature of the explo sion are cited. The use of a regenerator is referred to, and the original patent of Dr. Stirling is quoted in full in an appendix to the lecture. Modern hot-air engines of the Stirling type are mentioned and their construction described by the aid of diagrams. Attempts by Sir W. Siemens to apply the regenerator to internal-combustion engines are alluded to; and in this connection the author gives a short account of an experimental engine constructed by himself and Mr. A. C. Jameson, and of his own unsuccessful efforts to introduce gas-engines of a novel type. The author concludes by referring to the fact that Dowson gas allows the gas-engine even now to compete on favourable terms, as regards economy of fuel, with the steam-engine. 'Since this is the case now, and since theory shows that it is possible to increase the efficiency of the actual gas-engine two- or even three-fold, the conclusion seems irresistible that gas-engines will ultimately supplant the steam-engine. The steam-engine has been improved nearly as far as possible, but the internal-combustion gas-engine can undoubtedly be greatly improved, and must command a brilliant future.' The lecture is followed by a number of appendices giving (1) a more extended account of the theory of the subject: (2) examples of calculated theoretical indicator diagrams for air- and gas-engines of various types: (3) description of the engines referred to, and a full illustrated account of the operation of the valves of Otto's engine : (4) a Report by Dr. Slaby in which the Lenoir and Otto engines are compared and experimental data supplied : (5) an excerpt from Dr. F. Grashof's work on the Mechanical Theory of heat, giving physical constants relating to coal-gas and its combustion : (6) particulars of Mr. J. E. Dowson's cheap gas as fuel for gas engines : and (7) the Patent Specification of R. and J. Stirling for air engines (1827).

XXXIX. *Telpherage*, being the introductory address to the Class of Engineering in the University of Edinburgh for the Session 1883–4, October 30, 1883. 'The Electrician,' November 3, 1883 ; 'La Lumière Électrique,' November 10, 1883.

XL. *Telpherage*, being the substance of a lecture delivered at the School of Military Engineering, Chatham, on March 13, 1884.

XLI. *Telpherage*. A paper read before the Society of Arts, May 14, 1884. 'Journal of the Society of Arts,' May 16, 1884.

These papers refer to the invention the development of which formed Professor Jenkin's last work. Arts. XL. and XLI. are substantially identical. A reprint of the latter is given at p. 252.

LIST OF
PROFESSOR JENKIN'S BRITISH PATENTS.

1. TELEGRAPHIC COMMUNICATION (with Sir W. Thomson). No. 2,047, 1860.
2. BRIDGES. No. 667, 1861.
3. ELECTRIC TELL-TALE COMPASS. No. 1,553, 1863.
4. MACHINERY FOR MANUFACTURING TELEGRAPH CABLES. No. 2,155, 1865.
5. WINDING IN TELEGRAPH CABLES. No. 1,218, 1866.
6. APPARATUS FOR PRODUCING ELECTRIC LIGHT. No. 390, 1869.
7. BRIDGES. No. 3,071, 1869.
8. SUBMARINE TELEGRAPH CABLES. No. 3,236, 1869.
9. V-PULLEYS FOR THE TRANSMISSION OF POWER (with Mr. F. H. Ricketts). No. 1,886, 1873.
10. TELEGRAPHIC APPARATUS (with Sir W. Thomson). No. 2,086, 1873.
11. OBTAINING MOTIVE POWER. No. 2,441, 1874.
12. TELEGRAPHIC APPARATUS (with Sir W. Thomson). No. 1,095, 1876.
13. TRANSMITTING SOUND BY ELECTRICITY (with Mr. J. A. Ewing). No. 4,402, 1877.
14. CALORIC MOTOR ENGINES (with Mr. A. C. Jameson). No. 1,078, 1881.
15. CALORIC MOTOR ENGINES (with Mr. A. C. Jameson). No. 1,130, 1881.
16. CALORIC MOTOR ENGINES (with Mr. A. C. Jameson). No. 1,160, 1881.

MECHANISM FOR TRANSPORTING GOODS AND PASSENGERS BY MEANS OF ELECTRICITY. No. 1,830, 1882.

18. REGULATING SPEED IN MACHINERY DRIVEN BY ELECTRICITY. No. 3,007, 1882.
19. MECHANISM FOR TRANSPORTING GOODS AND PASSENGERS BY MEANS OF ELECTRICITY. No. 4,548, 1882.
20. MACHINERY FOR SPINNING AND WINDING (with Professor Ewing). No. 26, 1883.
21. DRIVING GEAR. No. 1,913

22. GEARING. No. 4,481, 1883.
23. DRIVING GEAR CALLED 'NEST GEAR.' No. 4,754, 1883.
24. GAS ENGINES. No. 2,635, 1884.
25. TELPHERAGE. No. 3,795, 1884.
26. TRUCKS AND LOCOMOTIVES FOR TELPHER LINES. No. 3,796, 1884.
27. CONTACT ARMS FOR TELPHER TRAINS. No. 4,167, 1884.
28. TELPHERAGE. No. 5,020, 1884.
29. TRANSPORTING GOODS AND PASSENGERS BY MEANS OF ELECTRICITY (with Mr. A. C. Elliott). No. 8,460, 1884.
30. TELPHER LOCOMOTIVE. No. 8,751, 1884.
31. REGULATION OF CURRENTS IN TELPHER AND OTHER ELECTRIC MOTORS. No. 8,906, 1884.
32. UNDERGROUND ELECTRIC HAULAGE. No. 10,907, 1884.
33. WATER MOTORS (with Mr. H. Darwin). No. 11,038, 1884.
34. MECHANISM FOR THE TRANSMISSION OF POWER (with Professor Ewing). No. 12,479, 1884.
35. GOVERNORS. No. 15,111, 1884.

THE END.

PRINTED BY
SPOTTISWOODE AND CO., NEW-STREET SQUARE
LONDON

Printed in the United States
By Bookmasters